Perspectives on Strategic Defense

About the Book and Editors

Bringing together proponents and opponents of the Strategic Defense Initiative, this book includes original essays by leading experts on every aspect of the issue. The collection provides a valuable introduction to the many complex questions involved in any serious consideration of the SDI. The contributors explore such issues as the strategic implications of the SDI, the technical feasibility of the proposal, its impact on U.S.-USSR relations and the arms control process, and the response of NATO allies to the program.

The book also includes a detailed history of the strategic defense debate, policy statements by the key political players, and petitions on its technical flaws and arms control implications that have been signed by U.S. scientists. A comprehensive glossary and index make this book a valuable reference tool.

Steven W. Guerrier is assistant professor of military and diplomatic history at the Virginia Military Institute. **Wayne C. Thompson** is associate professor of political science at the Virginia Military Institute and specializes in European politics.

Perspectives on Strategic Defense

edited by Steven W. Guerrier
and Wayne C. Thompson

Foreword by
Zbigniew Brzezinski

Routledge
Taylor & Francis Group
LONDON AND NEW YORK

First published 1987 by Westview Press, Inc.

Published 2019 by Routledge
52 Vanderbilt Avenue, New York, NY 10017
2 Park Square, Milton Park, Abingdon, Oxon OX14 4RN

Routledge is an imprint of the Taylor & Francis Group, an informa business

Library of Congress Cataloging-in-Publication Data
Perspectives on strategic defense.
(Westview special studies in national security and defense policy)
1. Strategic Defense Initiative. I. Guerrier, Steven W. II. Thompson, Wayne C., 1943- . III. Series.
UG743.P47 1987 358'.1754 86-22430

ISBN 13: 978-0-367-28275-2 (hbk)
ISBN 13: 978-0-367-29821-0 (pbk)

To
Nancy and Susie

Contents

PART TWO

US Government Statements

Statements by Allied Governments

Soviet Statements

Critical Statements by American Scientists

Foreword

Zbigniew Brzezinski

Four years have passed since President Reagan launched the Strategic Defense Initiative. The resulting debate has afforded us a unique opportunity. The United States is on the verge of a fundamental development in strategic weapons technology. In the past, technological development in the area of strategic weapons ran ahead of our ability to assimilate their implications for national strategy. The invention of nuclear weapons in the forties preceded serious discussion of the impact of nuclear weaponry on war and peace. For the next two decades, we engaged in a continuous debate about the role of nuclear weapons in our national strategy. Leaving aside for the moment the desirability of the subsequent development and deployment of a strategic defense, we are in a position that gives us the opportunity to engage in a kind of strategic re-evaluation, re-thinking, and re-development of our strategic doctrine that would make whatever decisions we eventually make about the SDI part of a coherent long-range strategy.

It is my view that the Strategic Defense Initiative provides the United States with the means of achieving mutual strategic security, either through a bilateral arms control agreement with the Soviet Union or through a unilateral deployment of a limited strategic defense. Our willingness to explore the possibility of building a strategic defense creates enormous leverage to induce the Soviets to accept in negotiations significant cuts in offensive nuclear forces, particularly those that are accurate and powerful enough to destroy hardened targets in a first strike. It is imperative that the United States continue with a vigorous research and development program for its negotiating position to appear credible. This program should be directed toward those defensive systems that can be developed in the near term to defend our retaliatory forces and command system. At the same time, this view implies that the United States must be willing *both* to limit the scope of its strategic defenses as part of an agreement and to take soon the initial deployment decisions in favor of a two-tier limited strategic defense system. Thus, if the negotiations succeed or fail to reach a stabilizing agreement, our strategic retaliatory forces will be more secure.

In this volume, Professors Steven Guerrier and Wayne Thompson have made a valuable contribution to the debate over strategic defense. Based on conferences at the Virginia Military Institute in the spring and fall of 1986, this book includes important insights on the SDI issue from prominent figures on both sides of the debate, as well as a transcript of a lengthy discussion session among these experts and a section that reproduces several official government policy statements. This debate, of course, will not end here, but our understanding of the fundamental issues involved with the SDI will surely be enhanced.

Preface

The debate over the Strategic Defense Initiative touches on the very survival of our nation and of mankind. How can we best be protected from atomic annihilation? Through defensive weapons, mutual vulnerability to offensive nuclear weapons, negotiated arms reductions, or some combination of these?

In a speech before the House of Commons on March 1, 1955, Winston S. Churchill said in reference to the policy of nuclear deterrence: "It may well be that we shall by a process of supreme irony have reached a stage in this story where safety will be the sturdy child of terror, and survival the twin brother of anihilation." Over four decades, nuclear deterrence has been the basis for keeping peace in the atomic age. Both sides have maintained swords, but no real shields. Is this approach now outdated? Is it time to attempt to render the nuclear threat impotent and obsolete by technical means? Would SDI undermine or strengthen the deterrence of war?

The Corps of Cadets at the Virginia Military Institute considered these questions to be so important that on April 7–8 and October 16–17, 1986, it organized two symposia on the subjects of SDI and the use of outer space. The participants, who represented widely divergent points of view, discussed the key elements of the debate over SDI: what it is and why it was proposed; its technical and financial feasibility; the response to it by the Soviet Union and America's allies; its impact on the United States's strategic doctrine and on arms control negotiations; and whether, in sum, SDI brings us hope or danger, stability or instability, peace or war. The presentations and discussions at those conferences provide the backbone of this book.

Part One of this volume is almost entirely composed of original and previously unpublished writings on SDI. Where a piece is an edited transcript of a speech presented at VMI, it is so noted. Most authors submitted papers which were used for this book. Part Two consists of important public pronouncements on strategic defense by American and non-American governmental leaders. These official statements have been the primary sources for the present public discussions concerning SDI. This part also contains critical opinions on SDI articulated by American and Soviet scientists. Finally, Part Two includes the text of the Anti-Ballistic Missile Treaty of 1972, which most participants in the public debate cite, but which far fewer have ever read.

The editors of this book wish to thank the Virginia Military Institute—particularly General Sam S. Walker, Superintendent; Brigadier General John W. Knapp, Dean of the Faculty; Colonel Leroy D. Hammond, Executive Assistant to the Superintendent; and Dr. Patrick M. Mayerchak, Senior Director of the VMI International Studies Program—for their total support of the symposia and for their encouragement to seek out spokesmen from all points of view. We wish to thank Dr. Willard M. Hays, Chairman of the Department of History and Politics, for generously providing the cadet staff with indispensable facilities and services and for tolerating our preoccupation with this project.

We are grateful to the VMI Foundation and the International Studies Program for a grant to support this publication. Dr. Edward J. DeLong, Audiovisual Librarian at VMI, provided crucial technical advice and assistance in taping the sessions and in producing the transcripts used by some of the authors. Cadets James Bradford Adams, Scott R. Harbula, Paul F. Hicks, Jr., Douglas M. Jacobsen, Michael R. Laban, Nicolas J. Lovelace, David J. Furness, Michael E. McGraw, A. B. Miller, and Mark A. Snedecor worked prodigiously and imaginatively to make the two symposia work; they have more than earned our gratitude and congratulations.

Professor T. Y. Greet and the VMI Symposium Committee gave us the benefit of their experience and guidance and prevented us from having to reinvent the wheel in planning our conferences. Mr. Richard McCormack of NASA was extremely generous with his time and contacts; much of the responsibility for the success of our second conference falls on his shoulders. Dr. Larry I. Bland, Marshall Papers Editor at the George C. Marshall Foundation, took the papers and documents gathered by two inexperienced editors and helped to transform those pieces into this book. Our appreciation for his part in this work matches the great extent to which we depended upon his meticulous efforts and professional experience.

We thank the BDM Corporation, which helped us bring a few of this book's authors to VMI. Of course, we thank the authors themselves whose ideas and arguments are the substance of this work. We can hardly express adequately our immense appreciation to Colonel Joseph M. Rougeau of the United States Air Force, who not only opened many doors for us, but who, more importantly, prodded and persuaded us to "think big." Finally, we owe very special gratitude to the VMI Corps of Cadets, without whose interest and initiative, this project would never have taken shape.

Steven W. Guerrier
Wayne C. Thompson

Contributors

James A. Abrahamson, General, USAF. Director, Strategic Defense Initiative Organization, Department of Defense

Barry M. Blechman, President, Defense Forecasts, Inc. Formerly: Senior Associate, The Brookings Institution; Assistant Director, US Arms Control and Disarmament Agency

Peter A. Clausen, Director of Research, Union of Concerned Scientists. Former Senior Arms Analyst, UCS

Jacquelyn K. Davis, Vice-president, The Institute for Foreign Policy Analysis, Cambridge, Massachusetts

Pierre M. Gallois, Brigadier General, French Air Force (Retired). Strategic adviser to General De Gaulle and a founder of the French nuclear force

Mikhail Gorbachev, General Secretary, Communist Party of the Soviet Union

Steven W. Guerrier, Assistant Professor of Military and Diplomatic History, The Virginia Military Institute

Michael Heseltine, Former British Minister of Defence

Geoffrey Howe, British Foreign Minister

Joseph T. Jockel, Director of Canadian Studies, St. Lawrence University. Council on Foreign Relations Fellow at the US State Department's Office of Canadian Affairs, 1984–85

Spurgeon M. Keeny, Jr., President, The Arms Control Association. Formerly: Deputy Director of the US Arms Control And Disarmament Agency; Head, US Delegation at Theater Nuclear Forces talks; Senior Staff Member, National Security Council

Sergei Kislyak, First Secretary, Soviet Embassy in Washington, D.C.

Steven A. Maaranen, Associate Director of the Center for National Security Studies at the Los Alamos National Laboratory

Louis Marquet, Director, Directed Energy Weapons, SDIO. Formerly Associate Head, Optics Division, Lincoln Laboratory, Massachusetts Institute of Technology

Jack Mendelsohn, Deputy Director, The Arms Control Association. Former Foreign Service Officer and member of the US Delegation, START talks

Brian Mulroney, Prime Minister of Canada

Paul H. Nitze, Special Adviser on arms control matters to the US President and Secretary of Defense

Richard N. Perle, Assistant Secretary of Defense for International Security Policy

George Rathjens, Professor of Political Science, Massachusetts Institute of Technology. Formerly: Chief Scientist and Deputy Director, DARPA; Special Assistant to the Director, US Arms Control and Disarmament Agency; Director, Systems Evaluation Division, IDA

Ronald Reagan, President of the United States

Joel J. Sokolsky, Assistant Professor of Political Science, Royal Military College of Canada

Gregory M. Suchan, Senior State Department Representative, Defense and Space Negotiation Group, US Delegation to the Negotiations on Nuclear and Space Arms with the Soviet Union

Edward Teller, Lawrence Livermore National Laboratory

Wayne C. Thompson, Associate Professor of Political Science, The Virginia Military Institute

Robert L. Walquist, Vice-president and SDI Program executive with TRW Electronics and Defense Sector. Formerly: Director, Voyager Spacecraft Study; Manager, Minuteman Program Office

PART ONE

1

Looking Back: Strategic Defense and US National Security

Steven W. Guerrier

TO many Americans, the search for an effective means of defense against the threat of nuclear war began with President Ronald Reagan's speech to the nation of March 23, 1983. In this address, the president called upon the American scientific community to "give us the means of rendering these nuclear weapons impotent and obsolete."[1]

The wisdom and feasibility of President Reagan's objective notwithstanding, his appeal serves as a dramatic illustration of the power of the presidency to shape the debate over issues of national and international importance. Over the next few months, as the organization to pursue the Strategic Defense Initiative (SDI) took form, it became clear that the framework for discussion of strategic nuclear weapons and their role in US national security policy had been significantly altered. Ballistic Missile Defense (BMD) quickly became the focal point of a great debate over nuclear strategy. All related issues took on a secondary status as SDI became the principal issue in the domestic debate over national defense—and the major obstacle to the conclusion of an arms control agreement with the Soviets.

The Strategic Defense Initiative has been billed as a "new vision," by which American technology would be harnessed to the search for a means to provide the United States with an effective defense against enemy attack. However, while the technology may be new, the "vision" is as old as the nation itself, and its fulfillment has been sought on many previous occasions. The present debate can be viewed as the latest episode in this long search for the best means of providing for the physical security of the United States.

1. Ronald Reagan, "Defense Spending and Defensive Technology," televised speech delivered March 23, 1983, "Weekly Compilation of Presidential Documents," March 28, 1983, pp. 423–66.

Traditional Approaches to Strategic Defense

The problem of strategic defense—that is the defense of the national homeland—is one that has, by definition, faced the United States since the beginning of the republic. Then, as now, the object has been to prevent the employment of an enemy nation's military capabilities directly against the territory of the United States, its population, and its productive resources. Throughout most of American history this has involved efforts aimed primarily at preventing adversaries from launching naval or amphibious attacks against the coasts and at establishing defense against cross-border invasions.[2]

During the nineteenth century, such defenses consisted primarily of fortifications, coastal batteries, and a navy that was in varying degrees capable of denying an enemy access to American waters. For most of the century, however, these were relatively low priority items. Except in times of crisis, appropriations were lacking and technological advances were haphazard at best. New approaches to strategic defenses were commonly the result of efforts to reduce defense expenditures and were rarely successful. President Thomas Jefferson's gunboats and the later use of ironclad monitors as harbor batteries are among the more entertaining examples. Overall, it was assumed that there would be sufficient time to mobilize to meet any threat that might jeopardize American security.

However, the American approach to strategic defense in the nineteenth century was not as unrealistic as this apparent lack of attention might indicate. After the War of 1812, foreign threats to the territory of the United States were minimal. Diplomatic efforts, such as the Rush-Bagot Agreement of 1817, led to the ultimate demilitarization of the border with Canada. The army was generally able to cope with the declining Spanish presence on the continent and with whatever threat might be posed by Mexico. Most importantly, after 1814 the growing commonality of interests between the United States and Great Britain—the world's greatest naval power—insured that the Atlantic would remain far more a defensive moat than a potential avenue of invasion. In any case, the American abstention from active involvement in world affairs during this period greatly lessened the chances of serious conflict.

By the end of the nineteenth century, with the United States assuming the role of a major power, the expanding navy became the true first line of defense for the homeland, as well as for the new American interests over-

2. General accounts of the traditional American approach to the problems of strategic defense can be found in such works as Russell F. Weigley, *The American Way of War: A History of United States Military Strategy and Policy* (Macmillan, 1972), and Allan R. Millett and Peter Maslowski, *For the Common Defense: A Military History of the United States of America* (The Free Press, 1984).

seas. Coastal and border fortifications grew irrelevant and were dismantled or converted into historic sites. The naval capabilities of the United States were believed to provide effective security against the threat of direct attack and would continue to give the nation time to mobilize in the event of a major conflict. As long as any attack on the American continent would necessarily come by sea, this would be adequate.

The rise of air power in the twentieth century, however, would change the fundamental assumptions upon which the American approach to strategic defense were based, albeit slowly.[3] During World War I, air power was primarily applied at a tactical level related to the battlefield situation below. The few strategic bombing raids against population and economic targets had little impact on the war or on military planning in the years that followed. Some authors, most notably Giulio Douhet, did argue that strategic bombing would render all other forms of military power obsolete, but most interwar efforts to translate such theory into policy met only resistance from military leaders.[4] In the United States, the foremost supporter of air power was Colonel Billy Mitchell, whose persistent advocacy and criticism of American inattention to the subject led to his court-martial in 1925. Only in the mid-1930s did the potential of air power, at both the tactical and strategic levels, begin to receive the serious consideration of military planners outside of the air services.

The experience of World War II removed all doubt that strategic air power would be an important part of any nation's military capability. It demonstrated that success in modern war would increasingly come to depend on an ability to strike at the industrial base which supplied the forces of the enemy. The Allied bombing campaigns against Germany and Japan, in particular, showed that this new weapon could provide that ability. Furthermore, there was continued hope that the bombing of population centers would demoralize an enemy and bring about its collapse. Postwar analyses of strategic bombing might challenge those advocates who argued that such campaigns alone would win future wars, but it was clear that strategic air power—and defense against it—would be important components of national security in the future.[5] For the United States, this meant that the traditional seaborne approach to strategic defense would soon be insufficient.

---

3. See Robin Higham, *Air Power: A Concise History* (St. Martin's Press, 1972), pp. 25–72. See also Trevor N. Dupuy, *The Evolution of Weapons and Warfare* (Jane's, 1982), pp. 240–44, and Weigley, *American Way of War*, pp. 223–41.

4. Giulio Douhet, *The Command of the Air* (Office of Air Force History, 1983). See also I. B. Holley, Jr., *Ideas and Weapons* (Office of Air Force History, 1983).

5. For example see Millett and Maslowski, *Common Defense*, pp. 435–40. See also United States Strategic Bombing Survey, *Summary Reports* (GPO, 1945–46).

Long-range bombers were not, however, the only new threat to be considered. The German development of the V-1 flying bomb—essentially a pilotless aircraft laden with high explosives—and, especially, the V-2 rocket were even more ominous. After all, the principles of defense against manned bombers were well understood. The development of radar, proximity fuzes, and jet aircraft offered the hope that strategic bombing compaigns might be defeated by attrition before they could have a decisive impact on the defender's capability and will to make war. The Battle of Britain, for example, demonstrated that an attrition rate of 10 percent was enough to force the abandonment of such a campaign by the Germans. But guided missiles were potentially far more challenging. They travelled too fast and flew too high to be intercepted reliably by a conventional antiaircraft defense. Also, they carried no crewmen who might be deterred by the prospect of flying into heavy fire.

Fortunately for the Allies, the weapons employed by the Germans during the war were relatively unreliable and inaccurate. The V-1 could deliver a ton of high explosive to a range of roughly two hundred miles, but it was only accurate to within about five miles of its target. Furthermore, with an average speed of only 350 miles per hour and an inability to take evasive action, it was highly vulnerable to a determined defense. Within three months of the initial use of V-1, the British were able to destroy over 80 percent of the flying bombs launched against them. While its speed and altitude made interception virtually impossible, the V-2 was not much more accurate than its predecessor. These rockets could deliver a slightly smaller warhead to within four miles of the intended target at a range of some two hundred miles. In any case, they were employed too late in the conflict to have a serious effect on the outcome.[6]

The potential of guided missiles, however, could not be ignored. As the war in Europe drew to a close, both the United States and the Soviet Union made determined efforts to capture as much as possible of the German rocket program and its personnel. The American effort, given the code name "Project Paperclip," was particularly successful. A number of German scientists, their files, and equipment (including three hundred railroad cars loaded with V-2 components) were soon on their way to the United States where they were integrated into the nascent American missile program at White Sands, New Mexico.[7] Although most experts believed that accurate missiles with a range greater than a few hundred miles were, at best, a decade away, it was clear that bombers would not be the only new challenge to strategic defense in the future.

---

6. P. M. S. Blackett, *Fear, War, and the Bomb: Military and Political Consequences of Atomic Energy* (McGraw-Hill, 1948), pp. 51–53.

7. See Clarence G. Lasby, *Project Paperclip: German Scientists and the Cold War* (Atheneum, 1975).

And yet, if the bombers and missiles of the next war were limited to the delivery of high explosives, the problem of strategic defense might still be manageable. The experience of the war had demonstrated the difficulties associated with mounting an effective strategic bombing campaign employing conventional explosives: the number of sorties required to have a serious impact on a war economy was huge, the cost of penetrating a determined defense was high and the delivery of bombs was often inaccurate. The damage done was difficult to assess and was often quickly overcome. While missiles were much harder to intercept, problems associated with their range, accuracy, and payload capacity gave hope that their impact might be limited as well. The problem, of course, was that the bombers and missiles of the future would not be limited to the delivery of conventional explosives.

The development of the atomic bomb changed everything. The tremendous increase in destructive power that came with the new weapon meant that now one bomber or missile could now cause a level of devastation that previously might have required hundreds of delivery vehicles. It also meant that accuracy would be less of a problem for the attacker. Since the radius of destruction of an atomic weapon was so much greater than that of a conventional bomb, pinpoint accuracy—though still desirable—was far less critical.

The challenge to strategic defense was clear. Against a long-term conventional bombing campaign, the defender could hope to survive through the attrition of the attackers. But the most effective sustained air defenses yet mounted were not more than 10 to 15 percent efficient. Against campaigns involving the thousands of sorties required by conventional explosives, this might be enough to degrade the attack before it could inflict an unacceptable level of damage. With atomic weapons, however, the successful penetration of only a handful of bombers could be disastrous. Even a defense of 90 percent efficiency would likely not be enough, and there are, of course, no perfect defenses. Furthermore, the prospect that atomic weapons might some day be adapted to long-range guided missiles seemed to eliminate whatever hope might remain for an effective strategic defense in the future. The American monopoly on atomic weapons offered some comfort—as did problems associated with the weight of the bombs and the ability to produce them in quantity—but few experts believed that this situation would last, and fewer still found it an acceptable basis for national security in the postwar era.[8]

---

8. Lawrence Freedman, *The Evolution of Nuclear Strategy* (St. Martin's Press, 1981), pp. 22–30.

Early Research into Missile-Based Defense

Attempts to deal with the new threats to the security of the United States were underway well before the end of World War II. Research continued in such areas as radar, antiaircraft artillery, fighter-interceptors, and other components of a conventional air defense. Additional programs sought to discover defensive applications of the same technologies which posed these new threats, but were generally accorded much lower priority than research into offensive weaponry. In part because of the apparent difficulty of creating an effective defense, the United States soon came to abide by the maxim that the best defense is the deterrent value of a good offense.

Although some academic strategists, such as Bernard Brodie, were arguing as early as 1946 that the only credible defense against atomic attack was through the creation of an effective deterrent force that would promise swift retaliation, the United States came to this policy mostly by default.[9] For three years following the war, the Truman administration gave surprisingly little attention to the implications of the atomic bomb for American policy. There was a general recognition that the new weapon was of great significance, but many in the administration resisted the notion that the atomic bomb had fundamentally altered the nature of warfare. Overall, it was thought that the next major war would be very much like the last. Atomic weapons would bring a new level of destruction, to be sure, but the basic outlines would remain the same.[10]

Only in 1948, during the Berlin crisis, did the president authorize the military to base its war plans on the assumption that atomic weapons would be used in a major conflict. Even then the decision was made for less than purely military reasons. With the wartime military demobilized and under

---

9. Bernard Brodie, ed., *The Absolute Weapon: Atomic Power and World Order* (Harcourt, Brace and Company, 1946).

10. For an excellent account of the early history of American nuclear strategy, see David A. Rosenberg, "The Origins of Overkill: Nuclear Weapons and American Strategy, 1945–1960," *International Security* 7(Spring 1983): 3–71. See also Gregg Herken, *The Winning Weapon: The Atomic Bomb in the Cold War, 1945–1950* (Alfred A. Knopf, 1981); and Harry R. Borowski, *A Hollow Threat: Strategic Air Power and Containment Before Korea* (Greenwood Press, 1982). The decision to authorize plans for the use of nuclear weapons is contained in NSC-30, "United States Policy on Nuclear Weapons," September 10, 1948, reprinted in Thomas H. Etzold and John Lewis Gaddis, *Containment: Documents on American Policy and Strategy, 1945–50* (Columbia University Press, 1978), pp. 339–42. Discussion of particular war plans is found in Etzold and Gaddis, pp. 302–38; Anthony Cave Brown, *Dropshot: The American Plan for World War III Against Russia in 1957* (The Dial Press, 1978); and Kenneth W. Condit, *The History of the Joint Chiefs of Staff: The Joint Chiefs of Staff and National Policy*, vol. 2, *1947–1949* (Michael Glazier, Inc., 1979), pp. 283–310.

pressure to curtail defense spending, the Truman administration increasingly turned to the establishment of a strategic bombing force, armed with atomic weapons, as the most cost-effective form of striking power. It was hoped that such a force would be potent enough to deter Soviet aggression in the first place. If it were not, however, atomic weapons would at least allow the United States to deliver a serious blow to the Soviet war economy at the outset of a conflict.

Thus, while a policy based upon the threat of atomic retaliation is more commonly associated with the Eisenhower administration, its origins can be found during the period of the American monopoly prior to the end of 1949. The development of atomic weapons by the Soviet Union only served to harden a policy already in existence and to make the deterrence of war the principal mission of American strategic forces. Meanwhile, however, the search continued for a more effective means of providing for the defense of the United States.

As early as 1944, the Army had contracted for two research and development studies which would have a bearing on the future of strategic defense. The first of these, Project Hermes, was an investigation of the basic operational problems involved in a rocket program and was ultimately aimed toward the development of long-range surface-to-surface missiles. The contract was awarded to General Electric Corporation, which carried out a series of tests at the White Sands Proving Ground in cooperation with the Army Ordnance Department. Hermes took over from ORDCIT, an earlier effort by the Ordnance Department and the California Institute of Technology that had led to the development of a series of test-bed missiles known as Private A and Private F. Building on the success of the Private series, which involved some forty-one firings to a range of twelve miles, Hermes moved on to the development of a longer-range missile known as the WAC (Without Attitude Control) Corporal. Again, the purpose was experimental. The Corporal would eventually be deployed in a modified form as a tactical surface-to-surface missile, and its basic design would be incorporated into the Nike-Ajax series of air defense missiles.[11]

Project Hermes concluded with a series of over sixty firings of V-2 rockets assembled from components captured during the war. Most of these were conducted at White Sands between March 1946 and June 1951 and involved high altitude research as well as the study of military applications. Subsidiary programs, such as Operation Sandy, looked to more novel applications for guided missiles—in this case the firing of a V-2 from the deck of the aircraft carrier *Midway* in 1947.[12] By its close, Project Hermes

---

11. For a discussion of Projects ORDCIT and Hermes, see Willy Ley, *Rockets, Missiles, and Men in Space* (Signet, 1969), pp. 284–94.
12. Ibid., pp. 291–93.

had produced a wealth of information that would provide the foundation for the subsequent development of rockets for both military and civilian use.

Although the most important in terms of basic research, Project Hermes was by no means the only program conducted in the early postwar era to have a bearing on prospects for an effective strategic defense. While Hermes was ultimately geared toward the development of offensive missiles, another Army program was begun in 1944 to focus on the defensive application of missile technology. Project Thumper involved basic research toward the development of a high altitude antiaircraft missile that would also be effective against rockets of the V-2 generation. The following year, Thumper would be merged with the Army Air Force's Ground-to-Air Pilotless-Aircraft program (GAPA) and then with a research program at the University of Michigan to become the Air Force's Project Wizard. This effort would ultimately lead to the two weapons with which the Air Force would contest the Army's claim for the mission of missile-based strategic defense—the Bomarc air defense missile and an antiballistic missile (ABM) program that would retain the code-name Wizard. Both, however, eventually lost out to the progeny of a third Army-funded research effort: Project Nike.[13]

Begun in February 1945, Project Nike was directed toward the actual production of a series of long-range surface-to-air missiles (SAMs) which could be employed against strategic bombers and, in later versions, against ballistic missiles. Development of the Nike series had, in fact, been initiated the previous year as a joint effort of Bell Telephone Laboratories and the Douglas Aircraft Company, under the direction of Western Electric as prime contractor for the Army. The establishment of the formal project reflected the growing potential accorded SAMs as a means of defense against strategic air attack.[14] This importance was enhanced by the results of a series of tests at the Army's Aberdeen Proving Ground, in 1947, which demonstrated that conventional antiaircraft artillery would be ineffective against jet aircraft.[15]

Building on the experience gained during the ongoing Project Hermes and employing the basic design of the WAC Corporal as a model, the Project Nike team completed work on a prototype by 1950.[16] Designated Nike-Ajax (MIM-3), the new weapon was a two-stage missile, with a range of twenty-five to thirty miles. Production was approved by the Truman administration, and the first units, armed with conventional warheads,

---

13. Projects Thumper and Wizard are discussed in Ernest J. Yanarella, *The Missile Defense Controversy: Strategy, Technology, and Politics, 1955–1972* (University Press of Kentucky, 1977), pp. 27, 32–33.

14. On the early history of Project Nike, see ibid., pp. 27–28.

15. Dupuy, *Evolution of Weapons and Warfare*, pp. 270–71.

16. Ibid., p. 271.

became operational in 1953. By the end of the decade, Nike-Ajax batteries were located around some thirty potential target areas in the United States. Ultimately, some fifteen thousand missiles would be produced under the Nike-Ajax program, with large numbers being transferred to allied governments.[17]

Once the Nike-Ajax became operational, the project team concentrated on the development of a second generation air defense missile that would bring increased range and accuracy, as well as the capability of delivering nuclear warheads. This phase of the project was spurred on by intelligence reports indicating a growing Soviet strategic bombing capability and led to the production of the Nike-Hercules (MIM-14/A/B/C).[18] Entering service in 1958, Nike-Hercules had a range of ninety miles and could be armed with either a nuclear or conventional warhead. Greater readiness was achieved by providing solid fuel engines for both stages, unlike the Nike-Ajax which employed liquid fuel in its second stage. At its peak deployment in 1963, some 134 Nike-Hercules batteries would be in operation. Overall, more than twenty-five thousand Nike-Hercules missiles were produced by Western Electric in the United States and under license by Mitsubishi in Japan. Of these, roughly thirty-eight hundred were transferred to NATO and other allies. Nike-Hercules remained in service in the United States until 1974.[19]

Well before the Nike-Hercules became operational, however, it was clear that the threat posed by Soviet strategic bombers was considerably less than had been anticipated by the US intelligence community. It became apparent that the Soviets, rather than investing in manned bombers, were directing their greatest effort toward the development of long-range ballistic missiles.[20] In response, research into defensive technologies shifted from attempts to defeat strategic bombers to the far more difficult problem of coping with an attack by Soviet ICBMs.[21]

US Policy Toward Continental Defense, 1945–1960

While Project Nike offered the hope that an effective missile-based defense might someday be constructed, the new threats to American security would seem to have required more immediate attention. Yet, despite the implications of the development of atomic weapons and long-range bombers

---

17. For specifications on Nike-Ajax, see Tom Gervasi, *Arsenal of Democracy II* (The Grove Press, 1981), p. 224; and Ley, *Rockets, Missiles, and Men in Space*, p. 628.

18. Yanarella, *Missile Defense Controversy*, p. 28.

19. For specifications on Nike-Hercules, see Gervasi, *Arsenal of Democracy II*, p. 224; and Ley, *Rockets, Missiles, and Men in Space*, p. 628.

20. See John Prados, *The Soviet Estimate: U.S. Intelligence Analysis and Russian Military Strength* (The Dial Press, 1982), pp. 45–50.

21. Yanarella, *Missile Defense Controversy*, p. 28.

for the defense of the United States, efforts to formulate a policy to deal with these threats were surprisingly slow in coming. Between 1945 and 1947, for example, there was little progress in the area of continental defense—as it was then called—beyond the revision of ABC-22, a US-Canadian plan drafted in 1941 for the joint defense of their coastal waters. In February 1947, this was superseded by the US-Canada Basic Security Plan. The new agreement listed the air defense of North America as the highest priority among eight "joint tasks," but little was done at the time to prepare for such a mission.[22]

There are several explanations for the apparently relaxed American attitude toward continental defense. In the first place, the pressure to dismantle the war-time military establishment and to cut defense spending greatly lessened the attractiveness of a program that would certainly be expensive. This reluctance increased as many government officials who were initially optimistic about the development of effective defenses came to understand the complexity of the problem.[23] Furthermore, there seemed to be no immediate threat to the security of the United States. Although it was widely acknowledged that the American monopoly of atomic weapons and the means to deliver them would not last, few expected that the Soviet Union would be capable of launching a major atomic attack until 1952 at the very earliest.[24] In any case, by the late 1940s, it was clear that the defense of the United States would indeed rest primarily on the development of a strategic nuclear force that would deter attack through the threat of certain retaliation.[25]

This is not to say that there was no recognition of the importance of continental defense. Administration officials frequently noted the need to meet any threat to the security of the United States and looked to the day when defensive systems would be deployed in strength. Air defense was listed among the highest priorities in every emergency war plan approved by the Joint Chiefs of Staff (JCS) since 1945, even when the means to carry out such plans were lacking.[26] Nevertheless, the administration continued to place almost exclusive emphasis on the development of offensive forces. By the fall of 1949, when the Soviets tested their first atomic bomb, active

22. James F. Schnabel, *The History of the Joint Chiefs of Staff: The Joint Chiefs of Staff and National Policy*, vol. 1, *1945–1947* (Michael Glazier, Inc., 1979), pp. 380–89. See also Joseph T. Jockel, *The United States and Canadian Efforts at Continental Air Defense, 1945–1957*. Ph. D. dissertation, Johns Hopkins University, 1978.

23. Freedman, *Evolution of Nuclear Strategy*, p. 30. On postwar demobilization, see Schnabel, *History of the Joint Chiefs of Staff*, pp. 380–89.

24. For example see Condit, *History of the Joint Chiefs of Staff*, p. 536.

25. Rosenberg, "Origins of Overkill," pp. 14–20.

26. Condit, *History of the Joint Chiefs of Staff*, p. 536.

defenses of the United States were virtually nonexistent—twenty-three fighter squadrons and thirty antiaircraft artillery battalions were assigned to provide air defense for the entire continental United States and Alaska. This prompted General Hoyt S. Vandenberg, Air Force chief of staff, to remark that "almost any number of Soviet bombers could cross our borders and fly to most targets in the United States without a shot being fired at them."[27]

Even after the end of the American atomic monopoly, resources devoted to continental defense remained limited. A Soviet attack was still thought unlikely for at least several years, and deterrence was still considered the most effective approach to the problem. One change that did come in the wake of the Soviet test was a decision by the Joint Chiefs to proceed with the construction of a temporary network of early warning radars, known as LASHUP. By the outbreak of the Korean War in June 1950, this system consisted of forty-nine stations in the continental United States and Alaska.[28]

American involvement in Korea brought with it the approval of a massive increase in defense spending that had been urged in a National Security Council report (NSC-68) prepared in April 1950.[29] This led to the expansion of conventional, as well as nuclear, forces. Appropriations for continental defense were also increased.[30] By November 1952, forty-six fighter squadrons and forty-five antiaircraft battalions, supported by eighty large radars, were deployed for the defense of American airspace. Still, the actual deployments continued to fall short of the requirements of American war plans through most of the 1950s.[31] Funding increased, more interceptor aircraft were assigned, and missiles—such as Nike-Ajax and Nike-Hercules—replaced antiaircraft artillery, but the emphasis of American strategy continued to be placed on the deterrent value of offensive nuclear arms.

The Truman administration had come to the conclusion that nothing even approaching a "perfect" defense was possible in the foreseeable future. A Soviet attack might be degraded, but the destructiveness of atomic wea-

---

27. Ibid., pp. 536–37.

28. Ibid., p. 541.

29. NSC-68, "United States Objectives and Programs for National Security," April 14, 1950, U.S. Department of State, *Foreign Relations of the United States, 1950,* vol. 1, *National Security Affairs; Foreign Economic Policy* (GPO, 1977), pp. 234–92. Series hereafter cited as *FRUS.*

30. For a discussion of the impact of NSC-68 on US defense programs, see Paul Y. Hammond, "NSC-68: Prologue to Rearmament," in Warner R. Schilling, Paul Y. Hammond, and Glenn H. Snyder, *Strategy, Politics, and Defense Budgets* (Columbia University Press, 1962), pp. 267–378. See also John Lewis Gaddis, *Strategies of Containment: A Critical Appraisal of Postwar American National Security Policy* (Oxford University Press, 1982), pp. 89–126; Steven W. Guerrier, *NSC-68 and the Truman Rearmament, 1950–1953* (forthcoming).

31. "Status of National Security Programs of the United States in Relation to Approved Objectives," November 5, 1952, *FRUS, 1952–54,* vol. 2, part 1, *National Security Affairs* (GPO, 1984), p. 168.

pons would insure that those bombers which did get through would do very grave damage. More than ever, deterrence seemed the only effective option. In September 1952, Truman told the National Security Council that, as far as he could see, "there wasn't much of a defense in prospect except for a vigorous offense."[32] This emphasis on the deterrent value of offensive forces increased with the Eisenhower administration's "New Look" defense policy—which emphasized the cost-effectiveness of nuclear weapons—and its threats of massive retaliation for any Soviet aggression.[33]

Nevertheless, the evolving idea of reliance on a deterrence-based defense, with its implication that little could be done to stop a determined Soviet nuclear attack on the United States, was unattractive to a number of individuals involved in defense policy. In April 1948, the magazine *Atlantic* published an article co-authored by James R. Killian, president of the Massachusetts Institute of Technology (MIT). Entitled "For a Continental Defense," the article served as a rallying point for supporters of a more vigorous effort to develop an effective air defense.[34] Backed by Secretary of the Air Force Thomas K. Finletter, these officials were able to convince President Truman, in the spring of 1950, to authorize an intensive study of the problems of continental air defense. Known as Project Charles, the investigation was conducted by some twenty physicists at MIT over the next year.[35]

One result of Project Charles was clear even before the study began. This was to demonstrate the extent to which opposition to a major air defense program had developed within the Air Force. Sensing a threat to its offensive role, the Air Force made repeated efforts to have the study killed; only the continued support of Secretary Finletter kept it alive. The Air Force might accept continental defense as an adjunct to its deterrent mission, but it would fight any attempt to change American strategy to one that was defense-dominated. As one Charles participant recalled, "it was an uphill battle to get the Air Force to think about defense at all."[36]

---

32. "Memorandum for the President of Discussion at the 122d Meeting of the National Security Council," ibid., p. 121.

33. Rosenberg, "Origins of Overkill," pp. 27–44. See also Glenn H. Snyder, "The New Look of 1953," in Schilling, Hammond, and Snyder, *Strategy, Politics, and Defense Budgets*, pp. 379–524; and Gaddis, *Strategies of Containment*, pp. 127–97.

34. James R. Killian and A. G. Hill, "For a Continental Defense," *Atlantic*, April 1948.

35. Gregg Herken, *Counsels of War* (Alfred A. Knopf, 1985), p. 61. Finletter had recently served as chairman of the President's Air Policy Commission, which called for the expansion of US strategic air power and air defense capabilities. See President's Air Policy Commission, *Survival in the Air Age* (GPO, 1948).

36. Herken, *Counsels of War*, p. 61.

In August 1951, the findings of the study were presented in a three-volume report, entitled "Problems of Air Defense." The report concluded that an effective continental defense was technically feasible and recommended that its development be undertaken immediately. It called for an increased number of fighters, rapid deployment of the Nike-Ajax, and increased funding for research on its successors. The report also called for the construction of a distant early warning radar network controlled by a system of computers, known as the Semi-Autonomous Ground Environment (SAGE), which was based on an MIT design for air traffic control.[37]

President Truman approved the study and soon authorized the establishment of a special research facility, the Lincoln Laboratory, at MIT to conduct further research on the problems of defense against bomber attack. Critics, however, quickly attacked the estimated two billion dollar cost of the Charles recommendations, and the Air Force charged that the report had grossly underestimated the difficulty of shooting down bombers. Of course, the Air Force by now had a considerable investment in bombers. It did not want their value called into question by discussions of how easily they might be destroyed.[38]

If the Air Force was unenthusiastic about continental defense, the same could not be said of the Army. The years since the end of World War II had been difficult for the Army. As American forces demobilized, it was the Army which suffered most from decreasing budgets. The National Security Act of 1947 had stripped it of the Army Air Force and, in establishing an independent United States Air Force, had created a powerful bureaucratic rival. The Key West and Newport Agreements, which assigned the services their basic roles and missions under the act, saw the Army bypassed as the principal strategic tasks were given to the Air Force and Navy. A major consequence of this was that, except during the Korean conflict, the Army's budget suffered well into the 1950s in comparison to its sister services. Therefore, in search of a lasting strategic mission, the Army increasingly came to place its hopes on continental defense.[39]

The Army's role in postwar strategic defense grew out of its responsibility for ground-based air defense during World War II. In the years after the war, the Army maintained antiaircraft artillery batteries for the point defense of a number of locations in the United States. But antiaircraft guns were of limited value against high-flying jet bombers. Therefore, as we have seen, the Army turned increasingly to research in missile technology. By the early 1950s, it was actively lobbying for operational control over missile-based continental defense programs.

---

37. Massachusetts Institute of Technology, *Problems of Air Defense,* August 1951, cited in ibid., p. 63.

38. Ibid., pp. 63–64.

39. Yanarella, *Missile Defense Controversy,* pp. 28–29.

One aspect of the Project Charles report did find support in the Air Force, as well as elsewhere within the defense community. This was the continuing need for a permanent early warning radar network. Air Force interest was spurred on by a 1952 Rand Corporation study which exposed the extreme vulnerability of Strategic Air Command (SAC) bases—at home and overseas—to a Soviet preemptive attack.[40] An early warning network would allow more time for SAC to get its bombers off the ground and thus help protect the retaliatory force. By the end of 1952, this broad agreement on the need for an early warning radar system led Truman to approve NSC-139, which called for the joint development with Canada of a system that, by the end of 1955, could provide at least three hours warning of a Soviet bomber attack.[41]

During the Eisenhower administration attention to the problem of continental defense grew. A series of National Security Council studies examined the status of air defense programs, as did special commissions such as the Killian and Gaither panels. The conclusions were almost always the same: the current program was judged inadequate to meet the Soviet threat, and new increases were recommended. Over time, however, this did lead to the creation of a powerful air defense force.[42]

By the early 1960s, three radar networks—the Distant Early Warning (DEW), Mid-Canada, and Pine Tree lines—provided early warning. Active defense was provided by sixty-seven regular and fifty-five Air National Guard fighter-interceptor squadrons, supported by batteries of Nike-Hercules missiles and the new Hawk low-altitude SAM. The only problem was that by then it had long been clear that the Soviets had foresaken the development of a large bomber force and had turned their attention instead to ballistic missiles. Consequently, programmed deployments were not completed, and many of these forces were soon assigned to other missions.

---

40. The SAC vulnerability study is discussed in Fred Kaplan, *The Wizards of Armageddon* (Simon and Schuster, 1983), pp. 101–10, 117–21.

41. NSC-139, "An Early Warning System," December 31, 1952, is discussed in "Memorandum by Paul H. Nitze and Carlton Savage of the Policy Planning Staff," *FRUS, 1952-1954*, vol. 2, part 1, pp. 318–23. NSC-139 is published in *FRUS, 1952-1954*, vol. 6.

42. For example see NSC-159/4, "Continental Defense," September 25, 1952, and NSC-5408, "Continental Defense," February 11, 1954, *FRUS, 1952-1954*, vol. 2, part 1, pp. 475–86, 609–24. See also NSC 5605, "Continental Defense," June 5, 1956. The report of the Killian Panel, "Meeting the Threat of Surprise Attack," February 1955, is discussed in Freedman, *Evolution of Nuclear Strategy*, pp. 158–60, and Kaplan, *Wizards of Armageddon*, pp. 130–31. The Gaither Report, "Deterrence and Survival in the Nuclear Age," is discussed in Freedman, *Evolution of Nuclear Strategy*, pp. 160–63, and Kaplan, *Wizards of Armageddon*, pp. 144–54.

Meanwhile, research efforts had been reoriented to meet the new threat of Soviet missiles.[43]

The Antiballistic Missile Debate

By the mid-1950s, as it became evident that the feared "bomber gap" would not materialize, the emphasis of research into air defense systems shifted to a search for means to counter long-range surface-to-surface missiles.[44] As early as 1953, the Army had begun to give serious consideration to that problem with the establishment of Project Plato, which studied the feasibility of tactical antiballistic missile systems designed to protect troops in the field. This particular project was cancelled in 1958, but by then it was clear that ABMs—or ballistic missile defense—could offer the Army the prominent strategic role it had been seeking.[45]

In 1955, evidence of Soviet advances in the development of long-range ballistic missiles prompted the Army to sponsor a feasibility study of potential means of defense against ICBMs. The results of the investigation, which was directed by Bell Telephone Laboratories, seemed promising and, in 1957, led to authorization for the development of the Nike-Zeus missile. Third in the Nike family, the Zeus was originally intended as a follow-on air defense missile that would compete with the Air Force's Bomarc.[46] It was to be a two-stage solid fuel missile that would have a range of approximately one hundred miles.[47] The Army now hoped that the nuclear-armed missile might be the key to a national ABM system that would provide effective defense against Soviet missile attack—a system that the Army would control.

---

43. John M. Collins, *American and Soviet Military Trends Since the Cuban Missile Crisis* (The Center for Strategic and International Studies, Georgetown University, 1978), pp. 133–37.

44. The most comprehensive discussion of the ABM debate is found in Yanarella, *Missile Defense Controversy*. Other useful accounts are found in Freedman, *Evolution of Nuclear Strategy;* Herken, *Counsels of War;* Kaplan, *Wizards of Armageddon;* and the following: Alain C. Enthoven and K. Wayne Smith, *How Much is Enough? Shaping the Defense Program, 1961–1969* (Harper Colophon Books, 1971); William R. Schneider, Jr., "Missile Defense Systems: Past, Present, and Future," in Johan J. Holst and William R. Schneider, Jr., eds, *Why ABM? Policy Issues in the Missile Defense Controversy* (Pergamon Press, 1969); David N. Schwartz, "Past and Present: The Historical Legacy," in Ashton B. Carter and David N. Schwartz, *Ballistic Missile Defense* (The Brookings Institution, 1984); U.S. Office of Technology Assessment (OTA), "Ballistic Missile Defense Technologies," published in *Strategic Defenses* (Princeton University Press, 1986); and Herbert F. York, *Race to Oblivion: A Participant's View of the Arms Race* (Simon and Schuster, 1970).

45. Yanarella, *Missile Defense Controversy*, p. 27.

46. Ibid., pp. 27–28.

47. Ley, *Rockets, Missiles, and Men in Space*, p. 628.

Missiles seemed to offer the Army the prospect of obtaining its long-sought strategic mission, but here too there was frustration. The development of missile technology led to a growing interest in strategic defense on the part of the Air Force. This reflected less an acceptance of the arguments for an active defense than it did a desire by the Air Force to gain control of all missile programs. By now it was clear that offensive missiles would become a major component of the American deterrent, and the Air Force wanted to control them. Defensive missiles were largely seen as part of the package.

Competition between the Army and Air Force for operational control over both offensive and defensive missile programs, along with other jurisdictional disputes among the services, led Secretary of Defense Charles Wilson to issue a memorandum on November 26, 1956, clarifying service roles. Again, the Army fared badly. The Air Force was given control over the development and deployment of intermediate-range ballistic missiles (IRBMs)—its Thor missile winning out over the Army's Jupiter. Jupiter was later saved when the Soviet launch of Sputnik prompted the Defense Department to proceed with both missiles; it went on to serve as the launch vehicle for America's first satellite. The Air Force was also given responsibility for the development of surface-to-air missiles for area defense. All that the Army could salvage at this time was control over SAMs for point defense.[48]

Despite its disappointment, the Army determined to make the most of its air defense mission and was soon devoting up to 15 percent of its budget to this purpose.[49] The memorandum had done little, however, to end competition between the services over defensive missiles because of its failure to define the terms "area defense" and "point defense." The Army, with its Nike-Zeus, and the Air Force, with its Wizard program, each continued to seek total control over ballistic missile defense. Even the Navy briefly entered the contest, making a short-lived case for the Talos-Terrier-Tartar family of shipboard SAMs that emerged from its Project Bumblebee.[50]

Final resolution of the jurisdictional dispute came on January 16, 1958, when Secretary of Defense Neil McElroy assigned to the Army sole responsibility for the development and operation of an ABM system. The Air Force was directed to support the development of communications and radar components for the proposed system and to step up construction of its Ballistic Missile Early Warning System (BMEWS), which had begun the previous year. There were a number of factors behind this decision. The Air Force was ambivalent about BMD despite its desire for control of the

---

48. Yanarella, *Missile Defense Controversy*, pp. 29–31.
49. Ibid., p. 31.
50. Ibid., pp. 27, 32–36.

program. Also, some defense officials hoped to buy off discontent within the Army over its continued exclusion from major strategic missions. Most important, however, was the clear superiority of the Nike-Zeus to any of the missiles under development by the Air Force.[51]

Also in 1958 came the creation of the Advanced Research Projects Agency (ARPA—later DARPA, as "Defense" was added to the name) to coordinate research and development within the Department of Defense. ARPA was given responsibility for ABM development beyond Nike-Zeus and soon began a long-term study of advanced BMD concepts, known as Project Defender. Over the years, it also funded a number of feasibility studies on missile defense systems conducted by the services.[52]

Development of the Nike-Zeus continued, spurred on by increased funding in the wake of Sputnik, but the decision giving the Army control over ABM programs did not mean that a system would actually be deployed. Even within the Department of Defense there was considerable skepticism about the technical feasibility of ballistic missile defense. In the spring of 1958, shortly after McElroy's directive, a Pentagon panel known as the Reentry Body Identification Group concluded that an ABM system could not be made to work in the foreseeable future. This view was supported by the President's Science Advisory Committee in a report issued in May 1959.[53]

Throughout this period, the Army lobbied for funding to begin production and deployment of the Nike-Zeus, but each time it was denied. Opposition came from the Air Force—which might be expected—and increasingly from the Office of the Secretary of Defense (OSD), where concerns about the effectiveness of the proposed system were mounting. Many had come to believe that the Nike-Zeus would not be able to stand up to the sort of heavy ICBM attack that the Soviets might be capable of launching by the late 1960s and particularly that its tracking and acquisition radars would be vulnerable to such potential countermeasures as decoys and chaff. Furthermore, with an election on the horizon, the Eisenhower White House was not enthusiastic about the program's estimated cost of $15 billion. Funding for research and development would continue, but that was to be all for the time being.[54]

With the inauguration of the Kennedy administration in 1961, the Army resumed its efforts to secure funding for deployment of an ABM system, and it found a growing number of allies in Congress. Kennedy's campaign

---

51. Ibid., pp. 40–41.
52. Ibid., pp. 40–41.
53. Kaplan, *Wizards of Armageddon*, pp. 343–44; Enthoven and Smith, *How Much is Enough*, pp. 184–85; Schwartz, "Past and Present," p. 333.
54. Yanarella, *Missile Defense Controversy*, pp. 60–61; Enthoven and Smith, *How Much is Enough*, pp. 185–86.

rhetoric about a supposed "missile gap" had prompted an increasing popular interest in defensive systems that would continue even after it was later demonstrated that any such gap was overwhelmingly in America's favor. In April, however, the administration elected to defer any such decision until the missile had completed its test program.[55]

Meanwhile, as doubts about the effectiveness and cost of the Nike-Zeus continued to grow within the Pentagon, Defense Secretary Robert McNamara increased funding for advanced ABM research. Project Defender, which had been examining such "exotic" BMD technologies as plasmas, chemical lasers, and X-ray lasers driven by nuclear explosions, concluded that none would be feasible until at least the 1980s. ARPA also sponsored a growing variety of service research projects. Among the more prominent of these were the Air Force's BAMBI (Ballistic Missile Booster Interceptor), which would have employed hundreds of missile-armed satellites to attack Soviet ICBMs; the Army's SAINT (Satellite Interceptor) and Field Army Ballistic Missile Defense System, an outgrowth of the Plato study; and a Navy program called Typhoon.[56]

None of these projects went beyond the basic research stage, but they did reflect a growing disenchantment with Nike-Zeus and the desire for a viable alternative. This sentiment continued, despite successful tests of the missile conducted in 1962. In mid-July, a Nike-Zeus intercepted a target Atlas ICBM over the Pacific Ocean. Two other successful tests followed in that year, including one involving the use of decoys.[57] Yet, while they provided valuable data, the tests did little to dispel doubts about the Nike-Zeus system. The interceptor was too slow—thus precluding the use of the atmosphere to filter out decoys—and the mechanically-steered radars could be too easily overwhelmed by a saturation attack.

On January 5, 1963, Secretary McNamara announced that the Nike-Zeus ABM system would not be deployed and that the program would be phased out at the conclusion of testing. In its place would come a program of research and development on a more advanced ABM system, to be known as Nike-X. McNamara also stated that the deployment of any future system should be accompanied by a massive program of civil defense as a matter of the highest priority. Without a sufficient number of fallout shelters, he argued, the effectiveness of BMD as a means of population defense would be severely limited. Similar arguments had been made in the 1950s in con-

55. Yanarella, *Missile Defense Controversy*, pp. 64–66.
56. Ibid., pp. 73–75.
57. Schneider, "Missile Defense Systems," p. 4; Yanarella, *Missile Defense Controversy*, p. 82.

nection with continental defense, but civil defense was always deemed too expensive to pursue on a large scale.[58]

The proposed Nike-X system addressed many of the failings of its predecessor. First of all, it would be directed by phased-array radars which would be steered electronically—projecting beams much as the electron gun of a television fires at the screen—rather than mechanically. It was hoped that this would enable the system to handle large numbers of targets simultaneously and prevent saturation. For terminal defense, the system would employ a new high-acceleration missile called Sprint. The speed of this nuclear-armed interceptor would allow it to use the atmosphere to filter out decoys (which are not hardened to withstand reentry) and attack only actual warheads. To provide area defense beyond the localities where Sprint was based, the Nike-X system would also field a long-range interceptor which would engage targets above the atmosphere. This missile, called Spartan, was to be derived from the Nike-Zeus and would also carry a nuclear warhead.[59]

The Army would control the new program and hoped that a decision on production and deployment might be reached at an early date. McNamara, however, was determined that no such decision would be made until the system had been fully tested. This angered ABM supporters in Congress, and over the next several years, the defense secretary was forced to fight off repeated attempts to appropriate funds for the deployment of Nike-Zeus and for the early production of Nike-X components. Congressional interest increased in 1964 with the circulation of unofficial reports that the Soviets were beginning to deploy ballistic missile defenses around Moscow.[60]

McNamara's desire to defer a decision on deployment of an ABM system was not based solely on issues of technical feasibility. Increasingly, he was coming to the view that ballistic missile defenses should not be deployed even if they could be built. McNamara's doubts were many and were largely supported by the results of studies carried out at the direction of OSD. Among these were the Threat Analysis Study (or Betts Report), a wide-ranging investigation of the strategic implications of ABM deployment begun in 1963, and an ongoing series of reports on specific problems prepared in the Office of Systems Analysis.[61]

---

58. Yanarella, *Missile Defense Controversy*, pp. 87–91; Enthoven and Smith, *How Much is Enough*, pp. 185–86.

59. Yanarella, *Missile Defense Controversy*, p. 91; Enthoven and Smith, *How Much is Enough*, p. 186. See also OTA, "Ballistic Missile Defense Technologies," p. 45; Schneider, "Missile Defense Systems," pp. 5–6; and Kaplan, *Wizards of Armageddon*, p. 345.

60. Yanarella, *Missile Defense Controversy*, pp. 92, 104–6; Enthoven and Smith, *How Much is Enough*, p. 187.

61. Yanarella, *Missile Defense Controversy*, pp. 111–13; Enthoven and Smith, *How Much is Enough*, pp. 187–89.

By 1965, McNamara had apparently reached the conclusion that a comprehensive ABM system directed against the Soviet Union should not be deployed. He was still concerned over the cost and effectiveness of such a system. Furthermore, he remained convinced that a major civil defense program was an essential adjunct to any BMD system, but he had come to realize that Congress was no more inclined to support the large-scale construction of fallout shelters at that time than it had been during the previous fifteen years.[62]

McNamara's major concern was his assessment of the likely Soviet response to American deployment of an ABM system. The Soviet Union, he reasoned, would not passively allow the United States to invalidate its retaliatory force by building defenses. To do so would leave it unable to deter an American first-strike attack. Yet, each of the likely Soviet responses would require a countermove by the United States that would simply send the arms race spiraling in costly new directions, with no real increase in security for either side. For example, the Soviet deployment of increasingly sophisticated penetration aids on their ICBMs would require the continual upgrading of the American ABM system. Or, it could attempt to flood the American system by building up its stockpile of strategic warheads, which would likewise require an expansion of the capabilities of that system. The Soviets would also surely deploy a comprehensive ABM system of their own, thus requiring the constant improvement of American offensive forces. The likely end result, in McNamara's view, was that each side would spend a great deal of money yet remain in roughly the same positions they occupied at the outset. With higher budgetary priorities, including a costly war in Vietnam, BMD did not strike the secretary as a good investment at that time.[63]

Despite these views, McNamara did not openly oppose the eventual deployment of a comprehensive ABM system. Instead, he simply attempted to defer a decision on the issue from year to year, citing the need for further research and development. This worked for a while, but by the end of 1966 events were moving against him. In October 1964, the People's Republic of China had detonated its first atomic bomb, bringing new calls for the implementation of defensive measures. The following year saw successful flight tests of the Sprint and Spartan missiles. These tests indicated that production could begin in the near future and brought growing pressure for a decision. Although McNamara had earlier expressed a willingness to consider the possibility of deploying a "thin" system which might serve as a counter to the minimal Chinese threat, he remained firmly opposed to a

62. Yanarella, *Missile Defense Controversy*, p. 112.
63. Enthoven and Smith, *How Much is Enough*, pp. 187–90.

larger system directed at the Soviets. But, by now, his appeared to be a losing cause.[64]

In early 1966, the Joint Chiefs of Staff (JCS), who had accepted McNamara's postponements up to that point, broke with the secretary and recommended an initial deployment of the Nike-X system as soon as possible. Specifically, they called for sufficient Spartan missiles to provide area defense for the entire United States and the deployment of Sprint missiles for point defense around twenty-five cities. The Sprint component would later be expanded to protect fifty-two cities. This action by the JCS encouraged congressional supporters of the ABM, who were successful in passing an appropriation of $167.8 million to begin production and deployment of the Nike-X system. The administration had fought the proposal and declined to spend the money, but a showdown was clearly in the offing. This became even more evident in November, when the Department of Defense announced that the Soviets were indeed in the process of deploying an ABM system of their own—the Galosh—around Moscow.[65]

On December 6, 1966, President Lyndon Johnson convened a meeting at Austin, Texas, to review the proposed defense budget for FY 1968. Among those present were McNamara, Deputy Secretary of Defense Cyrus Vance, the Joint Chiefs of Staff, and National Security Adviser Walt Rostow. When the discussion turned to the subject of ballistic missile defense, the Joint Chiefs were unanimous in their recommendation that the budget include funds for the production and deployment of the Nike-X. Their previous determination was strengthened by confirmation of the Soviet action. Rostow supported the Joint Chiefs, while Vance joined McNamara in his continued opposition. After stating the reasons for his position—and arguing that an expansion of American offensive forces, not deployment of ineffective defenses, was the proper response to the Soviet ABM—McNamara offered a compromise. He suggested that the administration request a small appropriation for ABM procurement, but state that none of the money would be spent and no decision would be made on deployment "until after we make every possible effort to negotiate an agreement with the Soviets" which would prohibit ballistic missile defenses and place limits on offensive forces.[66]

The president, who shared many of his secretary's doubts about the ABM, approved McNamara's proposal and shortly afterward instructed the Department of State to initiate negotiations with the Soviet Union. This

---

64. Yanarella, *Missile Defense Controversy*, pp. 114, 123.
65. Ibid., pp. 117–18; OTA, "Ballistic Missile Defense Technologies," p. 46.
66. Robert S. McNamara, *Blundering into Disaster: Surviving the First Century of the Nuclear Age* (Pantheon, 1986), pp. 55–56; Kaplan, *Wizards of Armageddon*, p. 346; and OTA, "Ballistic Missile Defense Technologies," p. 46.

policy was made public the next month in the State of the Union address.[67] Two weeks later, on January 23, 1967, McNamara sought to strengthen his position by arranging a special meeting at the White House between the president and every past and present director of Defense Research and Engineering and President's Science Adviser. The Joint Chiefs of Staff were also present but could not have been pleased when the scientists unanimously stated that an ABM system would not stop a Soviet attack.[68]

Despite McNamara's efforts, he was not able to delay a decision on deployment much longer. The Soviets showed no interest in negotiations on an ABM ban. In June, the subject was finally raised during the summit meeting between Johnson and Premier Aleksei Kosygin at Glassboro, New Jersey. Here, the secretary tried without success to explain that the deployment of defenses would only bring an increase in offensive weapons. Kosygin rejected McNamara's arguments and indicated that the Soviets would accept no limitation on the means to defend their homeland. He indicated that they would perhaps consider discussions on offensive arms, but not on ABMs.[69]

In effect, the failure of the attempt at negotiations ended the debate. The president was now committed to some sort of deployment, and congressional pressure to get on with it mounted over the summer. Indeed, some Republicans were already talking about an "ABM gap."[70] McNamara recognized his defeat, but still hoped—with the president's support—to shape the decision in favor of a limited deployment. In the words of one aide, the secretary's choice "was not a small ABM versus none at all, but rather a small ABM versus a big one." The rationale for the deployment of such a "thin" ABM system would be to provide defense against an attack by China, or other minor nuclear powers, and to protect against an accidental launch by the Soviets. Before his opposition to the ABM, the China argument had once appealed to McNamara, but by now he did not take it seriously. Nonetheless, it offered the best case for a limited deployment that might forestall something worse. As the secretary commented to his aide, Paul Warnke, "What else am I going to blame it on?"[71]

Following a final effort by Secretary of State Dean Rusk to interest the Soviets in negotiations on defensive systems, McNamara made public his decision.[72] The speech he delivered on September 18, 1967, before a conference of United Press International editors and publishers in San Francisco,

---

67. Yanarella, *Missile Defense Controversy*, p. 124.
68. Ibid., p. 124; Kaplan, *Wizards of Armageddon*, p. 346.
69. McNamara, *Blundering into Disaster*, p. 57; Kaplan, *Wizards of Armageddon*, p. 346.
70. Yanarella, *Missile Defense Controversy*, p. 125–26.
71. Ibid., p. 131; Kaplan, *Wizards of Armageddon*, p. 347.
72. Yanarella, *Missile Defense Controversy*, p. 138.

was quite remarkable. Speaking at some length, McNamara forcefully stated his case against the deployment of a comprehensive ABM system: it would be enormously expensive to build yet easily countered by an expansion of Soviet offensive forces to which the US, in turn, would be forced to respond. The only result, he continued, would be a dangerous acceleration of the arms race with no increase in security for anyone. This said, the secretary concluded by announcing that the United States would proceed with the deployment of a "thin" ABM system to provide defense against the threat of nuclear attack by China.[73]

The decision to deploy a limited ABM system of fifteen sites—using Nike-X components, but renamed Sentinel—did not end the debate. While McNamara hoped that the decision would vent some of the pressure for a comprehensive system, many congressional supporters of a "thick" deployment viewed this as the first step toward their goal. Somewhat ironically, the Army seemed satisfied with Sentinel. For the time being, at least, Vietnam provided it with the major mission it had so long sought, and its emphasis on full-scale defenses declined noticeably over the following months.[74]

Few critics were willing to accept even a "small ABM," but their hopes of preventing it initially appeared slight. Over the next year, however, there was a surprising growth in public opposition to the Sentinel program. Supporters of the ABM expected that any opposition would come from those cities not covered by proposed Sentinel deployments. To their dismay, it came instead from people in those cities that would be defended who objected to the presence of nuclear-armed interceptors in their neighborhoods.[75] Such sentiments were surely stimulated by the growing distrust of government that came with the stalemate in Vietnam, and many antiwar groups adopted the issue as a new cause.

Beginning in May 1968, the debate intensified as a number of scientists, many with long and distinguished careers in government service, made public their technical critiques of the proposed system. These opponents soon found allies in Congress, where a significant shift in opinion on the ABM was underway. Led by Senators John Sherman Cooper (R-Ky.) and

---

73. The speech is published in Robert S. McNamara, *The Essence of Security: Reflections in Office* (Harper and Row, 1968). The portion of the speech arguing against a comprehensive ABM is found in Chapter 4. The conclusion, which announces the deployment of a "thin" ABM system is found in Appendix I. See also Yanarella, *Missile Defense Controversy*, pp. 120–23; Kaplan, *Wizards of Armageddon*, p. 347–48; OTA, "Ballistic Missile Defense Technologies," p. 48; and Schwartz, "Past and Present," p. 339.

74. Yanarella, *Missile Defense Controversy*, p. 132.

75. OTA, "Ballistic Missile Defense Technologies," p. 48; Yanarella, *Missile Defense Controversy*, pp. 146–47; and Schwartz, "Past and Present," pp. 340–42.

Philip Hart (D-Mich.), a bipartisan coalition made several attempts to delay or defeat the Sentinel program. Congressional opponents also sponsored a series of hearings that would subject the ABM to a degree of public scrutiny unprecedented for a major weapons system.[76]

Meanwhile, the administration continued to seek negotiations with the Soviets that would lead to an agreement on both offensive and defensive arms. A favorable response finally came in late June 1968, and on July 1 President Johnson announced that the Soviets had agreed to initiate what would become known as the Strategic Arms Limitation Talks (SALT). Before the negotiations could begin, however, the Soviet invasion of Czechoslovakia compelled the United States to withdraw. Work on the Sentinel system continued through the end of the Johnson administration.[77]

Shortly after taking office in January 1969, President Richard Nixon directed an overall review of United States military posture, including an assessment of the still controversial ABM. Pending the outcome of the study, Defense Secretary Melvin Laird ordered, on February 6, a suspension of Sentinel deployment.[78] Two weeks later, a preliminary report on the strategic review offered four alternative approaches to the ABM question. The first three would continue to make use of the hardware developed for Sentinel. These included a "thick" system designed to protect the twenty-five largest cities in the nation; continuation of the Sentinel program to provide "thin" defense for some fifteen cities; and a variation on Sentinel, known as "Plan I-69," which would defend twelve ICBM bases and thus protect the American deterrent force from attack. The fourth alternative was to cancel the ABM program altogether.[79]

On March 5, the completed review strongly supported the I-69 option, and within four days the president had accepted this recommendation. Over the next week, it was decided that this new system would be deployed in phases, so that the costs of the program might be spread out. Work would begin on sites at two Air Force bases—Grand Forks in North Dakota and Malmstrom in Montana—in FY 1970. The timing of the other ten sites would be determined pending annual reviews of the program. On March 14, President Nixon announced his decision, naming the new program Safeguard.[80]

---

76. Kaplan, *Wizards of Armageddon*, pp. 349–50; OTA, "Ballistic Missile Defense Technologies," p. 48.

77. Ibid., p. 48; Yanarella, *Missile Defense Controversy*, p. 151.

78. Ibid., p. 144.

79. Ibid., pp. 170–71.

80. Ibid., pp. 173–74; OTA, "Ballistic Missile Defense Technologies," p. 48; Kaplan, *Wizards of Armageddon*, p. 350; and Schneider, "Missile Defense Systems," p. 9.

Nixon's action had two major implications for the future of the ABM. First, it made clear that there would be no rush to deploy a system of ballistic missile defenses. Second, the Safeguard decision signalled a clear shift in the emphasis of the ABM program. Unlike its predecessors, Safeguard was designed to provide only minimal defense of the American population, though there was still some talk of protection against Chinese or accidental Soviet launches. The real objective of Safeguard was to increase the security of the American strategic nuclear force by defending the nation's ICBMs—and some bombers based nearby—against any Soviet attempt at a disarming first strike. This, it was believed, would enhance the credibility of the American deterrent by giving added assurance that the Soviets would face certain retaliation for any attack on the United States.[81]

Safeguard's task was surely more managable than that of earlier systems. Leaving the bulk of the American population vulnerable to Soviet retaliation for an American attack, it was also less destabilizing. Many, however, were struck with the apparent irony of the program's attempt to defend missiles, instead of people, from nuclear attack.

In his announcement, Nixon explained his decision in terms reminiscent of McNamara:

> Although every instinct motivates me to provide the American people with complete protection against a major nuclear attack, it is not now within our power to do so. The heaviest defense system we considered, one designed to protect our major cities, still could not prevent a catastrophic level of US fatalities from a deliberate all-out Soviet attack. And it might look to an opponent like the prelude to an offensive strategy threatening the Soviet deterrent.[82]

The Safeguard decision did not end the domestic controversy over the ABM. Supporters of a comprehensive system were disappointed at this new limitation of the program, and critics still sought its termination. Congressional and public opposition to the ABM continued to mount. In August 1969, senators led by Cooper and Hart came within one vote of killing funds for Safeguard. Only the tie-breaking vote of the vice-president saved the program from a serious defeat. The following year's appropriation for continued deployment at the initial two Safeguard sites passed by a margin of five votes. In 1971 funding for these sites was approved by a wide margin, but there were indications that appropriations for the remaining ten sites would be more and more difficult to obtain, especially as cost estimates for Safeguard rose rapidly.[83]

---

81. Kaplan, *Wizards of Armageddon*, p. 350; Enthoven and Smith, *How Much is Enough*, p. 191.

82. OTA, "Ballistic Missile Defense Technologies," p. 48.

The president fought hard for his program, but he did so increasingly for reasons unrelated to its stated military value. Shortly after Nixon took office, the Soviets expressed interest in beginning the SALT negotiations that had been preempted by the invasion of Czechoslovakia. While receptive to the idea, Nixon deferred until his administration could prepare its negotiating position and begin the implementation of its strategic programs. In late January, he established a subcommittee of the National Security Council, known as the Verification Panel, to investigate the feasibility and implications of an arms control agreement.

The discussions that followed convinced Nixon that the sort of ABM system then available was largely irrelevant to American security. That security would rest, as it had since the 1940s, on the strength of the nation's nuclear retaliatory forces. In this view, an ABM might have marginal value in the face of a continued Soviet build-up of nuclear weapons, but it was also something that might be more profitably traded for an agreement limiting both defensive and offensive systems. Thus, Nixon came to view Safeguard primarily as an important bargaining chip in future arms control negotiations and fought for the program with this purpose in mind. If an agreement could not be achieved, the system could be either maintained or cancelled, as circumstances warranted.[84]

By the autumn of 1969, the administration was prepared to begin negotiations, and Nixon extended an invitation to the Soviet leadership. The Soviets, who by now had also come to the conclusion that antiballistic missiles were not promising, accepted on October 25. The Strategic Arms Limitation Talks began in Helsinki three weeks later.[85]

For two and a half years, the two sides struggled to reach a comprehensive arms control agreement that would address both offensive and defensive arms. Among the more difficult tasks they faced were finding the means to compare the strategic weapons in their respective arsenals equitably and to establish an agreed data base over which to negotiate. There was also considerable controversy over precisely what weapons should be considered as strategic. The United States placed only those weapons of intercontinental range in this category, while the Soviet view included any nuclear weapon that could fall within its borders, thus raising the question of American

---

83. Yanarella, *Missile Defense Controversy,* pp. 149–66. The debate is illustrated by such contemporary collections as Holst and Schneider, *Why ABM?;* Abram Chayes and Jerome Wiesner, eds., *ABM: An Evaluation of the Decision to Deploy an Anti-Ballistic Missile System* (Signet, 1969); and William R. Kintner, ed., *Safeguard: Why the ABM Makes Sense* (Hawthorn Books, 1969).

84. Yanarella, *Missile Defense Controversy,* pp. 178–79, 183–84.

85. Ibid., pp. 183–84. On the SALT I negotiations, see John Newhouse, *Cold Dawn: The Story of SALT* (Holt, Rhinehart and Winston, 1973); and Gerard Smith, *Doubletalk: The Story of the First Strategic Arms Limitation Talks* (Doubleday, 1980).

nuclear forces in Europe. The Soviets finally relented, but the talks nearly deadlocked on this point. Throughout the negotiations, defensive weapons caused far fewer problems. Since both sides had already questioned their feasibility, an agreement limiting ABM deployment was achieved with much less difficulty than one covering offensive arms.

On May 26, 1972, the SALT I Agreement was signed by President Nixon and Soviet General Secretary Leonid Brezhnev at a ceremony in Moscow. SALT I consisted of two parts: the Interim Agreement on Strategic Offensive Arms and the Treaty on the Limitation of Anti-Ballistic Missile Systems, more commonly known as the ABM Treaty.[86]

The Interim Agreement froze ICBM deployments at their current levels of 1,054 for the United States and 1,618 for the Soviet Union. The agreement also limited the number of submarine-launched ballistic missile (SLBM) deployments to 656 American and 740 Soviet weapons. The desparity in force levels was explained by the technical superiority of the American systems and by the exemption of strategic bombers—in which the United States had a substantial lead—and of those American nuclear forces in Europe which were capable of striking the Soviet Union. Furthermore, the American stockpile of nuclear warheads was three times the size of the Soviet arsenal. The Interim Agreement was to last for five years, during which it was understood that negotiations for more stringent limitations would proceed.

The ABM Treaty was far more comprehensive. Under its provisions, the United States and the Soviet Union pledged not to deploy antiballistic missile systems for the defense of their national territory or any portion thereof. The treaty contained two strictly limited exceptions to this otherwise total ban. One allowed each side to deploy an ABM system of up to one hundred launchers—with one missile each—and six radar complexes within a radius of 150 kilometers around its national capital. The other exception permitted each nation a similar deployment of launchers and missiles, though with more radars (including two large phased-array radars), in defense of an area in which ICBM silos were located. The two deployment sites were to be separated by at least 1,300 kilometers, and all existing systems beyond these limits were to be destroyed.

To support the ban on ABM systems, the treaty included a number of supplementary prohibitions. The parties agreed to forego the development of ABM systems that were based in the air, at sea, or in space, as well as land-based systems which were mobile or which could be rapidly reloaded. Furthermore, the two sides agreed that they would neither transfer ABM systems to other nations nor interfere with each others' means of verifying

86. The texts of the SALT I Interim Offensive Agreement and the ABM Treaty are found in US Arms Control and Disarmament Agency (ACDA), *Arms Control Agreements: Texts and Histories of Negotiations* (GPO, 1982.)

compliance with the treaty. The ABM Treaty also created a Standing Consultative Commission to resolve any issues relating to observance of the agreement that might arise in the future, including the development of ABM systems based upon "other physical principles" which might be "capable of substituting for ABM interceptor missiles."

On August 3, 1972, the United States Senate ratified the ABM Treaty by a vote of eighty-eight to two. Approval of the Interim Agreement followed on September 25 by the same margin. The SALT I accords entered into force with the formal exchange of ratifications on October 3, 1972. Seven weeks later, negotiations for a SALT II treaty that would bring further limitations on offensive weapons began in Geneva.[87] Such an agreement took on increasing importance as it became clear that one of the major results of the search for effective defenses was the expansion of nuclear stockpiles on both sides, which McNamara had predicted. The development of multiple-warhead technology, first by the United States, brought substantial increases in the number of deployed weapons despite the SALT I limitations on launchers.

With the deployment of anything approaching an effective defensive system prohibited by the ABM Treaty, the maintenance of those sites that were permitted increasingly seemed to be an expensive and pointless luxury. Even if they worked, the two hundred interceptors allowed to each side could be easily overwhelmed by the growing number of warheads available to an attacker. Therefore, on July 3, 1974, the two nations agreed to a further limitation on defenses. In a protocol to the ABM Treaty, nominally designed to encourage progress on SALT II, Nixon and Brezhnev cut the number of allowed deployments to a single system of one hundred missiles on each side. The United States chose to continue its Safeguard site at Grand Forks, while the Soviets elected to maintain their Galosh system around Moscow. The protocol was Nixon's last diplomatic accomplishment. Five weeks later, he resigned in the wake of the Watergate scandal.[88]

Neither side, however, completed even this minimal deployment. The Soviets built only sixty-four launchers and later reduced this number to thirty-two (although they have added to this number in recent years). For the United States, the story of the ABM ended when Congress voted in 1975 to cancel funds for the Safeguard program.[89] The Grand Forks site was deactivated the following year.[90] For the time being, the strategic defense of

87. See "SALT Chronology," *Countdown on SALT II* (The Arms Control Association, 1985), p. 32.

88. OTA, "Ballistic Missile Defense Technologies," p. 50. The text of the protocol, which entered into force on May 24, 1976, is found in ibid., pp. 281–82; and ACDA, *Arms Control Agreements*, pp. 161–63.

89. Gervasi, *Arsenal of Democracy II*, p. 30.

90. OTA, "Ballistic Missile Defense Technologies," p. 50.

the United States against attack by Soviet missiles would rely exclusively on deterrence through the threat of certain retaliation.

The Strategic Defense Initiative

The conclusion of the ABM Treaty did not mean that either party had abandoned interest in ballistic missile defense.[91] In the years following the agreement, both the United States and the Soviet Union maintained active, though modest, research programs. The purpose of such research was two-fold. On the one hand, it allowed each side to continue the development of BMD-related technology, particularly in tracking and acquisition, up to the limits imposed by the treaty. More importantly, this research had the purpose of informing decision-makers on each side of what was possible under the current state of the art so that they would not be surprised by any sudden technical advances by the adversary. This, it was hoped, would deny the other side any chance of achieving the ability to "break-out" of the limits imposed by the ABM Treaty by rapidly deploying a defensive system for which there would be no ready counter.

In the United States, concern over a potential Soviet break-out drove most BMD-related research in the years following the conclusion of the agreement. While early research into such "exotic" technologies as directed energy indicated that there might be some long-term promise in the defensive application of weapons based on lasers and particle beams, the main emphasis through most of this period was on more conventional systems that could be rapidly deployed in response to a Soviet abandonment of the ABM Treaty. Much of this was associated with the search for a means of basing and protecting the MX missile, the next generation of American ICBM, and with concern over a potential "window of vulnerability" brought on by the expansion of the Soviet strategic arsenal.[92]

---

91. The relationship between the ABM Treaty and continuing BMD research is discussed in George Schneiter, "The ABM Treaty Today," in Carter and Schwartz, *Ballistic Missile Defense,* pp. 221–50; Sidney D. Drell, Philip J. Farley, and David Holloway, *The Reagan Strategic Defense Initiative: A Technical, Political, and Arms Control Assessment* (International Strategic Institute at Stanford, 1985), pp. 7–38. See also OTA, "Ballistic Missile Defense Technologies," pp. 51–55; Albert Carnesale, "Special Supplement: The Strategic Defense Initiative," in George Hudson and Joseph Kruzel, *American Defense Annual, 1985–1986* (Lexington Books, 1985), pp. 198–202. See also Alan B. Sherr, "Sound Legal Reasoning or Policy Expedient? The 'New Interpretation' of the ABM Treaty," *International Security* 11(Winter 1986–87): 71–93.

92. OTA, "Ballistic Missile Defense Technologies," p. 57. See also Jonathan B. Stein, *From H-Bomb to Star Wars: The Politics of Strategic Decision Making* (Lexington Books, 1984), pp. 54–56; Donald M. Snow, "Ballistic Missile Defense: The Strategic Defense Initiative," in Stephen J. Cimbala, *The Reagan Defense Program: An Interim Assessment* (Scholarly Resources, 1986), p. 147.

During the late 1970s and early 1980s, the Army Ballistic Missile Defense Program Office worked on the development of components for its Low Altitude Defense System (LoADS). Conceived as a successor to the Sprint portion of the Safeguard system, LoADS would employ five hundred high-acceleration nuclear-armed Sentry missiles and small mobile phased-array radars to provide terminal defense to hardened ICBM silos. Not all American ICBMs would be protected by this system; the objective, rather, was to insure that a significant portion of the nation's retaliatory forces would survive a Soviet first-strike. In addition, the Army conducted research on a nonnuclear endoatmospheric interceptor missile to be incorporated into the LoADS.[93]

Despite considerable interest within the Army, the LoADS program was cancelled in February 1983 as a result of a shift in emphasis toward more advanced BMD technologies. The development of components for the Sentry missile was continued at a slower pace for possible future deployment, perhaps utilizing the old Sprint warheads, but more likely using a new enhanced-radiation warhead. Such systems are not given a high priority under the Strategic Defense Initiative and are of significance largely because they could be deployed as a relatively rapid response to an attempted Soviet break-out. Throughout this period, the Army also conducted research on an exoatmospheric kinetic interceptor of the type successfully tested against a target warhead in the Homing Overlay Experiment of 1984 as part of a program still underway.[94]

While the initial emphasis of the BMD research in the years following the ABM Treaty was placed on conventional systems involving missile interceptors, work continued on the development of more advanced weapons based on directed energy, particularly lasers and particle beams. In itself, this was nothing new. Military interest in the defensive applications of lasers dated back to the early 1960s. A decade later, a number of research projects, coordinated through the Defense Department's High Energy Laser Group, were showing promise. In 1973 the Air Force conducted the first successful test of a laser weapon against a drone aircraft. Five years later a laser weapon destroyed in quick succession three TOW antitank missiles traveling at five hundred miles per hour. Research begun during the 1970s continues in programs operated by the various services and by DARPA. Among these are DARPA's "Talon Gold" project, which will employ the space shuttle for experiments in the acquisition and tracking of objects in space; "Alpha," a program to build hydrogen-flouride chemical lasers;

93. OTA, "Ballistic Missile Defense Technologies," pp. 57–58; Thomas B. Cochran, William M. Arkin, and Milton Hoenig, *Nuclear Weapons Databook*, vol. 1, *U.S. Nuclear Forces and Capabilities* (Ballinger, 1984), pp. 163–67.

94. OTA, "Ballistic Missile Defense Technologies,"; Cochran, Arkin, and Hoenig, *Nuclear Weapons Databook*, p. 165.

"Lamp," which is exploring the feasibility of aiming lasers with orbiting mirrors; and "Dauphin," a program to develop X-ray laser weapons being conducted at the Lawrence Livermore National Laboratory. The Air Force, meanwhile, has been investigating the possibility of employing lasers for defense against SAMs and air-to-air missiles with its Airborne Laser Laboratory (ALL).[95]

Similarly, the Pentagon has had a long interest in the potential weapons applications of particle beams. In 1958 DARPA initiated a program named "Seesaw," which explored the possibility of employing such beams of atomic and subatomic particles in ballistic missile defense. Seesaw was terminated in 1972 when it was concluded that such weapons would be costly and very difficult to engineer. Two years later, however, in response to reported Soviet efforts in this area, the United States resumed research on particle beams. The Navy initiated a project titled "Chair Heritage" to investigate the feasibility of using electron beams for shipboard air defense. The Army followed with a program called "Sipapu," which again looked at the applicability of particle beams for ground-based BMD. The Air Force later began its own program of basic research.[96]

Despite proponents' enthusiasm for these various technologies, policymakers showed little interest in ballistic missile defense in the decade following the signing of the ABM Treaty. Funding for research remained at a relatively low level, and few officials expressed any confidence that an effective BMD system could be developed in the foreseeable future. The prevailing view in the Departments of State and Defense—and among outside experts—was that deterrence through Mutual Assured Destruction (MAD) was the only way to maintain a stable balance between the superpowers. A move to a strategy based on ballistic missile defense, it was believed, would be destabilizing and would only serve to spread the arms race into an area that had been closed off by the treaty.[97] Furthermore, abrogation of the ABM Treaty, which would be required by the deployment of any new BMD system, would surely end all hope of further arms control agreements for many years to come. Even in the early years of the Reagan administration, such views were not uncommon within the government. As late as February 1983, senior Defense Department officials stated that effective BMD systems would not be feasible in the near future and expressed little interest in them beyond a continuation of basic research to prevent a Soviet break-out.[98]

---

95. James Canan, *War in Space* (Harper and Row, 1982), pp. 149–51; Wolfgang K. H. Panofsky, "The Strategic Defense Initiative: Perception vs. Reality," *Physics Today*, June 1985, p. 26; William J. Broad, "Reagan's 'Star Wars' Bid: Many Ideas Converging," *New York Times*, March 4, 1985, p. A-8.

96. Canan, *War in Space*, pp. 156–58.

97. Stein, *From H-Bomb to Star Wars*, , pp. 54–56.

98. Snow, "Ballistic Missile Defense," pp. 147–48.

In this context, the shift in American defense policy that came with President Reagan's "Star Wars" speech of March 23, 1983, is quite remarkable. Calling on the nation's scientific community to find the means to deliver the United States from the threat of nuclear attack, Reagan's Strategic Defense Initiative made BMD-related research, particularly in space-based systems, one of the administration's highest priorities.

The shift in policy did not come overnight and was foreshadowed by developments during the Carter administration. Concerned with expanding Soviet research in antisatellite (ASAT) weaponry during the previous year, President Jimmy Carter signed Presidential Decision Memorandum 37 on May 13, 1978. This document committed the United States to "activities in space in support of its right to self-defense, thereby strengthening national security, the deterrence of attack, and arms control agreements." Carter's principal concern was with the protection of American satellites, but the policy embodied in PDM-37 marked a significant shift from his previous determination to keep weapons out of space.[99] Funding for research in technologies of relevance to space weaponry was increased somewhat. Also, the Directed Energy Transfer Office was established in the Department of Defense to oversee projects involving lasers and particle beams.[100] In August 1980, Carter's adoption of a counterforce nuclear targeting strategy in Presidential Directive 59 again raised questions concerning the desirability of providing terminal defenses for American ICBMs.[101] In the political sphere, the Soviet invasion of Afghanistan and the subsequent withdrawal of the unratified SALT II agreement from Senate consideration marked a downward shift in Soviet-American relations.[102] These events, along with the frustrations associated with the ongoing hostage situation in Iran, brought increases in defense expenditures and a growing national desire for a reassertion of American power. Despite these developments, there was little thought during the Carter administration of a major program in strategic defense, and the president remained committed to the ABM Treaty.

In 1981 Ronald Reagan came to the presidency committed to a major expansion of American military capabilities—nuclear and conventional—in

---

99. Canan, *War in Space*, p. 23. For an excellent discussion of US military space policy, see Paul B. Stares, *The Militarization of Space: U.S. Policy, 1945*–1984 (Cornell University Press, 1985).

100. Snow, "Ballistic Missile Defense," p. 147.

101. Stein, *From H-Bomb to Star Wars*, pp. 66–67; Snow, "Ballistic Missile Defense," p. 145.

102. On the decline of the détente as it relates to arms control and BMD, see OTA, "Ballistic Missile Defense Technologies," pp. 61–62. The best discussion of arms control policy in the Reagan administration is Strobe Talbott, *Deadly Gambits: The Reagan Administration and the Stalemate in Nuclear Arms Control* (Alfred A. Knopf, 1984).

the hope that clear superiority over the Soviets could be reestablished.[103] He also came with a long interest in the possibility of employing American technology to provide a shield against Soviet nuclear weapons. Reagan had little background, or interest, in the details of nuclear strategy, but he believed instinctively that deterrence based on Mutual Assured Destruction was immoral and that a workable defense could be achieved.[104]

This belief was strengthened over the years through contacts with a number of prominent supporters of ballistic missile defense. In 1967, as the newly elected governor of California, Reagan toured the Lawrence Livermore National Laboratory. While there, he had an extended meeting with Edward Teller, the laboratory's founder and a long-time proponent of BMD. Reagan came away clearly impressed with what Teller had told him about the potential for using American technological superiority to provide a defense against nuclear attack.[105]

Reagan's interest in moving away from MAD would continue over the years. During his run for the presidency in 1976, he frequently described the nuclear balance in terms of two men pointing cocked pistols at each other and stated that "there's got to be a better way." These sentiments were shared by his campaign's defense adviser, Lieutenant General Daniel O. Graham, who had recently retired as the director of the Defense Intelligence Agency, but there were no specific proposals put forth at that time. Graham returned as defense adviser in the 1980 campaign.[106] In late 1979 Reagan's interest in BMD increased when he toured the North American Air (now Aerospace) Defense Command facility at Cheyenne Mountain, Wyoming, which monitors objects in space and would coordinate early warning of a Soviet attack. Reagan was impressed with the system's capabilities, but he was clearly disturbed by the inability of the United States to do anything about an attack once it was detected, save to respond in kind.[107] As he later stated, "I think the thing that struck me was the irony that here, with this great technology of ours, we can do all of this, yet we cannot stop any of the weapons that are coming at us."[108] Again, there were no specific proposals,

---

103. Broad, "Reagan's 'Star Wars' Bid" p. A-8; Stein, *From H-Bomb to Star Wars,* p. 82. For an excellent discussion of the Reagan administration's search for military superiority, see Tom Gervasi, *The Myth of Soviet Military Supremacy* (Harper and Row, 1986).

104. Broad, "Reagan's 'Star Wars' Bid," p. A-1.

105. Ibid., pp. A-1, A-8; Nova/Frontline Special Report, "Visions of Star Wars," (WGBH Educational Foundation, 1986), transcript of broadcast on April 22, 1986, p. 3. Hereafter cited as "Nova."

106. Ibid., pp. 4–5.

107. Broad, "Reagan's 'Star Wars' Bid," p. A-1; Nova, "Visions of Star Wars," p. 3.

108. Robert Scheer, *With Enough Shovels: Reagan, Bush and Nuclear War* (Vintage Books, 1982), p. 104. Scheer is an excellent source for early Reagan administration thinking on strategic issues.

but Reagan's continuing interest in the subject helped to inspire Graham's Project High Frontier, a private study of defensive technologies completed in 1982 under the auspices of the Heritage Foundation.

Meanwhile, there was growing interest within the Republican party in the new defensive technologies. In 1979 Senator Malcolm Wallop (R-Wyo.) published an article in *Strategic Review* in which he criticized the Pentagon's emphasis on the new MX missile and called for a system of space-based laser battle stations to protect the United States against Soviet missiles and bombers. Wallop's claim that such a system could be developed as reasonable cost and with technology already in existence met strong opposition in the Defense Department, but his lobbying efforts paid off when an increase in laser funding was approved by the Senate in 1981.[109] Interest in strategic defense was also reflected in the 1980 Republican platform, adopted on July 15, 1980. The platform called for "vigorous research and development of an effective anti-ballistic missile system, such as is already in hand in the Soviet Union, as well as more modern ABM technologies." It further stated that the objective of American policy should be "overall military and technological superiority over the Soviet Union."[110]

The Republican platform clearly reflected the views of candidate Reagan. Despite this, the new administration did not rush headlong into strategic defense upon taking office. President-elect Reagan discussed the subject in several meetings following the election, including one with Senator Harrison Schmitt (R-NM) which dealt largely with the potential of space-based lasers.[111] Still, there were to be no major departures in policy during the first two years of his presidency. Even in the Defense Department, BMD was viewed as a concern for the future, although the president's call for the modernization of strategic nuclear forces in October 1981 did increase interest in the subject.[112] Nonetheless, Secretary of Defense Caspar Weinberger's first *Annual Report*, for FY 1983, stated: "For the future, we are not sure how well ballistic missile defenses will work; what they will cost; whether they would require changes in the ABM Treaty; and how additional Soviet ballistic missile defenses—which would almost certainly be deployed in response to any U.S. BMD system—would affect U.S. and allied offensive capabilities."[113]

---

109. Stein, *From H-Bomb to Star Wars*, p. 58; Angelo M. Codevilla, "Strategic Defenses: Technical Success and Failure of Policy," paper presented to a conference of the American Political Science Association, August 31, 1985, pp. 61–75.

110. Broad, "Reagan's 'Star Wars' Bid," p. A-8; Stein, *From H-Bomb to Star Wars*, p. 82.

111. Broad, "Reagan's 'Star Wars' Bid," p. A-8.

112. Snow, "Ballistic Missile Defense," p. 145.

113. Caspar W. Weinberger, *Annual Report to Congress for FY 1983* (GPO, 1982), p. III-65.

During the first year of the Reagan presidency, momentum in favor of an active program to create an effective BMD system began to build. In May 1981 Dr. George A. Keyworth II, an associate of Teller's and a strong supporter of directed energy weapons, was appointed as the President's Science Adviser. Later in the year, the Heritage Foundation, a conservative think-tank based in Washington, began sponsoring a series of meetings to develop plans for the establishment of a defensive system to protect the United States from nuclear attack. Among the participants in the study group were Edward Teller, his Lawrence Livermore associate Lowell Wood, and several members of Reagan's "kitchen cabinet," including businessmen Joseph Coors and Justin Dart. The group was chaired by Karl R. Bendetsen, a former under-secretary of the Army. Its vice-chairman was Lieutenant General Graham, who helped to initiate the meetings and arranged for support from the administration in the form of access to classified information.[114]

By the end of 1981, the group—which agreed on the urgent need for a program of strategic defense—had split over the best means of achieving that objective. Bendetsen, Teller, and the members of the "kitchen cabinet" came to favor a system based on exotic technologies which would require extensive research over a number of years prior to deployment. Among these were the "third generation" nuclear-driven X-ray lasers under development at Livermore. The other faction, led by General Graham, supported a program based largely upon technologies already in existence which could be deployed rapidly.[115]

In the bureaucratic struggle that followed the split, the Bendetsen-Teller group won. In their first meeting with the president, in January 1982, they recommended the adoption of an increased program of advanced research in BMD-related technologies, particularly lasers and particle beams. They were also critical of the Pentagon's apparent lack of serious interest in such weapons. Reagan was receptive to the group's presentation and met with them on two more occasions over the next year.[116]

Meanwhile, Graham's group, operating as Projcct High Frontier, continued to develop its case for a BMD system employing "off the shelf" technologies. In February 1982, after being denied the presidential access enjoyed by his competitors, Graham published his group's findings in a book entitled *High Frontier: A New National Strategy.* Graham argued for the replacement of MAD with a new doctrine of Mutual Assured Survival which would eliminate the Soviet threat to the territory of the United States. The High Frontier proposal held that technologies currently availa-

114. Broad, "Reagan's 'Star Wars' Bid," p. A-8.
115. Ibid.
116. Ibid.

ble could be deployed in a layered defense consisting of some four hundred space-based weapons platforms firing nonnuclear "kill vehicles" and a network of ground-based terminal defenses employing high-velocity interceptor missiles. These, in turn, would be supplemented by increased attention to passive civil defense of the American population. The entire system, it was argued, could be in place by 1990 at a cost of approximately $40 billion.[117]

Following publication of the High Frontier report, Graham and his associates conducted numerous briefings for defense officials and members of Congress. The reception was mixed, at best. Those actively involved in BMD-related programs were generally supportive, since High Frontier would employ systems they were currently developing. The vast majority, however, were skeptical of High Frontier's technical ability to provide an effective defense. Many within the Departments of State and Defense argued that a shift away from MAD would be dangerous and destabilizing. Even those who supported some sort of ballistic missile defense saw the proposed system as vulnerable to a wide variety of Soviet countermeasures. As Teller commented: "It is very do-able. And very easily destroyed. I do not think it is the right way to go. It is much too simple to be effective in the way in which it has been proposed." These views were shared by various senior Defense Department officials, the Joint Chiefs of Staff (with the possible exception of the chairman, General John Vessey), and Science Adviser Keyworth.[118]

While the specific proposals embodied in High Frontier were not convincing, the underlying notion of a shift to a strategy of defense was well received by conservatives, particularly those interested in attaining military superiority over the Soviets. High Frontier received considerable publicity and helped serve as a doctrinal stimulus in the months leading up to the president's launching of the Strategic Defense Initiative.[119]

Other factors, as well, contributed to the growing momentum within the White House in favor of change. Arms control negotiations with the Soviets were going nowhere, and there was growing concern that the Soviets themselves were making significant advances in BMD research. The Soviets were also thought to be continuing the modernization of their offensive forces at a rapid pace. Furthermore, the administration was worried that the popularity of the Nuclear Freeze Movement and the drafting of the Catholic

117. Daniel O. Graham, *High Frontier: A New National Strategy* (Heritage Foundation, 1982). See also Stein, *From H-Bomb to Star Wars*, pp. 57–59; Broad, "Reagan's 'Star Wars' Bid," p. A-8; Nova, "Visions of Star Wars," p. 5; OTA, "Ballistic Missile Defense Technologies," pp. 294–96.

118. Stein, *From H-Bomb to Star Wars*, pp. 58–60; Broad, "Reagan's 'Star Wars' Bid," p. A-8; Nova, "Visions of Star Wars," p. 7.

119. Stein, *From H-Bomb to Star Wars*, pp. 58–59.

Bishops' letter of November 1982, which denounced nuclear weapons, signalled a loss of political will to continue with its own defense build-up.[120]

On September 14, 1982, Edward Teller held a private meeting with President Reagan at the White House. The invitation was issued following a comment made by Teller in a June interview on the PBS television program "Firing Line," in which he complained about his lack of access to the president. In fact, Teller had already met with Reagan at least twice that year as part of the Bendetsen group. Now he desired a less formal opportunity to make a case for his "third-generation" weapons and expanded BMD research in general.[121]

Again, the president expressed interest in the proposal. However, William Clark, Reagan's National Security Adviser, also attended the meeting. Clark was skeptical and played the part of devil's advocate. Furthermore, he was reportedly fearful of the known opposition to BMD within State and Defense. A major shift in policy, he believed, would be difficult to achieve and might provoke paralyzing opposition within the bureaucracy. Within weeks, though, Clark's deputy, Robert McFarlane, began to move in favor of a shift in strategy. Apparently, this was out of concern over the Nuclear Freeze Movement and the lack of progress in arms control negotiations. According to Laurence Barrett of *Time,* "McFarlane reasoned that a defensive research program would attract bipartisan support at home and might someday be useful in Geneva." It is unclear whether McFarlane supported at this point the eventual deployment of a defensive system, or simply saw a research program as a useful bargaining chip. In any case, he discussed his views with the president and was soon able to convince Clark that a shift in policy should be considered. McFarlane also found an ally in Admiral James Watkins, the chief of naval operations, and the two consulted frequently in the closing months of 1982. The result of their meetings was a paper which Watkins presented to the Joint Chiefs of Staff, who in turn approved a briefing for the president. Meanwhile, Reagan held his third meeting with the Bendetsen group as a follow-up to his September discussion with Teller.[122]

The JCS met with the president over lunch on February 11, 1983. Although the agenda called for an initial disscussion of the MX missile, Watkins soon turned the conversation to the subject of defense. According to Barrett:

> Watkins took the opportunity to talk about the growing threat of instability. Then he made his pitch: the advances in defensive technol-

---

120. Lawrence J. Barrett, "How Reagan Became a Believer," *Time,* March 11, 1985, p. 16; Nova, "Visions of Star Wars," p. 7.

121. Broad, "Reagan's 'Star Wars' Bid," p. A-8; Barrett, "How Reagan Became a Believer," p. 16.

122. Ibid.

> ogy were so promising that the president should throw his weight behind a major research effort. McFarlane interjected: Are you saying that over time this could lead to deployable systems? Exactly, Watkins replied. McFarlane then polled the other four military leaders around the table. None dissented.
>
> Reagan promptly seized on Watkins' argument. It validated his conviction that there had to be a way out of the MAD trap and played on his often stated faith in US science and industry. Reagan said that he wanted the ideas pursued promptly.[123]

In effect, McFarlane and his colleagues succeeded in making an end-run around the bureaucratic opponents of strategic defense. Had the normal procedure of developing the proposal through departmental and interdepartmental channels been followed, it surely would have been stalled and perhaps killed outright. Obtaining the full support of the president at the outset at least guaranteed that the proposal would not be stillborn.[124]

Over the next six weeks, the ideas discussed during the meeting on February 11 were fleshed out by a small number of officials in the White House on a highly secret basis. Secretary of State George Shultz and Defense Secretary Caspar Weinberger were not informed of the project until a week before the president's speech inaugurating it. Science Adviser George Keyworth II was told only five days ahead of time, and, by some reports, the Joint Chiefs were not informed until the day before that the proposal would actually be made.[125]

The timing of the president's speech was apparently subject to some debate within the White House. McFarlane, who is believed to have written much of it, hoped to go slowly, but Reagan's political advisers "wanted the president to express a large, fresh idea in his next defense policy speech."[126]

Therefore, on March 23, 1983, in the closing moments of a televised address which began with a call for increased defense spending, President Reagan launched his Strategic Defense Initiative. Speaking in terms of a "vision of the future which offers hope," Reagan proposed to move away from the policy of deterrence through retaliation and to "embark on a program to counter the awesome Soviet missile threat with measures that are defensive." He concluded by calling upon the nation's scientific community, "those who gave us nuclear weapons, to turn their great talents now to the cause of mankind and world peace, to give us the means of rendering

---

123. Broad, "Reagan's 'Star Wars' Bid," p. A-8; Barrett, "How Reagan Became a Believer," p. 16.
124. Ibid.; Nova, "Visions of Star Wars," p. 7.
125. Barrett, "How Reagan Became a Believer," p. 16.
126. Ibid.; Broad, "Reagan's 'Star Wars' Bid," p. A-8.

these nuclear weapons impotent and obsolete."[127]

The speech caught the nation—and most people in government—by surprise. The initial public reaction was generally favorable despite the near instant labeling of the proposal as "Star Wars" (after the film of that name) because of its heavy reliance on space-based systems. Many defense experts outside of government were critical, and most in government were simply caught off guard. Ashton B. Carter, one of the nation's leading experts on ballistic missile defense, recalled: "Having left the Pentagon shortly before the speech was given, my first throught was, 'What is going on here? I never heard anything about this.'"[128] Several days after the speech, Deputy Defense Secretary Paul Thayer remarked, "This is truly a vision, and that's about all it is at this stage of the game."[129] A month later, the final report of the President's Commission on Strategic Forces (the Scowcroft Commission) stated that current technology offered "no real promise of being able to defend the United States against massive nuclear attack in this century."[130] Confusion, it seems, was the price of McFarlane's end run around the bureaucracy, but the program was established.

As time passed, the president's obvious enthusiasm for the program—and the career implications of that enthusiasm for recalcitrant bureaucrats—was enough to rally support among the majority of defense officials in the government. In the public debate, the SDI was one of the most controversial military programs in American history, with a growing number of critics attacking the technical feasibility and strategic wisdom of the president's "vision." The proposal would quickly become a major political issue, as well, and it remains so.

On March 25, the White House began to give substance to the SDI. On that day, the president signed National Security Decision Directive (NSDD) 6-83, "Defense Against Ballistic Missiles," which ordered "the development of an intensive effort to define a long-term research and development program" leading to an effective system of ballistic missile defense. The approach, clearly, would be that of the Bendetsen-Teller group; there was no indication of support for Graham's view that a BMD system could be

---

127. Reagan, "Defense Spending and Defense Technology." Although the president's long interest in BMD is well documented, and he clearly played an important role in launching the new program, he has frequently seemed unaware of the history of the BMD debate. For example, on March 18, 1985, he told *Newsweek:* "It kind of amuses me that everybody is so sure I must have heard about it, that I never thought of it myself. The truth is, I did." For this and similar statements, see The Arms Control Association, *Star Wars Quotes* (The Arms Control Association, 1986), p. 26.

128. Nova, "Visions of Star Wars," p. 7.

129. Stein, *From H-Bomb to Star Wars,* p. 61.

130. President's Commission on Strategic Forces, "Report of the President's Commission on Strategic Forces," (GPO, 1983), pp. 5, 9. See also Stein, *From H-Bomb to Star Wars,* p. 61.

quickly deployed with technologies taken "off the shelf."[131]

The directive led to the commissioning of three study panels in June to investigate the feasibility and implications of the president's proposal. The panels' final reports were submitted in October and, although they remain classified, summaries of two of them have been made public. In addition, the magazine *Aviation Week and Space Technology* obtained copies of the full reports of two of the panels and used them as the basis of a series of articles published between October and December 1983.[132]

The Defensive Technologies Study Team—more commonly known as the Fletcher Panel, after its director, the once and future NASA Administrator James C. Fletcher—looked into the technical feasibility of a comprehensive ballistic missile defense system. Its sixty members were drawn from the national weapons laboratories, private industry, universities, research institutions, and the military. The Fletcher Panel reported to Richard DeLaurer, under-secretary of defense for research and engineering.[133]

The Fletcher Report, which included some twelve volumes of classified technical studies, was cautiously optimistic and concluded that a number of emerging technologies offered hope that an effective BMD system could be made to work within a reasonable time. Among these were infrared chemical lasers, ground-based excimer lasers, short-wavelength chemical lasers, neutral- and charged-particle beams, electromagnetic railguns, and X-ray lasers. According to the report, a vigorous research effort might yield "meaningful levels of defense" during the 1990s with the initial deployment of the terminal and mid-course layers of a projected multi-tiered defense. A complete system, including the capability for boost-phase intercept, might be in place by the year 2000 at a cost of some $90 billion. The panel recommended an initial research and development of $18 to $27 billion between FY 1985 and FY 1989.[134]

---

131. Stein, *From H-Bomb to Star Wars,* p. 60; Stares, *Militarization of Space,* p. 226; Broad, "Reagan's 'Star Wars' Bid," p. A-8.

132. Stein, *From H-Bomb to Star Wars,* pp. 61–66; David L. Haffner, "Assessing the President's Vision: The Fletcher, Miller, and Hoffman Panels," in Franklin A. Long, David Haffner, and Jeffrey Boutwell, *Weapons in Space* (W. W. Norton and Company, 1986), pp. 91–107. See also Clarence A. Robinson, Jr., "Panel Urges Defense Technology," *Aviation Week and Space Technology* (hereafter cited as *AWST*), October 17, 1983, pp. 16–18; "Scientific Canvas Locates Innovative Defense Ideas," *AWST,* October 17, 1983, p. 19; Clarence A. Robinson, Jr., "Study Urges Exploiting of Technologies," *AWST,* October 24, 1983, pp. 50–56; Clarence A. Robinson, Jr., "Shuttle May Aid in Space Weapons Test," *AWST,* October 31, 1983, pp. 74–78; and Clarence A. Robinson, Jr., "Panel Urges Boost-Phase Intercepts," *AWST,* December 5, 1983, pp. 50–61. These articles focus on the Fletcher Report, but also contain useful information regarding the Hoffman Report.

133. Haffner, "Assessing the President's Vision," p. 106.

134. Ibid., pp. 91–96, 106; Stein, *From H-Bomb to Star Wars,* pp. 61–66; Stares, *Militarization of Space,* pp. 226–27.

The panel was far less optimistic that any such system could totally eliminate the threat of nuclear attack on the United States. While stating that the "technological challenges of a strategic defense initiative are great but not insurmountable," the panel concluded that the president's hope for a leakproof defense was "not technically credible."[135] Quite simply, some warheads would always get through. As Major General John Toomay, a member of the panel, later commented: "No imaginable set of defenses can prevent a determined and resourceful enemy from detonating nuclear weapons in our country."[136]

The other two panels investigated the strategic implications of ballistic missile defense. The Miller Panel was an interagency review group chaired by Franklin C. Miller, director of Strategic Forces Policy in the Department of Defense. Its report remains classified and no summary has been made public.[137]

The Future Security Strategy Study Team, chaired by Fred S. Hoffman of the defense consulting firm Pan Heuristics, did release a twelve-page summary that is reported to be virtually identical to its final report. The Hoffman Panel consisted of twelve members drawn from the same sources as the Fletcher Panel, plus a nine-member senior review group, and reported to Franklin Miller. Its report was skeptical that a "near perfect" defense could ever be achieved, but it argued that even a partial BMD system would be worthwhile in that it would strengthen retaliatory deterrence—a goal quite different from the comprehensive population defense that the president had in mind. The Hoffman Report called for the phased deployment of missile defenses, beginning with anti-tactical ballistic missile defenses (ATBM) in Europe. This would be followed by terminal and mid-course defenses designed to protect military assets in the United States and, finally, by systems for boost-phase intercept. The report also concluded that BMD research would not necessarily violate the ABM Treaty, but it implied that any US effort would be met by a strong Soviet program to develop countermeasures.[138]

Defense officials spent over a month digesting the contents of the panel reports. The consensus was that a BMD progam was worth pursuing, but that the cost and difficulty of the task should not be underestimated. Shortly after receiving the reports, Richard DeLaurer testified at a Senate hearing that a viable defense system was twenty years away and would be very costly since each of the proposed technologies would require an effort

---

135. Haffner, "Assessing the President's Vision," p. 94; Stein, *From H-Bomb to Star Wars,* p. 62.

136. McNamara, *Blundering into Disaster,* p. 94.

137. Haffner, "Assessing the President's Vision," pp. 96, 106.

138. Ibid., pp. 96–98; Stein, *From H-Bomb to Star Wars,* p. 83; Stares, *Militarization of Space,* p. 227.

equivalent to or greater than the Manhattan Project. "When the time comes that you deploy any one of these technologies," he stated, "you'll be staggered at the cost that they will involve."[139]

In late November 1983, Defense Secretary Weinberger announced that, on the basis of the study team reports, the administration would proceed with a long-term research and development program to create an effective ballistic missile defense system for the United States. Weinberger acknowledged the magnitude of the task and the difficulties involved, but he expressed confidence that the required technologies could be achieved. Given continued support, he concluded, a BMD system might be deployable by the end of this century or the beginning of the next.[140]

The findings of the study panels were formally approved by President Reagan in National Security Decision Directive 119, issued on January 6, 1984. This document, which marked the official birth of the Strategic Defense Initiative, served as policy guidance for the Department of Defense in the preparation of its FY 1985 budget requests for BMD research. That budget, submitted to Congress on February 1, called for expenditures of $1.77 billion in research and development funding for the SDI and $26 million for related space systems research in areas such as antisatellite weaponry.[141]

A month later, in March 1984, the Strategic Defense Initiative Organization (SDIO) was created within the Department of Defense, under the direction of Lieutenant General—now General—James A. Abrahamson, USAF. The SDIO was established to provide central direction for the various BMD programs already underway and to initiate research in such new areas as might appear promising.[142]

Conclusion

With the Strategic Defense Initiative, the United States was, once again, committed to the search for an effective means of protecting the national homeland from direct enemy attack. Given the state of offensive strategic weapons technology—and certain advances in this realm—the objective is clearly an ambitious one that will require many years of research before a decision on deployment can be contemplated by a future president.

In the years since the program was launched, funding has steadily increased, with over $5 billion being requested for research and development

---

139. Stein, *From H-Bomb to Star Wars,* pp. 68–69; McNamara, *Blundering into Disaster,* p. 94.

140. Stein, *From H-Bomb to Star Wars,* p. 68.

141. Snow, "Ballistic Missile Defense," p. 148; Stein, *From H-Bomb to Star Wars,* p. 69; Stares, *Militarization of Space,* p. 228.

142. Colin S. Gray, "Strategic Forces," in Joseph Kruzel, ed., *American Defense Annual, 1986–1987* (Lexington Books, 1986), p. 79.

in FY 1988, and the SDIO has reported significant progress in several of the technologies under investigation.[143] The Reagan administration remains fully committed to the program, but in recent years there has been somewhat less discussion of population defenses and leak-proof shields. Despite talk of defensive strategies, the United States continues an unprecedented peacetime military buildup. This may still be in search of the military superiority over the Soviet Union that was discussed in the 1980 Republican platform, but the administration has lately been more cautious in its statements on this point. American security remains based on a strategy of deterrence through retaliation, with no significant change likely in the near future.

As the SDI program continues, so, too, does the controversy over the feasibility and wisdom of its objectives. In domestic politics and in allied relations the Strategic Defense Initiative remains the subject of often intense debate. In Soviet-American relations, while the SDI may deserve the credit which supporters give it for bringing the Soviets back to the negotiating table in Geneva, it has clearly become the principal obstacle to progress toward a new arms control agreement—as illustrated in October 1986 at the Reykjavik summit meeting.[144]

---

143. Appropriations and budget requests for SDI through FY 1987 are noted in ibid., p. 80.

144. In addition to the works cited above, the current debate over BMD is illustrated by the following: William J. Broad, *Star Warriors* (Simon and Schuster, 1985); Alun Chalfont, *Star Wars: Suicide or Survival* (Little, Brown and Company, 1985); Stephen J. Cimbala, ed., *The Strategic Defense Initiative: Technology, Strategy, and Politics* (Westview Press, 1986); Dorinda G. Dallmeyer, ed., *The Strategic Defense Initiative: New Perspectives on Deterrence* (Westview, 1986); William J. Durch, ed., *National Interests and the Military Use of Space* (Ballinger, 1984); Colin S. Gray, *American Military Space Policy: Information Systems, Weapon Systems and Arms Control* (Abt Books, 1981); P. Edward Haley and Jack Merritt, *Strategic Defense: Folly or Future?* (Westview Press, 1986); Jeff Hecht, *Beam Weapons: The Next Arms Race* (Plenum Press, 1984); David Hobbs, *Space Warfare: Star Wars Technology Diagrammed and Explained* (Prentice Hall Press, 1986); Thomas Karas, *The New High Ground: Systems and Weapons of Space Age War* (Simon and Schuster, 1983); James Everett Katz, ed., *People in Space: Policy Perspectives for a "Star Wars" Century* (Transaction Books, 1985); Robert M. Lawrence and Sally M. Reynolds, *The Strategic Defense Initiative: Bibliography and Reference Guide* (Westview Press, 1986); Keith B. Payne, *Strategic Defense: "Star Wars" in Perspective* (Hamilton Press, 1986); Curtis Peebles, *Battle for Space* (Beaufort Books, Inc., 1983); G. Harry Stine, *Confrontation in Space* (Prentice-Hall, 1981); Union of Concerned Scientists, *The Fallacy of Star Wars* (Vintage Books, 1983); John Tirman, ed., *Empty Promise: The Growing Case Against Star Wars* (Beacon Press, 1986). See also The Arms Control Association, *Arms Control Today* (The Arms Control Association, monthly); American Academy of Arts and Sciences, "Weapons in Space, Vol. I: Concepts and Technologies," *Daedalus* 114(Summer 1985); and American Academy of Arts and Sciences, "Weapons in Space, Vol. II: Implications for Security," *Daedalus* 114(Summer 1985). Articles on the subject also appear regularly in such journals as *Foreign Affairs; Foreign Policy; International Defense Review; International Security;* and *Scientific American.* This list is by no means exhaustive, but it does indicate the breadth of the current debate over strategic defense.

The arguments of opponents, however, have had little impact on the administration, and its commitment to the Strategic Defense Initiative remains as firm as ever. Nevertheless, several developments during 1986 promised new difficulties for the program. First of all, the tragic loss of the space shuttle *Challenger* and the subsequent halt in shuttle flights brought a delay in a number of SDI-related experiments which required the orbiting of large payloads or human operation. Second, the failure of the administration to find a solution to the problem of the record federal deficit promised to have a significant impact on the long-term viability of the SDI. Although the Fletcher Panel spoke in terms of $95 billion for a deployed system, the actual cost might be ten times that, or more—a cost that would be very difficult to sustain. Third, the extent of the Democratic victory in the November 1986 elections meant growing difficulty for the administration in achieving its objectives, including increased funding for the SDI. Finally, by the beginning of 1987 the implications of the "Iran-Contra" scandal indicated that the administration could become so paralyzed that it might be unable to mount a viable defense of a program as controversial as the Strategic Defense Initiative. In any case, a new president would take office in 1989 with an agenda that may—or may not—include ballistic missile defense.

Yet, despite these difficulties for the Reagan administration and regardless of the particular fate of the SDI, it is clear that the problem of strategic defense of the national homeland will always be of concern to the United States, as it will with all nations. Even a successful ballistic missile defense system would not remove all threats to American security, and debate would continue over means of dealing with other challenges. On the other hand, if the current program is cancelled, the attractiveness of the basic idea—regardless of its wisdom in this peculiar nuclear age—and the technological momentum spurred by the SDI will insure that, at some point, the debate will reemerge. As long as the threat of war remains, this much is certain.

2

SDI: A Personal Vision

James A. Abrahamson

IN pursuing the Strategic Defense Initiative, I believe that, in Thomas Jefferson's words, "we act not just for ourselves alone, but for the whole human race." When Jefferson spoke those words, our nation was weak in arms but rich in spirit. Today we are the richest nation on earth. We are one of the world's two superpowers, with interests in every continent. But our growth in wealth and strength has not guaranteed us peace.

Over the last two hundred years, but especially in this century, war has touched every family in America. For all of us, war has been a learning experience. For some war is remembered for moments of glory; for others it is distinguished by unforgettable examples of self-sacrifice and valor. But for all, war is an experience that is more often tragic than glorious. For all, war is an experience that teaches us, above all, that there is no security in weakness. General George Catlett Marshall, the architect of the greatest military force in history and a man whom the world may better remember as a statesman than as a soldier, best expressed this truth when, in 1945, he stated: "We have tried since the birth of our nation to promote our love of peace by a display of weakness. This course has failed us utterly."

Carrying out the tradition of General Marshall and being strong also requires us to dare to try to create a better way of conducting our national security than to adhere to the current process of holding our nation and our people hostage to an implied ballistic missile threat; hostage to the threat of nuclear annihilation; hostage to a process that we now call Mutual Assured Destruction or MAD.

We have to pull together to create a better way. Our arms and our wealth alone will not do it. General Marshall understood this very well. "Without . . . heart, . . . spirit, . . . and soul," he said, "you have nothing." President Reagan agrees, stating in his first Inaugural Address that "no weapon in the arsenals of the world is so formidable as the will and moral courage of free men and women." The president's initiative, which established SDI, was an enlightened and extraordinary act of historical significance. We are working very hard to create a more secure way of life, and any discussion of SDI must cover more than ballistic missiles, strategy, and doctrine. The discussion must recognize, as before, that there is no security in weakness. It must highlight the fact that we now have the collective will, wisdom, and technological potential to free us all from the tyranny of the nuclear balance of terror with which we have had to live.

We must balance our armed strength with our spirit and our compassion. We must also recognize, however, the harsh reality that, over the last three and one-half decades, the Soviet Union has achieved greater advances in military power than any other nation in the world. The Soviets have made tremendous investments in armament, and there can be no question that defense programs today enjoy an absolute primacy in the economic and industrial development of the Soviet Union.

The USSR and its allies have continued to expand and modernize their armed forces. They have capable, sophisticated weapons designed for the entire spectrum of conflict. They have begun to deploy a new generation of nuclear-armed, air-launched and sea-launched cruise missiles. They have massive conventional forces. They have positioned short-range ballistic missiles in Eastern Europe, and they have increased the number of ballistic missile submarines on patrol in the Atlantic and Pacific oceans.

In contrast to the Western Allies, the Soviet Union has also built and maintained a strong defensive capability, especially against nuclear delivery systems. This is not criticism—this is fact! The Soviets have the world's only functioning antiballistic missile—or ABM—system around Moscow. It has an active civil defense program and a very effective air defense network that employs approximately 4,000 fighter aircraft, 6,300 radars, and nearly 10,000 surface-to-air missile launchers. Clearly the challenge from the East is great, very great. One hundred fifty years ago, Alexis de Tocqueville had remarkable insight into the world of the 1980s when he wrote:

> There are at the present time two great nations in the world, which started from different points, but seem to tend towards the same end. I allude to the Russians and the Americans. . . . The American struggles against the obstacles that nature opposes to him; the adversaries of the Russians are men. The former combats the wilderness and savage life; the latter, civilization with all its arms.
>
> The conquests of the American are therefore gained by the plowshare; those of the Russian by the sword. . . . Their starting point is different and their courses are not the same; yet each of them seems marked out by the will of heaven to sway the destinies of half the globe.

De Tocqueville's observations are still valid. The Soviet challenge will continue. I am confident of our ability to respond to the Soviet challenge. But my vision and hope is that in adapting the SDI technologies, we can change the direction of this challenge to one where we both cooperate and strive to bring the maximum returns of the new space renaissance to all of the peoples of the world and for the enduring benefit of mankind.

I do not want to cast this presentation in terms of "we" verses "them." I do not like to do that. But, as a minimum, we must recognize that we face a strong and determined foe: a foe that currently holds a military edge.

Under Mikhail Gorbachev there have been some encouraging signals from the Soviet Union. But the signals are a result more of our determination and resolve than of Mr. Gorbachev's personal initiatives and desires. The signals are the result of our addressing the imbalance that existed vis-à-vis the Soviet Union, a process begun in the last part of the Carter administration and pursued even more vigorously by the Reagan administration. In this respect, President Reagan would, in 1983, state with confidence that "we have put in place a defense program that redeems the neglect of the past decade. We have developed a realistic military strategy to deter threats to peace and to protect freedom if deterrence fails."

As part of the program to restore our neglected defenses, the president took a giant step on March 23, 1983, when he directed that we undertake "a comprehensive and intensive effort to define a long-term research and development program to begin to achieve our ultimate goal of eliminating the threat" posed by ballistic missiles. He described this program as "a vision of the future which offers hope." This vision is now called the Strategic Defense Initiative or SDI. It is an exciting, motivating program, but more important, it is a commitment to all people, toward a better, more peaceful world.

In proposing SDI, President Reagan did not suggest that we abandon deterrence or arms control efforts. Instead he challenged the American nation and the entire world to determine if the means might not be discovered to save lives, not avenge them. He challenged our scientists, our engineers, our strategists, and our policy makers to find these means—and to do it within the context of existing arms control agreements.

There was some skepticism that the president's vision could be realized. The skepticism ranged even among members of the distinguished group of scientists convened to determine if strategic defense was, in fact, feasible. Close on its heels came cynicism in the glib and shallow-minded shorthand of the press's sobriquet "Star Wars." However, the skepticism turned to optimism when members of the scientific team rigorously examined our capabilities and concluded that not only were the technologies necessary for a Strategic Defense Initiative attainable, but that many of the technologies were already available. The scientists recommended that the US embark on an in-depth research program that would increase our options for the future. Their examination validated a fact which I had long felt to be true: the unparalleled leadership of the United States of America in the development and application of technology. And that is what we are doing. We are conducting research into those technologies that will enable the development of defensive systems capable of intercepting and destroying ballistic missiles after they have been launched and, consequently, preventing these missiles from hitting their targets. I like to call SDI the "Strategy of Hope."

SDI research is focusing on defenses against ballistic missiles of all ranges, including intermediate-range (IRBM), intercontinental (ICBM), and submarine-launched ballistic missiles (SLBM). Our emerging technologies have the promise of overcoming previous obstacles, thereby making possible a layered defense at each stage of a ballistic missile's flight: the boost (or launch) phase when the first- and second-stage rocket motors are burning and an intense infrared signature is created; the post-boost phase when multiple warheads and penetration aids are deployed; the midcourse phase when the warheads and penetration aids travel on ballistic trajectories above the earth's atmosphere; and the terminal phase when the warheads and penetration aids reenter the atmosphere. Each phase offers different opportunities for a defense system; each poses different challenges. A highly effective counter to a massive missile attack on the US will require multiple tiers of defense, each designed to significantly reduce the number of incoming warheads until their number is actually too small to be of any military utility.

Countering the potential offense is not going to be easy. It will not, however, as some have intimated, be impossible. At each tier the defense system must carry out a set of essential functions: surveillance, acquisition, and discrimination; pointing and tracking of weapons, be they space-based lasers or ground-based lasers, neutral particle beams, or hyper-velocity launchers; the interception and destruction of missiles and warheads; and, of course, battle management to orchestrate the entire defense scenario.

While our technological capabilities are not yet adequate to mount a robust defense against a plausible Soviet threat and potential countermeasures, they are advancing rapidly. Current technologies offer new opportunities for active defense against ballistic missiles, opportunities that did not exist just slightly more than a decade ago. A decade ago there were no means available or even envisioned that were capable of intercepting a missile during the boost phase. Midcourse intercept would also have proven very difficult, since an effective way of differentiating between warheads and penetration aids did not exist. Even intercept in the terminal phase would have posed significant problems: the computer hardware, software, and signal-processing equipment of the late 1960s and early 1970s were not sufficiently advanced to cope with multiple tiers of defense or with a massive attack. We have come a long way in addressing and overcoming the shortfalls of the past. We are confident in our belief that, with a continuing effort, a future administration and a future Congress, if they so choose, will have the very real option to design, build, and deploy an effective defense against ballistic missiles.

Our progress has been dynamic. Our spirits were given a big boost, for example, when the Army's Ballistic Missile Command demonstrated the practicality of terminal defense by intercepting and destroying a dummy Minuteman re-entry vehicle off Kwajalein, approximately four thousand miles from the Minuteman's launch point at Vandenberg Air Force Base. We equated this success to "hitting a bullet with a bullet." We have also burned holes through remotely piloted vehicles and other laboratory test articles with high-energy lasers. In September 1985 we blew a stationary Titan I booster apart.

With the help of the space shuttle, we demonstrated that we could overcome the atmosphere's distorting effect on a laser beam. Some members of the scientific community had thought this problem to be insurmountable. They do not think that it is insurmountable now. We have also made very good progress in reducing the cost associated with laser-beam output, almost in orders of magnitude. We have greatly improved our technology and techniques for discriminating between warheads and penetration aids. The work on our particle-beam weapons and hypervelocity (or railgun) launchers is proceeding very satisfactorily. It is truly amazing how big a hole can be made in a cast aluminum block by a tiny plastic projectile travelling several thousand miles an hour! Perhaps more important, we are making great strides in terms of energy storage, high speed computational capabilities, and sensor development. Over the last twenty years, electronic information hardware costs have decreased by a factor of three; size has decreased by a factor of three; and speed has increased by a factor as much as ten every three or four years! Indications are that we will improve upon this record!

In short, SDI like this great nation is alive and well. Our program is every day proving that the scientific panel was correct in asserting that not only are the technologies needed for strategic defense attainable, many of them are, in fact, already on hand. We are proceeding deliberately, keeping in mind that a strategic defense system must be highly survivable. An aggressor must know that he cannot destroy our strategic defense system and expect his own capabilities to remain intact. We are also mindful of the criteria that it must be as cheap or cheaper for us to increase our defensive capability as for an aggressor to add nuclear warheads to his arsenal.

With SDI, I believe that our generation has crossed the threshold of a new era in human history. In building a better way, we have an awesome responsibility: a responsibility made possible by the continuing evolution of our science and technology. In the months ahead, we will have many additional technical achievements to report to the nation. Nonetheless, these achievements will only be the tip of the iceberg. There will be no way in which the spectrum of benefits—for defense, in direct stimulation of the economy, and for enhanced productivity—can be accurately measured.

And, of course, there will be no way to accurately assess the sense of pride, hope, and optimism that the SDI program, and its celebration of scientific and technological discovery, will give to each and every American.

With SDI, we have explicitly recognized our strength in the development and application of technology, and we have chosen this as the gateway to the third century of our existence as a nation. Survival and peace are the business of us all, and SDI, in my opinion, will significantly reduce the chances of war. In the long run, I am convinced that SDI will be a positive catalyst in promoting more cooperation on earth and in space with all nations.

I also consider it a natural corollary that our investments in SDI technologies will help foster new growth industries and build a more robust economy. This is not a unique idea. Our nation's military and civilian aeronautics and space programs directly, or indirectly, have made us all beneficiaries of this process. The process has been going on for years, on a wide range of matters, including electronics, air transport, and data automation. Military investment has traditionally sparked investment in the civil sector, and it has been a supportive foundation for the flexible and innovative elements of our industry.

Our primary objective in SDI is to enhance our national security posture, and of this, there can be no doubt. However, we are also mindful of the opportunities for increased commercial enterprise based on our technologies. New products and new services can be introduced; productivity improvements can be achieved. The prospects are exciting and, apart from national security considerations, their applications are limited only by our imagination and our resourcefulness.

The basic goals of the national space policy of July 4, 1982, include strengthening the security of the United States, maintaining US space leadership, and obtaining economic and scientific benefits through the exploitation of space. The SDI program gives us an opportunity to satisfy aspects of these goals by creating the potential for tapping the significant contributions that can be made by the private sector.

In this regard, I like to think of SDI as part of the new space renaissance. The science and technology which developed in the seventeenth century were the tools which Renaissance man needed to complete his emancipation from the Middle Ages. In the twentieth century, another Renaissance was brought about by the space program. Our work in space opened new opportunities for expanding our understanding of the universe and for improving the quality of human life. The space renaissance brought to us much of our microelectronics capability, as well as more accurate forecasting of weather and natural disasters, low-cost global communication and navigation, and the surveillance from space which has proven so vital for our national security.

In conjunction with other programs and in league with other agencies, SDI could be the nucleus of yet another renaissance: the Twenty-first Century Renaissance. It will spur the growth of much new technology and, to the degree permitted by our national security requirements, the SDI program will assist in identifying new commercial applications and techniques and promoting joint endeavors that have economic benefits for the nation. I have created an Office of Educational and Civil Applications to carry out this aspect of our program.

Clearly, the potential for an economic payoff from investment in the SDI technologies is great. The introduction of new, advanced technologies historically enhances a society's rate of growth. Europe and the US, for example, have benefited immensely as the result of the aerospace technologies that they have developed. I am confident that most of the emerging SDI technologies can, over time, readily be adapted for other space activities and for the consumer market. The whole field of enhanced solid-state electronics, including more durable integrated circuits and power switches vital for our hypervelocity launchers, has considerable promise for application in areas and programs other than merely stretegic defense. Computer, propulsion, and laser technologies also have attractive spinoff possibilities. They could serve as the vehicles to introduce our civil applications program as well as to alleviate technological problems in related fields. Their adaptation could help the SDI program pay for itself, and, in the process, they could help other civil and defense programs. For example, we have the potential of harnessing laser technology to create communication lines of almost unlimited bandwidth. For military applications, laser communication will be difficult to intercept and counter. For commercial applications, line rates (the number of circuits) could go up exponentially.

With a lot of patience and a little creativity and innovation, the SDI program can create technological and economic opportunities that could be among the most important in our history. We are working in conjunction with the National Aeronautical and Space Administration (NASA) and other government agencies to make this happen. I also think that SDI, with its spark of new technology, will grow to be recognized as more than just a vital defense activity. Even at $26–30 billion spread over a five- to six-year period, the research program is not really expensive: it averages about $20.00 per year per American. I consider that a pretty good investment, especially given the human and moral dimensions of the program.

Admittedly, many scientists have opposed our program. Some have opposed it on technical grounds based on the shaky premise that SDI, or rather some aspects of if, will never work. How on earth can they be so positive until we at least try? When we do try, I am convinced that these opponents will join the legions who said that man would never fly, who said that man would never go to the moon. As one who, in 1969, thrilled at watching man standing on the moon, I am not discouraged by these criticisms. I might add that neither is my friend, Sir Frank Whittle. In 1940 the National Academy of Sciences concluded that "the gas turbine could hardly be considered a feasible application to airplanes." Whittle first became aware of this opinion in 1976, long after his invention was commonplace. Not bad for something that was not feasible!

I frankly think that many of our critics fail to appreciate the dynamic character of science. New discoveries are constantly changing technology and adding to its potential for new processes and new products. I remain convinced that there is little that the alliance of American science and technology cannot provide for our armed forces and for our nation! I remain convinced that anything that is theoretically possible will be achieved in practice, especially if we want it.

In the 1930s, one of my favorite authors and a great aviator, Antoine de St. Exupery, decried the so-called moralists who viewed technology as "the enemy of spiritual civilization." He stated that technological accomplishments "have the single aim of bringing men together." In my vision, so does SDI.

I have always had a fascination with the future. After all, that is where we will spend the rest of our lives; so I think my interest is understandable. I am convinced that, through SDI, we are charting a better course and creating a better way. In my vision, SDI holds unlimited opportunities for strengthening our defense; for amassing new scientific knowledge; for inspiring our young; for reinvigorating our economy; and for promoting peace and good will. In my vision, SDI will benefit the whole human race.

There have been some astonishing conversions among some of our critics. Some formerly fierce opponents of Mutual Assured Destruction now act to defend the concept. What was formerly called a crime against humanity is now considered by some as an indispensable instrument for maintaining peace. I am not questioning their motives or their sincerity. But frankly, I fail to understand or appreciate their change of heart and mind. I also feel uncomfortable—and I think our critics should also feel uncomfortable—with a security concept premised on the threat of annihilating millions of people on each side.

SDI is trying to create a new security concept: a concept built on the will and moral courage of free men and women and premised on the fact that we now have the wisdom and technological potential to free us all from the nuclear balance of terror with which we have had to live. The political and military interests of the Free World will be well served by SDI. SDI will be completely defensive. It will not take a single life, but it could indeed serve to save hundreds of millions of people from nuclear annihilation. As President Reagan noted early in 1985, SDI is not about war, "it's about peace. It isn't about retaliation, it's about prevention. It isn't about fear, it's about hope."

On October 11 and 12, 1986, President Ronald Reagan and General Secretary Mikhail Gorbachev met in Reykjavik, Iceland, to discuss arms reductions. The Strategic Defense Initiative was a major topic in those conversations. On October 16, 1986, General Abrahamson returned to the Virginia Military Institute to discuss SDI in the post-Reykjavik period.

The first uses of space were for weapons. Now we are attempting to find ways to take the weapons that were considered omnipotent, weapons that indeed could obliterate mankind, and to find a way to make, as the president said, those ballistic missiles "impotent and obsolete."

We are challenging every frontier in order to do that. We are challenging the technical frontier. I think everybody understands that; that is straightforward. We are challenging the scientific frontier because we do not even know enough about our world to be able to do that, to do it reliably. We are challenging the frontiers of military strategy. How can a defensive system, one that is made up of weapons which will destroy weapons, as opposed to weapons which will destroy people, be used to make our world safer and to prevent war? Again, I want to emphasize, that is what SDI is truly about.

Finally, we are challenging the very foundations and the very idea of arms control as it has been attempted to be exercised over the past several decades, the decades of the nuclear age. In that area, arms control for nuclear weapons and ballistic missiles has been an exercise in counting warheads and counting ballistic missiles themselves. But as their technology is changing, it is not at all clear that merely counting those kinds of systems can give us a great new dimension and a new means of achieving greater safety through effective arms control or arms reductions.

What SDI can and will be able to provide in the future is an entirely new strategy, one in which the uncertainty is so large that, in fact, a targeteer on the other side will say: "It isn't worth it. Do not strike under any conditions, even in a condition of crisis, because the military objective cannot be achieved." That is the level at which we are trying to ensure that SDI effectiveness will be able to operate.

But as the president dramatically stated and laid out for us in Iceland, it is also a lever, a lever in a whole new arena of arms reduction to try indeed to move the whole arsenals of ballistic missiles dramatically from partial reductions down to total elimination. That is what was presented. Unfortunately it has been presented now in the media as a failed attempt, and I think that is exactly the wrong way to look at what happened. What really happened is that dramatic progress was made on both sides in terms of this new regime for arms control. The stage is now set, I believe, to continue to use the lever of strategic defense to try to nail down those kinds of varifiable and usable arms reductions which, in turn, make the concept of strategic defense easier. And then strategic defense can be used, as the president again outlined, as insurance—insurance to ensure that whatever agreements are achieved, will endure and will provide modified behavior so that it will not be merely a signed piece of paper but really and truly will provide increased security for all of us.

The military uses of space go through all of those regimes. I am pleased that we are going back full circle to try to take away the first things that appeared to be viable and the things that appeared to be the ultimate weapon, because there is no such thing as an ultimate weapon, ever. Our final solution is a human solution. But we must be dedicated to using whatever our advantages are in space, on the ground, at sea, and in the air—everywhere—to find the way to ensure that we endure long enough and safely enough so that human solutions can finally be found.

Thus, there is a military role in space. And unlike what the critics say—that we are taking war to the heavens—the most important thing is to be able to preserve peace here on earth and to use these new frontiers of knowledge, these new frontiers of human endeavor, to ensure peace on earth.

QUESTION: *You have said that we need a "Marshall Plan for space." What is meant by that?*

ABRAHAMSON: I believe we have the ingredients today for a marvelous new step forward—a human step forward as well as a technical step forward—in space, and one in which we can, in fact, share and cooperate at every level and in every kind of endeavor. Clearly, we have already been doing much of that in our civil space program. We are offering, in our commercial space program, the opportunity to share both in the potential for space business, but also the benefits for all mankind.

Finally, perhaps the most important thing of all: the president has even outlined a possibility, in conjunction with a shift to a more defensive orientation and a safer world associated with SDI, a means of sharing for the whole world the benefits of this space frontier that I mentioned; and I think that is where it really is. We must grasp the fact that space is a frontier for mankind—for any level of effort. In fact, we must find a way to ensure that that potential will actually be realized for all of us.

QUESTION: *If enemy missiles cannot be intercepted all of the time, what is the value of SDI?*

ABRAHAMSON: The reason that we try to emphasize that is because very often the critics say that if one or two or five or ten ever get through in a major attack, then it is useless. None of us is promising a perfect weapon of any kind. Nothing that mankind ever does is perfect. However, I think it is important to underscore that the president's challenge to us was not to go halfway to the moon and not to go nine-tenths of the way to the moon, but in fact to be creative and to find ways to deal with the weapons so that we can have a thoroughly reliable defense. I just refuse and cannot responsibly say that it is 99.99 percent perfect. However, all of our studies, all of our war games, all of our analyses of the problem indicate that, in fact, you achieve deterrence because what we are really trying to do is to prevent war.

Remember that you achieve effective defense deterrence at much lower levels than a perfect defense. You only need the perfect defense in the terrible circumstance where deterrence fails. And then, of course, we would want to be as effective as possible. If we have a small number to deal with, because we have been successful in achieving arms reductions, we will probably be very, very effective. But the kind of planning that we are doing is not to rely on that but, in fact, to examine the possibility of very profound and very robust kinds of attacks much more intense than we are faced with today.

By the way, as we examine the ability to counter those kinds of attacks, we think that is the best way of ensuring that, in fact, arms reductions can be made possible. This is because if the other side and we both understand that that can be done, I think we will never have to deploy all of that capability. We will only have to deploy sufficient to be an insurance policy, as the president indicated in Iceland.

Some of the critics will say, "After all, if only fifteen or twenty get through, it is a tragedy," because, they imply, the other side will know which fifteen or twenty get through; and they will be the fifteen or twenty that are targetted at our major population centers or whatever. I can assure you that the kind of layered defense that we are working on is one that will guarantee that they will never be able to have assurance that they can know which ones will get through. That is the way you prevent war. Obviously, if the terrible tragedy, if major attack would occur, that would be the way in which you, in fact, save as many lives as possible. But I reject the arithmetic that says if a few get through, it is useless, because that is equating a few to the eight thousand warheads that we are faced with today. And I do not understand that particular arithmetic.

Let me go back one more time. Our aim is to have the technology so that we could build something which could be just as perfect as possible. But I do not think we can credibly say today that that is what we can do. That is why we are engaged in the research program. And none of us claims that anything in this world is perfect.

QUESTION: *Do you have any particular concerns about the SDI program?*

ABRAHAMSON: I think they fall into several categories. The first concern is that until the president announced that we as a nation set a goal to try to establish a defensive posture and a defensive kind of deterrent, the Russians had a set of plans, both in offense and in defense, that they could carry through at their own pace and not be challenged. So that I think that the fact that we have challenged their plan is a very, very important one. That is one major area of concern.

There is a second area of concern that is a technical area of concern that goes beyond the question: "Can we do such a thing, can we create an SDI?" They know that in the process of doing that, we will be creating such new technology and such a major step forward that they will have difficulty dealing with that and that there will be many benefits, both for our security and for civil things as well. And I think that has been expressed very simply and clearly in public statements by key Soviet officials.

3

The Case Against SDI

Spurgeon M. Keeny, Jr.

IN 1983 President Ronald Reagan announced his goal of eliminating the threat posed by strategic nuclear missiles and called on the scientific community to make these nuclear weapons "impotent and obsolete." Secretary of Defense Caspar Weinberger elaborated that the administration sought a "thoroughly reliable and total" defense system that would render all attacking missiles, including low-flying cruise missiles, "impotent." He asserted that the envisaged system "would guarantee that there would be no longer any danger from nuclear weapons." As time went on, the secretary of defense offered to extend the coverage of the shield to our allies as well as the United States.

With these incredibly ambitious and confident statements, the government of the United States shattered the long standing and almost universally supported consensus that the defense of populations and urban society was not possible in the nuclear age. The world was told that henceforth the United States would seek to change its long held offense-dominated nuclear strategy to a defense-dominated strategy. The Strategic Defense Initiative was created to discover and develop the impenetrable defensive shield necessary to implement this radical new doctrine. This program is well on its way to becoming by far the most ambitious research and development project in human history. While it is still much too early to project actual systems costs, crude estimates ranging from several hundred billion to more than a trillion dollars have been advanced.

As the scope of the SDI program has expanded, the goal of the program has become less clear. Some spokesmen have retreated to the more modest goal of a less-than-perfect defense that would strengthen classical deterrence by increasing the survivability of otherwise vulnerable retaliatory forces; others avoid the issue of goals entirely and explain SDI as merely a program to explore the possible technical options that might be available. I think that it is fair to say that the president and his secretary of defense have held to the original vision of an essential impregnable shield against nuclear weapons. The president has given this message repeatedly to the nation, to the world at the United Nations, and apparently to Soviet Secretary General Mikhail Gorbachev personally at the Geneva summit. It is certainly the picture that the American people have of the program. Therefore, I will initially focus on this concept.

What, one may well ask, could be wrong with a defense-dominant strategy that seeks to protect our population and society from the horrors of nuclear war? Such a military posture certainly has more popular and emotional appeal than a strategy based on deterrence of nuclear attack by the threat of assured nuclear retaliation in the event of a nuclear strike. The resulting threat of Mutual Assured Destruction (MAD) was not arrived at as a preferred strategy but rather as a grim fact of life in the nuclear age.

For thirty years, almost since the first deployment of nuclear weapons, the technical community has explored the possibilities of defensive systems. In the past such systems have been found to be technically inadequate, and I am confident they will continue to be found wanting in the future for several fundamental reasons:

—such systems are technically unattainable against a sophisticated and determined adversary because of the power of offensive nuclear weapons;

—moreover, the attempt to achieve such a defense would simply accelerate the nuclear arms race and increase the danger to both sides;

—finally, the deployment of such a system, whatever its capabilities, would tend to decrease strategic stability and increase the possibility of a nuclear exchange in the event of a major US-Soviet crisis.

The fundamental barrier to developing a highly effective nationwide ballistic missile defense (BMD) is the incredible destructive power of thermonuclear weapons when matched against the extreme vulnerability of populations and urban society. A single nuclear warhead can have more than a million times the yield of a comparably sized conventional high-explosive weapon. Today a single such thermonuclear warhead can destroy a major city.

Let me be a bit more specific as to what this really means. A megaton explosion on a major city would destroy multistory concrete buildings out to a distance of three miles with almost complete loss of life within that zone; out to five miles, there would be spontaneous ignition of combustibles and high immediate loss of life; out to nine miles there would be severe damage to frame buildings and second degree burns to exposed people. The innumerable fires ignited by the blast would either merge into an outward-moving conflagration or more likely create a giant firestorm of the type experienced on a much smaller scale in Hamburg and Dresden during World War II. The resulting uncontrollable fire would destroy the remnants of all structures in the region and kill a large fraction of those surviving the initial explosion. If the fireball of the explosion reached the ground, the resulting radioactive debris would produce lethal fallout far beyond the destroyed city.

This would be the result of a *single* nuclear weapon. But we live in a world of fifty thousand nuclear weapons. The United States and the Soviet Union each have approximately ten thousand thermonuclear weapons dedicated to strategic missions. The total consequences of an exchange of these arsenals are impossible to comprehend. In addition to the local impact of the multiple explosions, nuclear fallout would cover most of the country with lethal levels of radioactivity.

When General David Jones, former chairman of the Joint Chiefs of Staff, was asked at a hearing in the Senate Foreign Relations Committee in November 1981, while he was still on active duty, what the consequences would be of a major nuclear exchange, he replied as follows:

> We have examined that over many, many years. There are many assumptions that you have as to where the weapons are targeted. Clearly, the casualties in the northern hemisphere could be, under the worst conditions, into the hundreds of millions of fatalities. It is not to the extent that there would be no life in the northern hemisphere, but if all weapons were targeted in such a way as to give maximum damage to urban and industrial areas, you are talking about the greatest catastrophe in history by many orders of magnitude.[1]

One might understandably ask, if the consequences of nuclear war are so catastrophic, should defense not be our first priority? I would agree to this if there were only a small number of such weapons and if they were extraordinarily expensive or difficult to deliver. But the fact is that we are faced with ten thousand Soviet strategic weapons, and the Soviet Union could easily produce many more since they are relatively inexpensive and easy to deliver.

The power and number of nuclear weapons has fundamentally changed the nature of war and the relationship of offensive and defensive forces. A nationwide defense would have to be very effective at all points since leakage anywhere would be catastrophic to the defended targets. This requirement can be appreciated by comparing defensive against conventional and nuclear weapons. In the past, a defense that destroyed 10 percent of the attacking aircraft per sortie would be considered very successful. After ten such attacks the bomber force would be reduced to less than one-third of its initial size. The offense could normally not accept that level of attrition given the limited destructive power of conventional weapons. But in defending urban areas against nuclear weapons, destruction of 90 percent or even

1. US Congress, Senate, Committee on Foreign Relations, *Strategic Weapons Proposals Hearings. First Session on the Foreign Policy and arms Control Implications of President Reagan's Strategic Weapons Proposals* (Washington, D.C.: GPO, 1981), p. 44

99 percent of incoming warheads would be ineffective since a single penetrating weapon would destroy a target vastly more valuable than ten or a hundred warheads.

SDI would seek to achieve very high levels of attrition by a layered defense. This would involve a complex network of detectors, assessors, trackers, and exotic kill mechanisms operating from earth and space stations. The system would have to be linked and operated essentially automatically by computers dependent on unprecedentedly large and complex software programs. The system would have to work flawlessly when used for the first time in a very hostile nuclear environment. This is a staggering technical assignment even under idealized conditions. But the concept is "fatally flawed" since in the real world the system would be operating against a sophisticated adversary using every technique at his disposal to defeat the system. And there is a wide range of very effective countermeasures against such systems that the Soviet Union could develop at a fraction of the cost of the defenses that the SDI is contemplating.

Vulnerability

First, the offense could always attack vulnerable, critical points in the system. Extremely costly space-based battle stations or giant optically perfect mirrors or other critical space assets would be particularly vulnerable and lucrative targets. Such space-based systems components could be attacked at the outset of hostilities. They could also be attacked either clandestinely or openly during peacetime. It is most unlikely that space would remain a sanctuary if either side deploys major space-based weapons systems that the other side considers seriously threatening.

The operation of a multilayered defense system would require the interaction of many vulnerable space and earth-based components. The specific failure modes of such a system would depend upon the details of its ultimate design. But one can be sure that there would be easily identifiable vulnerable modes, destruction of which could easily result in the collapse of major portions of the system. It is instructive to recall that one of the principal reasons the United States rejected the early Sentinal-Safeguard ABM systems as serious candidates for nationwide defense systems against a concerted Soviet attack was the extreme vulnerability of the prime radars, on which the system depended, to direct attack or to blinding by nuclear blasts.

Firepower

Second, the offense could overwhelm the defense with increased firepower. This could be done by the brute force approach of increasing the number of attacking missiles or increasing the number of MIRVed (multiple independently targetable reentry vehicle) warheads on each missile. For example, without the constraints of the unratified second Strategic Arms

Limitation Talks (SALT II) Treaty, the Soviet Union could rapidly increase the total size of its missile force. The large SS-18 missile, which is now limited to ten warheads by SALT II, could carry as many as thirty warheads. The same objective could be accomplished by launching large numbers of balloons or other decoys designed to simulate real reentry vehicles. In principle, false warheads could be identified by multiple sensors; in practice this will prove very difficult to accomplish with confidence on a real-time basis. In addition, the offense has the tactical option of concentrating its firepower against the targets of its choice, while the defense must be prepared to defend all valuable targets against such a selective attack.

Countermeasures

Third, the offense would have the option of adopting technical modifications that would nullify or diminish the performance of the defense's sensors or kill mechanisms. A particularly striking example would be the development of a "fast burn" booster which could release its post-boost delivery vehicle in a small fraction of the time required by current missile boosters. By significantly shortening the time that the highly vulnerable missile boosters would be subject to attack, fast-burn boosters could eliminate or greatly reduce the effectiveness of a defense system that depended on attacking the boost phase. Failure of the high-leverage boost phase layer defense would be far more significant than simply the loss of one of several independent barriers since the entire weight of the attack would then have to be handled by subsequent layers. Unless they had been sized to meet this contingency, the system would be immediately overwhelmed.

Technical ingenuity will produce a host of ideas that could be introduced at a fraction of the cost of the defense system and on a much shorter time scale. Such ideas include spinning boosters, shields, and coatings to reduce, at relatively low cost, the effectiveness of directed energy systems against the vulnerable booster. Sensors could be blinded by precursor nuclear explosions or more delicately incapacitated or confused by a variety of jamming and spoofing techniques.

Circumvention

Finally, the offense has the option of circumventing the defense system by introducing an entirely different mode of attack. In the case of the SDI approach, this could be accomplished by air-breathing systems that fly under the shield. For example, low-flying cruise missiles launched from submarines close to the US coast would completely defeat the missile defense system and constitute a new offensive threat as difficult to defend against as ballistic missiles. In this connection, former Secretary of Defense James Schlesinger has noted that the addition of a serious air defense component would add at least another $50 billion annually to the cost of a ballistic missile defense. But without a very effective air defense, the defensive shield could hardly be considered to have rendered nuclear weapons "impotent and obsolete."

So far I have emphasized the fundamental technical difficulties in developing an effective defensive shield. I believe that most independent scientists who are familiar with the program would agree with this assessment. But some clearly do not agree, and one can never be certain about the future. Why then, one might ask, would it not be prudent to develop such a system on the chance that it might work despite the tremendous technological barriers to be overcome.

One answer is the immense expense of the undertaking both in dollars and in our national high technology capabilities. These dollars and brains would be diverted from other military and civilian activities where they are badly needed to maintain a balanced military force and an internationally competitive economy. SDI funding levels are already said to be as large as the combined services' technology-based research and development programs. The actual development of such a system, involving many hundreds of billions or even trillions of dollars, would certainly impact seriously on the ability of future military budgets to sustain and strengthen our conventional forces. Such a reorientation of priorities is impossible to justify given the very low probability that the program could achieve its avowed goals.

The decisive argument against this crash program, however, is the extremely dangerous impact it would have on the stability of the US-Soviet strategic military relationship. The goal of the program simply put is to deprive the Soviet Union of its nuclear deterrent against the United States. Not surprisingly, neither the Soviet Union nor the United States is going to allow the other side to eliminate the deterrent value of its strategic offensive forces by precluding the possibility of retaliation. Fear of the possibility of such nuclear "impotence" would arise long before the systems are deployed. And the fear would persist, even if the other side judges that the system ultimately would probably not prove to be very effective.

The attempt to achieve a highly effective ballistic missile defense system would also be seen by the other side as a direct offensive threat. Today both the United States and the Soviet Union know that neither could initiate a strategic nuclear attack against the other side without *themselves* being destroyed as a society by the inevitable retaliation. But if either the United States or the Soviet Union develops, or is perceived by the other side as developing, a highly effective ballistic missile defense, the other side would fear that its adversary is positioning itself to be able to conduct a preemptive strike. Even if his defense were not perfect, the attacker might believe that it would be able to handle a reduced and disorganized "ragged" retaliatory strike. In these circumstances, we would not accept Soviet protestations of good intentions. And there is no reason to believe that the Soviet Union would accept ours.

The intensity of these fears is best reflected in statements by US officials. For example, Secretary of Defense Caspar Weinberger has said that *Soviet* development of an effective defense "would be one of the most frightening prospects I could imagine."[2] The White House paper on the president's Strategic Defense Initiative issued last year contains the alarming statement that *if* the Soviet Union deployed a nationwide antiballistic missile (ABM) defensive system, "deterrence would collapse, and we would have no choice between surrender and suicide."[3] There is no reason to think that Soviet leaders have a more charitable view of a nationwide American SDI defense system.

The fact of the matter is that the deployment of a nationwide ABM system by either the Soviet Union or the United States would not lead to the "surrender or suicide" of the other side. What it would do would be to lead to a major acceleration of the nuclear arms race. The side that perceived itself threatened would move rapidly to increase the size or quality of its strategic offensive forces. It would make sure that the defensive systems would not result in the "collapse" of its deterrent capabilities. As I have outlined earlier, there is a broad spectrum of technical options to accomplish this. Secretary Weinberger described this situation clearly in a memorandum to the president in November 1985 in which he stated that "even a probable [Soviet] territorial defense would require us to increase the number of our offensive forces and their ability to penetrate Soviet defenses to assure that our operational plans could be executed."[4]

I should add that the Soviet reaction to SDI would not be limited to strategic offensive forces. The Soviet Union would also certainly accelerate its own strategic defensive efforts. No matter what technical advice they might get from their scientists, Soviet politicians and military leaders would not take the chance of lagging too far behind the US military technology they so admire and fear. In response the United States would have to build up its offensive forces.

It is not idle theoretical speculation to conclude that a buildup in strategic defenses would lead to a dangerous acceleration in the strategic arms race. In the past, when faced with Soviet strategic defense programs, the United States has *always* acted to insure its retaliatory capabilities. In the late 1960s the United States was concerned that the Soviet Union might be undertaking a nationwide ballistic defense system. In response, the United States

---

2. Caspar Weinberger, speech delivered to the Atlantic Institute, Paris, France, December 2, 1983.

3. The White House, *The President's Strategic Defense Initiative* (Washington, D.C.: The White House, January 1985), p. 4.

4. Caspar Weinberger, memorandum for the president on "Responding to Soviet Violations Policy (RSVP) Study," November 13, 1985, quoted in the *Washington Post*, November 18, 1985, p. A-4.

introduced MIRVs on the Minuteman III and Poseidon missiles in order to increase the firepower of our ballistic missiles and to insure their ability to penetrate the projected Soviet defenses and maintain existing target coverage. There was no suggestion that we abandon or reduce our reliance on ballistic missiles.

In the same manner, even though the Soviet Union has invested vast sums on its air defense system over the past thirty years, there has never been any suggestion that the United States should abandon the air-breathing leg of its triad of retaliatory forces. On the contrary, the United States has continuously upgraded the ability of its strategic bombers to penetrate Soviet air defenses. These qualitative improvements included upgrading the performance of the B-52 itself, electronic countermeasures against Soviet air defense radars, and short-range nuclear missiles to suppress Soviet air defenses. As Soviet air defenses improved, the US introduced the air-launched cruise missiles whose low altitude flight profile and small radar cross-section greatly complicate the problem of the defense. Then came the higher performance B-1B bomber. And today we are pursuing stealth technology which would make it even harder in the future for Soviet defense radars and other sensors to detect penetrating aircraft and cruise missiles. Throughout this period the US Air Force has remained confident of the ability of its aircraft and cruise missiles to penetrate Soviet defenses.

As I indicated earlier, the actual goal of the SDI program has become somewhat blurred. Some supporters now assert that the objective is not to protect the population from nuclear weapons, but to provide a much more modest level of defense. Such a defense, we are told, might serve to complicate enemy war plans and contribute to the survival of our strategic deterrent. Such a goal is probably attainable, particularly since our deterrent is at present quite secure in any event. But a multi-layered ballistic missile defense of the type being considered by SDI would appear to be an incredibly expensive way to seek additional reinsurance for the future survival of our deterrent forces. This requirement can be achieved at much less cost and with higher confidence by such straight-forward means as replacing vulnerable high-value fixed land-based MIRVed ICBMs with low value mobile single-warhead missiles such as the Midgetman or with further sea-based deployments.

Above all, by simply lowering the *initial* paper specifications for the system, we would not alter the Soviet perception of the ultimate intended purpose of the system—a nationwide population defense. Faced with such an over-designed system, the Soviet Union would no more accept our statement of intentions that we would the Soviet Union's in similar circumstances.

These are not new problems. They are the same problems that we struggled with in 1972 when the ABM Treaty was negotiated with the Soviet Union. The ABM Treaty, which the Senate ratified by a vote of eighty-eight to two, committed the United States and the Soviet Union not to *deploy* or provide a *base* for a nationwide ballistic missile defense system. The treaty sought to prevent an open-ended arms race that would follow deployment of these systems and to provide a basis for future reductions in strategic offensive arms. The ABM Treaty is, of course, in direct conflict with the goals of the SDI whether they be comprehensive or limited. The treaty would have to be abrogated or openly violated if the SDI program goes beyond the present research stage.

President Reagan has called for a 50 percent reduction in strategic offensive nuclear weapons. Soviet General Secretary Gorbachev has agreed to this objective. But Gorbachev has made it clear that the Soviet Union will not agree to such reductions if the United States embarks on a strategic defense program to eliminate the Soviet deterrent. In fact, he has promised to take steps to assure that this will not happen.

I believe the American people will soon have to make a clear choice that will affect their security for decades to come. They can pursue the Strategic Defense Initiative in the vain hope of finding a technical means of eliminating the threat of nuclear weapons. This would lead to a very dangerous and extremely expensive nuclear arms race with the Soviet Union. Or they can limit the SDI program for the foreseeable future to the research allowed under the ABM Treaty and seek major reductions in the level of strategic nuclear arms.

We cannot have it both ways.

4

SDI: Why It Is Important

Edward Teller

I have lived through a part of history in which technology has brought revolutionary changes. From the time that fission was discovered, as a surprise to all of us, to the day where the first atomic bomb exploded, the time passed was less than seven years. It happened that I was involved in practically every phase of that development. A democracy was in control of the new weapon and, therefore, the freedom of the world was preserved. Think of what would have happened if the first who got hold of nuclear weapons had been Hitler or Stalin?

Every one of us has to try and guide the development as best we can; guide it by listening to each other, by answering questions, by clarifying our minds. There is hope that we might be able to get away from weapons of mass destruction and move toward the use of technology for defense against weapons of mass destruction. Unfortunately, we in the United States have neglected to move in that direction. The Soviets have done what they could in developing methods of defense. And because they have worked for more than two decades they are in fact ahead of us today. In a few minutes I cannot describe to you the whole evidence, but I can give you some illustrations.

In Siberia, on the shores of Lake Balkash, on a military test site the Soviets have deployed a big laser. They can shoot with light velocity and destroy with precision a satellite or maybe an incoming missile—this is just one of the examples. We work in the laboratory; they are deploying. Another example: twenty years ago a great and wonderful Soviet scientist, Andrei Sakharov, published a paper about a magnetic field of twenty-five million gauss. This is a magnetic energy density thirty times higher than you get in a high explosive. I do not believe that this could have been done

---

Dr. Teller's speech was delivered at a conference on SDI at the Virginia Military Institute, April 7, 1986.

otherwise than with the help of a nuclear explosion. To this day we do not know how to convert nuclear explosive energy into electromagnetic energy to parallel Sakharov's feat.

Sakharov is a courageous man who dares to differ from his very single-minded government. He suffered and his family suffered. I do not know how many of us could match his courage. I never met the man. He was the one to develop the hydrogen bomb in the Soviet Union and his fate is one of the few points in which all American scientists are interested and on which all of us agree. But back to SDI.

We have a new possibility of lasers not of light but of X-rays. The one truly novel thing in the SDI development is this X-ray laser. I cannot tell you how effective it will be. I cannot tell you because the work is secret. But while I do not know whether it will work really effectively, I also have to tell you that it is more significant than any other approach to defense. In other words, it is something on which research must proceed because ignorance has to be replaced by knowledge. We know that the Soviets invented the principles on which the X-ray laser is based. They are trying to impede by every means our work on that subject.

The technology involved in defense is breaking new ground. President Reagan's efforts toward international cooperation are even more innovative. Indeed, our government has taken unprecedented steps to work with everyone who is willing to work with us. There may be cooperation from private enterprise to private enterprise, from government to government, from our government to private enterprise abroad, whatever our partners choose. The British, the Germans, the Israelis are already working, and I hope many others will.

We should work on defense of every kind, not only on the important antimissile missile. Defense against incoming missiles is already on its way, whether they are long-range or short-range. I believe that defense against the short-range missiles is the easiest because short-range missiles soon rise above the horizon, and while they are accelerating they are very vulnerable. That is when they might best be attacked by antimissiles or lasers.

In the early phases of the Strategic Defense Initiative we had several successes; but every time we are succeeding I have to ask myself, how far along this road have the Soviets traveled? They have worked on neutral particle beams, on electron beams, on high velocity objects, on lasers, on varieties of nonnuclear and nuclear defenses. As long as these are used against weapons of mass destruction, the Russians are justified to work on them, and so are we.

There is one question that the opponents of SDI ask again and again: What will happen to our negotiations? Actually we cannot negotiate if we are weak. If we want to negotiate, we must have the strength to resist. We may bring about cooperation first with our friends so as to create unity. We may bring about cooperation among those nations which have a deep and abiding respect for the individual. But in the end we have to negotiate with everyone. The president has said we should share with the Soviets. Only if there is universal agreement on some measures of security, can we get away from the dreadful balance of terror in which there is today more terror than balance. We must replace terror with knowledge and with some real program for assured survival—survival for everyone.

Our hope is that by joint use of technology, defense can become easier and less expensive than the development and deployment of means of aggression. This is the aim of SDI.

If we succeed in defensive technology, we shall be prepared to make the last step which I shall now mention. Because the last step is not just technical knowledge but human understanding resulting in reliable agreement. This is the aim of negotiations. Human understanding without technical knowledge will not suffice. Technical knowledge without human understanding will not lead to real peace. Technical knowledge and human understanding jointly, that is the two-sided unit on which peace on our globe can be founded in a secure manner.

5

The SDI Debate: A Critic's Perspective

Peter A. Clausen

ALMOST four years after President Reagan initiated a major US commitment to the development of ballistic missile defenses, two things are apparent. First, the Strategic Defense Initiative (SDI) dominates the landscape of strategic and arms control policy as few issues have done before. Second, we still lack a coherent statement of the objectives of the program, its strategic rationale, and its relationship to other elements of American national security policy. This remarkable situation accounts for the unfocused and often unproductive nature of the debate that has raged around the SDI since its inception. Proponents and critics of the program have argued past each other, seldom agreeing even on what the questions are, let alone the answers. We have yet to sort out the real from the false choices and to structure the issue of strategic defense so as to encourage intelligent decision-making.

As has often been pointed out, ballistic missile defense raises two kinds of questions—those of feasibility and those of desirability. Neither can be addressed without first carefully specifying the technical objectives and strategic functions of the defense, a condition the SDI debate has notably failed to satisfy.

Instead, that debate goes on at two very different levels. At the level of public perceptions and presidential rhetoric, it is about protecting American society from nuclear attack. The goal set forth by President Reagan is a defense so effective as to make nuclear weapons "impotent and obsolete," replacing the policy of deterrence whereby the US and the Soviet Union hold each other hostage to the threat of devastating retaliation. At the level of practical policy, however, the SDI is something much less radical. As described by SDI program managers, the objective is to use missile defenses

This is a revised version of an article that appeared originally in *The Fletcher Forum,* Winter 1986.

to increase the uncertainty and complexity of a Soviet first strike. In essence, this means defending US nuclear forces, especially vulnerable land-based missiles, to preclude a successful preemptive attack against them. The role of defenses here is not to shield American cities but to ensure the survival of nuclear retaliatory capabilities—the classic requirement of deterrence.

The disparity between these two versions of the SDI—one revolutionary, the other mundane—is striking. The first has compelling appeal but is almost certainly unattainable. The second is much less demanding technically, but may or may not be a prudent way to deal with the problem of deterring a Soviet attack.

Although it continues to provide the basis for public support of the SDI program, the president's vision of an invulnerable America is understood by virtually all technical specialists, both in and out of government, to be unattainable. In order to deny the Soviet Union the physical capability to devastate American society, we would need a defense approaching perfection against all means of delivering nuclear weapons onto US territory—not only ballistic missiles but bombers, low-flying cruise missiles, and all the conceivable ways that bombs might be smuggled into the country. Near-perfection is required simply because of the immense destruction that even a handful of nuclear weapons (of the approximately ten thousand currently targetted by the Soviet Union on the United States) would cause. A "leakage" through the SDI shield of only 1 percent of a full-scale Soviet attack would leave the US exposed to historically unprecedented levels of death and destruction.

The shield that could fundamentally alter our condition of vulnerability is a technological fantasy. The misplaced optimism of some SDI advocates on this issue, as reflected in analogies to the Apollo moon landing or the Manhattan Project, ignores two critical distinctions. The first is between an individual device and a working weapon system. The task of SDI is not simply to build, for example, a laser beam or rocket that can destroy a ballistic missile. It is to build a system of hundreds or thousands of such weapons and their associated sensors and control systems; deploy them survivably; and operate them with almost perfect reliability, without previous testing, the first time they are called upon. Moreover, the system must operate while defending itself in a hostile environment, against a determined effort to defeat it.

Hence the second critical distinction: Nuclear defense is not a problem against nature, but against an active adversary who will attempt to deny us a highly effective shield. Those who believe technology can accomplish anything should bear in mind that technology works for both offense and defense, and that a vast array of offensive countermeasures exists to exploit the weaknesses and vulnerabilities of missile defenses. Moreover, since the defense must be nearly perfect if the goal is to protect cities, the offense need only get a small fraction of its missiles through to frustrate that objective. This basic asymmetry in performance requirements gives the offense an inherent and overwhelming advantage.

In these circumstances, vulnerability to nuclear attack will remain a fact of life for both superpowers whether SDI is pursued or not. No foreseeable technical development can alter the high probability that nuclear war would be suicidal for both countries. Recognition of this fact is at the heart of nuclear deterrence and undoubtedly is a major reason for the lack of armed conflict between the US and Soviet Union during the postwar period. "Mutual assured destruction" is not, as advocates of strategic defense often assert, a perverse policy of intentional vulnerability, but rather a description of the likely outcome of a nuclear war. As such it is a reminder that the only hope of protection is through the avoidance of nuclear war. In the absence of nuclear disarmament or a political reconciliation of the superpowers such that war between them becomes unthinkable, this means we will continue to rely on deterrence.

The real issue raised by SDI, then, is not, "Can we escape from our vulnerability to nuclear destruction?" It is, "Would a defense against Soviet missiles, of uncertain but less-than-perfect effectiveness, improve nuclear deterrence?" This is only in part a technical question; the critical issues have to do with strategy and politics and the dynamics of the superpower arms competition. A serious debate on the merits of the question has scarcely begun.

This debate would be well served by the acceptance of two ground rules. First, advocates of SDI should drop the claim of moral superiority they often make for defenses. A shield corresponding to President Reagan's utopian goal would of course be morally preferable to the existing balance of terror, but the same cannot be said for deterrence-enhancing defenses. As a way of deterring a Soviet attack, defenses are no more or less moral than other methods (such as hardened missile silos or submarine basing) for ensuring the survival of nuclear forces against a preemptive strike. In each case, the objective is to maintain the ability to do mortal damage to Soviet society. The "deterrence by denial" favored by SDI supporters is at bottom simply another way of underwriting the existing strategy of "deterrence by retaliation," and fully shares the latter's moral ambiguity.

Second, there is a need for realistic net assessment of the strategic consequences of deploying ballistic missile defenses. The choice is not between the present world and one in which the US is defended, but between the present and a world in which both superpowers are simultaneously deploying defenses. The judgment underlying the 1972 Anti-Ballistic Missile Treaty, which strictly limits missile defenses, was that the second world was likely to be considerably less stable, and less congenial to American interests, than the first. While it is appropriate that this judgment be reassessed as technical and political conditions change, the burden of proof is on those who would set aside the ABM Treaty—surely the most significant constraint the superpowers have managed to place on the arms race—to allow SDI to proceed.

However, advocates of SDI have for the most part not supported their case with realistic strategic analysis. Instead, they have made dubious assumptions about Soviet reactions to SDI and have displayed a perplexing blind spot toward the contribution of the ABM Treaty to their professed objectives of stable deterrence and arms control. For example, those who argue for defenses to strengthen US deterrence of a Soviet first strike have consistently missed the point that the ABM Treaty itself protects US retaliatory capabilities by limiting Soviet defenses. In a world of defenses, we would sacrifice this protection. As a result, while more American missiles might survive a Soviet attack, fewer would reach their targets in the Soviet Union.

The outcome of this trade-off would depend on the relative effectiveness and coverage of US and Soviet defenses. On *a priori* grounds, however, the US would seem to have more to lose than to gain, since the majority of American nuclear warheads (those based on submarines) are already invulnerable to preemptive attack. In exchange for protecting the vulnerable, land-based minority, we would be subjecting all of our missile warheads to the attrition that a Soviet defense would extract. If that defense were to achieve even a modest 50 percent effectiveness, the US would suffer a net loss in retaliatory strength.

Of course it is hardly likely that the US would in practice allow this situation to develop. Instead, it would invest in the new offensive weapons and countermeasures necessary to offset Soviet defenses and prevent the erosion of American deterrence. Thus, Secretary of Defense Caspar Weinberger has stated that the threat of a Soviet territorial defense "would require us to increase the number of our offensive forces and their ability to penetrate Soviet defenses to assure that our operational plans could be executed." Nor is it likely that the Soviet Union would acquiesce in the erosion of its own deterrent that an American defense would cause. Yet SDI proponents seem to expect the Soviets to do just that, and indeed to accelerate the process by agreeing to cuts in offensive forces in the face of a concerted American effort to develop effective defenses.

The notion that SDI is a catalyst to arms control, despite pervasive evidence that it is the key obstacle to progress at the Geneva talks, is perhaps the least plausible of the arguments used to support SDI. This is not to deny that a US agreement to *curtail* SDI might well lead to an arms control breakthrough. But the president has so far rejected the idea of using the program as a "bargaining chip," and has presented SDI as the cornerstone of a whole new arms control framework. The problem with this vision is that it assumes a level of cooperation, trust, and mutuality of interests between the superpowers that would be difficult to imagine even in the best of circumstances, and that is directly at odds with the Reagan administration's own premises about US-Soviet relations.

This incongruity points to a basic tension between two different models of the SDI's role in superpower relations. The first model posits that the transition to a "defense-dominant" world is in the mutual interests of the US and Soviet Union, and should be pursued as a cooperative endeavor with arms control playing a central role. The second sees SDI as a way of improving America's strategic position relative to the Soviet Union by shifting the arms competition to areas of US technological advantage. In this view, SDI should be used as a lever to seek arms control on US terms; failing that, it should be deployed unilaterally.

These two models coexist uneasily within US policy. The first is more prominent in the public rhetoric of the administration, but there is ample evidence that the second is closer to its ideological heart and informs its thinking about the strategic and military role of SDI. Needless to say, the second also defines Soviet perceptions of the program.

Ironically, the first—"idealistic"—approach to SDI is in one sense the more realistic of the two: A stable defensive transition could be accomplished, if at all, only with the closest superpower cooperation. If attempted competitively, defensive deployments would result not in unilateral American gains but in an accelerated arms race and a steep decline in strategic stability. In practice, however, the political ingredients for a cooperative transition are not in sight. In effect, the superpowers would have to agree to base their mutual security on something other than the threat to destroy each other. And the transformation of superpower relations that would make such an agreement possible would at the same time make SDI unnecessary.

6

SDI's Impact on US-Soviet Relations

Gregory M. Suchan

I wish to discuss the impact of the Strategic Defense Initiative on US-Soviet relations from two viewpoints. The first is arms control, which some people think is the sum total of the US-Soviet relationship; it is not. The second is the broader context of the political and security relationship between the two superpowers.

Arms Control

One could argue that SDI has already had a beneficial impact on the arms control process, in that it may have been a key element in the Soviet decision to resume the negotiations on strategic and intermediate-range nuclear forces, which the Soviet Union broke off in 1983. Clearly the Soviets have expressed great concern over SDI, and the unique difference between the current negotiations and those conducted prior to late 1983 is the addition of a third Negotiating Group dealing with defense and space issues, including SDI.

That said, SDI is clearly a major issue separating the two sides in the Geneva negotiations. The United States believes that the US and the USSR should begin now to implement deep reductions in their strategic and intermediate-range nuclear forces. At the same time, the United States is committed to pursue, within the limitations of existing arms control agreements, the possibility that new technologies may permit an effective defense against nuclear weapons. If effective defenses prove feasible, the United States hopes to cooperate with the Soviet Union in the joint management of a transition to a new strategic regime in which both sides rely increasingly on defenses for their security and less on their ability to threaten the other with nuclear weapons.

For its part, the Soviet Union argues that SDI is a violation of the 1972 ABM Treaty, a charge which I doubt that informed Soviets themselves take seriously. Secondly, the Soviet Union charges that the prospect of effective defenses against nuclear weapons makes agreement on the limitation and reduction of nuclear weapons impossible.

Regarding the issue of SDI's compliance with arms control commitments, I must first point out that the president has directed that the Strategic Defense Initiative be conducted in a manner consistent with the ABM Treaty and other US legal obligations. This injunction is, of course, being taken seriously, and the SDI program is adhering strictly to treaty limits. The Soviets themselves, who have an active ABM research, development, testing, and deployment program, recognize that extensive activities in the ABM field are permitted under the 1972 treaty. Indeed, in 1972 the defense minister of the Soviet Union, Marshal Grechko, informed the Supreme Soviet that the ABM Treaty "imposes no limitations on the performance of research and experimental work aimed at resolving the problem of defending the country against nuclear missile-attack."[1]

The US Congress conducts a constant review of SDI activities. Although the Legislative Branch contains quite a few opponents of SDI, no one in Congress argues that SDI has gone beyond the limits of the ABM Treaty. In addition, SDI's prominent and vocal critics outside the government, such as scientist Dick Garwin and the former Ambassador Gerard Smith, concede that current SDI activities are consistent with the treaty.

This should be contrasted, incidently, with the Soviet Union's own record under the ABM Treaty. At Krasnoyarsk in central Siberia, the USSR is constructing a large phased-array radar which is essentially identical to other radars which the Soviets themselves have identified as being for early warning purposes. Article VI of the ABM Treaty prohibits the construction of new early-warning radars except on the periphery of the national territory and oriented outward. The Krasnoyarsk radar, however, is located 750 kilometers from the nearest Soviet border and is oriented in a direction where the Soviet border is 4000 kilometers away. Again, even critics of SDI such as Ambassador Smith and Senator Edward Kennedy have agreed that the Krasnoyarsk radar is a violation and a threat to the existence of the ABM Treaty.

Let us turn now to the more serious issue of whether effective defenses against nuclear weapons are consistent with efforts to negotiate the reduction of such weapons. The Soviets have argued that reductions in strategic offensive arms are impossible unless SDI is banned. In the United States, former Ambassador Smith and other prominent SDI critics have written that "it is possible to reach good agreements, or possible to insist on the Star Wars program as it stands, but wholly impossible to do both."[2] This point

1. *Pravda*, September 30, 1972.

2. McGeorge Bundy, George F. Kennan, Robert S. McNamara, and Gerard Smith, "The President's Choice: Star Wars or Arms Control," *Foreign Affairs* 63(Winter 1984–85), 277.

of view is an important part of the conventional wisdom associated with the ABM Treaty. The underlying rationale for this view (at least in the United States) is that, in order to deter war, both sides must be convinced of their ability to destroy the other under any circumstances, including after having absorbed a first strike. If one side's defenses reduce the other's confidence in its capability to inflict such damage, the latter would have no alternative but to increase its offense to compensate for the other's defenses.

But there is an alternative way of approaching deterrence. Deterrence is based on convincing a potential aggressor that he cannot achieve his objectives at an acceptable cost. An offensive-reliant strategic regime, such as what we have at present, focuses on increasing the potential cost to the aggressor. Alternatively, however, it would be possible to base deterrence on the ability to deny an aggressor his ability to achieve his objectives, that is, by establishing an ability to defend against a potential aggressor.

At this point I should point out that the Soviet Union, despite its attacks on SDI, clearly understands the importance of strategic defenses. The Soviet Union retains, around Moscow, the one ABM deployment site permitted under the ABM Treaty, which has recently been upgraded; the US ABM site was dismantled years ago. The Soviet Union also maintains the world's largest strategic air defense system, including more than twelve hundred dedicated interceptor aircraft, nine thousand strategic surface-to-air missile launchers, and some ten thousand air defense radars. Neither the Moscow ABM system nor the mammoth Soviet investment in strategic air defense are consistent with the view held by many in the United States that mutual vulnerability to the other side's nuclear forces is not only an inevitable but a desirable fact of life.

Nor can we assume that Soviet strategic defense efforts are confined to these more traditional concepts. We know that thousands of Soviet scientists and engineers are working on many of the same advanced technologies under investigation in SDI and have been doing so for years. In the context of what we know about Soviet military doctrine and Soviet investment in other areas of strategic defense, it would be unreasonable and imprudent for us to assume that a major part of Soviet research into advanced laser and particle-beam technology, for example, is not devoted to ballistic missile defense applications.

Returning to the issue of whether the prospect of effective defenses against nuclear weapons is consistent with negotiated reductions in such weapons, the US government clearly believes that it is, provided truly effective defenses prove feasible. Why is this so?

The first reason is that one of the key criteria we have set for judging whether future defenses are "effective" is "cost-effectiveness at the margin." This criterion refers to creating disincentives to attempt to overwhelm defenses through increases in offensive capabilities. If this criterion is met, then it would not make sense for one side to proliferate its offensive weapons but to channel its resources into defending against the other's offenses. Defenses which meet this criterion, therefore, would be entirely consistent with the limitation and reduction of offensive forces, because expanding one's offenses would essentially be a waste of effort.

In addition, by diminishing the military utility of offensive nuclear forces (and particularly ballistic missiles), effective defenses would facilitate the reduction of these forces beyond the 50 percent cuts agreed in principle by President Reagan and General Secretary Gorbachev in Geneva. Also, if negotiated reductions in offensive nuclear forces reach very low levels or zero, effective defenses would help underwrite security against problems which would be more threatening than at present. These problems include cheating, sudden abrogation of treaty limits, or the acquisition of nuclear weapons by other states.

In conclusion, therefore, regarding SDI's impact on arms control, the Strategic Defense Initiative is a major issue separating the two sides in the Geneva negotiations. In the view of the US government, it need not be. We believe that effective defenses are consistent with and conducive to reductions in offensive nuclear forces, both now and in the future. I might also point out that, even with these considerations aside, one should not take too seriously the claims that SDI is all that is blocking an arms control agreement requiring deep reductions in offensive nuclear forces. As Jack Mendelsohn well knows, we were engaged in START negotiations before there was an SDI, and there was no meeting of the minds between the United States and the Soviet Union. The removal of SDI as an "obstacle," therefore, would be no guarantee of an acceptable START agreement.

US-Soviet Strategic Relationship

Let us turn now to the impact of SDI on the broader US-Soviet security and political relationship, which involves much more than arms control. Unquestionably, this relationship could stand improvement. However, this was also true before there was an SDI. Indeed, it is clear that SDI is irrelevant to most of the broader issues separating the United States and the Soviet Union, such as regional conflicts like Afghanistan and Central America and the important issue of human rights. An important point to bear in mind is that military programs, whether SDI or a new tank, do not cause international antagonism; rather, military programs reflect problems which already exist. Nevertheless, if the SDI program proves that effective defenses are feasible, then it *should* have a significant impact on the US-Soviet relationship. And despite what Soviet spokesmen say, it is not at all clear that the result would be an increase in tensions.

As I mentioned earlier, the current offense-reliant strategic regime is based on the assumption (at least in the United States) that if each side can maintain its ability to threaten nuclear retaliation against any attack from the other, it can impose on the aggressor costs which outweigh the potential gains to the aggressor. But there are certain limitations to this approach. The first is that, from what we know of Soviet strategic doctrine, the ability to preempt in a crisis is a key element of reducing damage to the Soviet Union in a future nuclear conflict. A nagging concern of US strategic planners for many years has been that, in a future crisis, the Soviet Union might be tempted to use its nuclear forces first to limit the damage from a US retaliatory response or possibly to deter the United States from responding at all.

Another problem is the amount of uncertainty involved. Back in October 1964, a Soviet general named N. Talenskiy wrote in the Soviet journal *International Affairs* the following statements:

> When the security of a state is based only on mutual deterrence with the aid of powerful nuclear missiles, it is directly dependent on the goodwill and designs of the other side, which is a highly subjective and indefinite factor. . . . The creation of an effective anti-missile system enables the state to make its defenses dependent chiefly on its own possibilities and not only on mutual deterrence.

We should ask ourselves how well we understand the way the Soviets think and what risks they might be willing to take in some circumstances. Conversely, we must ask ourselves how well the Soviets understand us and our determination to defend a society which is so very different from their own. There are sufficient reasons to believe that our respective wisdom on this topic is not infallible. It is clear, from hindsight, that both sides miscalculated greatly during the Cuban Missile Crisis—the United States by failing to anticipate Khrushchev's bold gamble to put the missiles in, and the Soviets by misjudging Kennedy's willingness to accept a crisis to get the missiles out. Stability in the current strategic regime, therefore, requires that East and West not misread the intentions and capabilities of the other in a crisis. We should therefore question, given the high stakes involved, whether it is necessary or prudent for us to rely indefinitely on what General Talenskiy called "a highly subjective and indefinite factor."

The prospect of effective defenses against nuclear weapons offers the prospect of altering the US-Soviet strategic relationship in a way in which both sides would rely more on their own ability to defend against the other's forces, and less on their ability to influence what they perceive to be the other's goals, objectives, and willingness to take risks and accept costs. To me this sounds like a better and more stable basis for US-Soviet relations.

Let me underscore again that the United States is attempting to assure the Soviet Union that this—and not the achievement of US superiority—is the goal of the Strategic Defense Initiative. We are conducting current SDI activities within the limitations of the ABM Treaty. We have proposed discussion now, years before we will even know whether effective defenses will be feasible, of the relationship between offensive and defensive forces and of how we might, in the future, carry out a transition to a more stable environment. And finally, if defenses prove feasible, we would seek to cooperate with the Soviet Union in jointly managing the transition to a defense-reliant strategic regime, so that both sides would know that their interests were being protected.

7

Five Fallacies of SDI

Jack Mendelsohn

IT is difficult to imagine a more disruptive issue in US-Soviet relations than the Strategic Defense Initiative. SDI has paralyzed arms control discussions, buffetted US-Allied relations, mobilized the Soviet military and propaganda apparatus, and jeopardized the already slim chances for improving bilateral relations between the superpowers.

It is also difficult to determine which aspect of SDI is the most disruptive: the falsely-placed hopes for a leak-proof astrodome defense; the hype surrounding the abolition of nuclear weapons; the waste involved in a "top-down" driven technology hunt; the futility of attempting to delegislate mutual deterrence; or the lack of logic which envelops the entire program. While SDI certainly provides a vast field for critical comment, this paper will concentrate on five major fallacies of SDI and their impact on US-Soviet relations.

SDI is predicated on five major fallacies which have affected the US-Soviet relationship: the cooperative fallacy; the technological fallacy; the exhaustion fallacy; the last-move fallacy; and the strategic fallacy.

The Cooperative Fallacy

In response to the knotty problem of how to assure the Soviet Union of the basically benign intent of the SDI program and how to restrain the Soviet Union from any rash acts before SDI comes on line, the president has several times stated that the US would share its SDI technology with the Soviet Union. Most recently, in an interview before the Geneva summit conference with a group of Soviet journalists, the president said: "If such a weapon is possible, and our research reveals that, then our move would be to say to all the world, 'here, it is available.' . . . And we make that offer now. It will be available for the Soviet Union, as well as ourselves."[1]

1. President Reagan, interview with Soviet journalists, October 31, 1985.

In addition, when faced with the problem of how to untangle the Soviets from their devotion to the existing offensive-dominated world of mutual deterrence, Paul Nitze, speaking for the administration, has described the need for a "cooperative" shift to a defense-dominated world. In a speech celebrated for at last creating a coherent rationale for SDI, Nitze noted that "what we have in mind is a jointly managed transition, one in which the US and the Soviet Union would together phase in new defenses in a controlled manner while continuing to reduce offensive nuclear arms. We recognize that the transition period . . . could be tricky. We would have to avoid a mix of offensive and defensive systems that, in a crisis, would give one side or the other incentives to strike first. That is exactly why we would seek to make the transition a cooperative endeavor with the Soviets."[2]

As regards the first cooperative proposal, is it conceivable that the US would consider transfering its most advanced technology to the Soviet Union and thereby aiding an adversary either to obtain or defeat our own defenses? Unless the Soviet leadership has totally changed their spots, handing over US defense plans and technology—which, incidentally, we are not prepared to do with even our closest allies—is an ultimate folly. As a recent Office of Technology Assessment (OTA) study on ballistic missile defense (BMD) notes: "If BMD plans or devices are transferred, potential adversaries might be able to study them to discover vulnerabilities, enabling them to circumvent or destroy our own such components. . . . Furthermore, many BMD-relevant technologies have applications in other military areas that we may not want to help the Soviets develop."[3]

As for the second cooperative aspect, why should the US have any reason whatsoever to trust or expect the Soviets to cooperate in the implementation of a transition to a defense-dominated world? Everything this administration and critics of the Soviet Union have said about the "evil empire" would lead one to just the opposite conclusion: that the Soviet Union would do everything within its power *not* to cooperate with the United States. And everything we intuitively believe about how nations interact, whether correct or not, would lead one to conclude that the Soviet Union, as a major adversary, cannot be trusted to cooperate.

A much more realistic assessment of SDI's impact on US-Soviet relations is that it will trigger a reaction in the Soviet Union, the general outlines if not the exact details of which are clear:

---

2. Paul Nitze, address to the North Atlantic Assembly, October 15, 1985. General James Abrahamson has said the same thing: We "would hope the Soviets would recognize the inevitability of the emergence of defensive systems and cooperate in the establishment of mutual deployment arrangements and mutual offensive force reductions."

3. Office of Technology Assessment, *Ballistic Missile Defense Technologies*, September 1985.

—the Soviet Union will assume that the US has no intention of cooperating in the development and implementation of SDI;[4]
—the Soviets will resist or respond to any effort to overturn offense-dominated deterrence;
—the Soviets will attempt to develop and obtain SDI technology, both overtly and clandestinely, and to subvert US efforts to develop a workable system; and
—the Soviets will exploit any changes or instabilities in the strategic balance created by SDI.

The Technological Fallacy

This fallacy also has two aspects to it. One aspect is the notion that since the US is technologically advanced it will somehow humble the Soviet Union in the race to develop SDI. Former Science Advisor George Keyworth put it this way: "I see this shift [from offensive to defensive weapons] as a decided advantage to the West in maintaining a stable peace. The reason stems from the superiority we and other Western countries have over the Eastern bloc in terms of industrial capacity and industrial base. . . . [The Soviets] have to play catch up when it comes to advanced technology. . . . In that way, by the expedient of always staying several steps ahead, we can thwart even the most aggressive attempts by our adversaries to keep up."[5]

Secondly, and conversely, the case is also made that the Soviets are so far ahead of the US in SDI-type research that we are compelled to catch up with them. For example, the administration has made clear its concern about a ground-based laser facility at the Sary Shagan test site, and President Reagan has said that the Soviets have "been conducting research in this sort of thing for a long time. And they already have far beyond anything we have."[6] Secretary of Defense Caspar Weinberger has recently reported that "the Soviets [are] ahead of us today in the development and deployment of strategic defenses."[7]

But Donald A. Hicks, the undersecretary of defense for research and development, notes in his annual technical report to Congress that the US leads the Soviet Union in fourteen basic technologies and trails in none, with the two nations about even in six areas. The Soviet Union has worked hard to redress the balance, the report adds, but with little success. As a

---

4. Administration spokesmen (e.g., General Abrahamson) have already made it quite clear that "it is imperative that we have a much more effective defense than they have." *New York Times,* December 17, 1985.

5. *Science,* July 1, 1983. General Abrahamson agrees with this: "In the key technologies needed for a broader defense—such as data processing and computer software—we are far, far ahead."

6. Remarks to the Committee for the Free World, London, March 19, 1985.

7. Department of Defense, Annual Report, Fiscal Year 1987.

matter of fact, this same report shows that since 1982 the relative US position in basic technologies vis-à-vis the Soviet Union has improved, from two areas where the Soviets were superior to the US and five where they were equal to none where they are superior and six where they are equal.[8]

The fact of the matter is that US technology in those areas likely to be critical to SDI is five to ten years ahead of comparable Soviet programs. For example, in target acquisition and computer technologies, the Soviets are six to ten years behind the US. And Soviet infrared tracking and homing tests have failed thirteen out of thirteen times. Even so, it is a fundamental mistake to attribute to a Soviet SDI program, or to a Soviet response to a US SDI program, the requirement that it reach the comparable US technological level. It probably will not do so, nor does it need to.[9]

The simplest and most immediate Soviet response to a high-tech SDI program may be a low-tech offense (proliferation and deception) and mid-tech counter-measures (hardening and antisatellite [ASAT] systems). Coupled with this is the fact that the technology to produce an effective ASAT weapon (for use against space-based defenses) in most instances is less stressing than the technology to produce an effective ABM weapon. This means that the demands on Soviet technology to defeat SDI will actually be less than those on the US to develop it. As Harold Brown has noted, "everything that works well as a defense also works somewhat better as a defense suppressor."[10]

But the Soviet Union cannot allow itself to be perceived as technologically inferior to the United States. Therefore, we can expect some sort of parallel effort to whatever the US undertakes in the area of SDI. The obvious mid-term Soviet response will, of course, be to develop those aspects of SDI which they can successfully master, for example ground-or space-based laser systems. Additionally, Soviet scientists claim that certain SDI counter-measures are already "on-the-shelf" and will be put on-line as the exact nature of the American SDI threat becomes clearer. And while there is no question that the complexity and "sweetness" of American SDI technology will by far outshine that of the Soviets, if they are not running, a Mercedes affords no better transportation than a Pontiac.

The Exhaustion Fallacy

This fallacy maintains that the US can, through its technological dominance, use SDI to race the Soviet Union into the ground economically, if not militarily. Again, it is unclear why those who distrust the intentions of

8. Undersecretary of Defense, Research and Engineering, Annual Report, Fiscal Year 1987. Radar sensors have moved from US-USSR equal to US superior.

9. Council on Economic Priorities Newsletter, December 1985.

10. "Is SDI Technically Feasible?" *Foreign Affairs, America and the World, 1985.*

the Soviet Union or who see the Soviets as engaged in the largest military build-up in history, refuse to heed Soviet leaders when they declare that "all attempts at achieving military superiority over the USSR are futile. The Soviet Union will never allow them to succeed. It will never be caught defenseless by any threat. Let there be no mistake about this in Washington."[11]

The Soviet nation is accustomed to making the economic sacrifices needed to meet a real or even imagined challenge, perhaps even more so than is the US. Even Secretary of Defense Weinberger, in the recent edition of *Soviet Military Power* (*SMP*), admits that "the Soviet leadership can devote a large percentage of the national income to defense programs—a cost no Western nation is willing to pay nor need incur in times of peace."[12] Moreover, external security threats are a mobilizing factor to the Soviet leadership. The American SDI program—which the Soviets characterize as an offensive threat—gives the Soviet leadership a convenient excuse to demand further sacrifices from its citizens while exhorting them to produce even more to meet the threat of US "aggression."

In other ways, "exhausting" the Soviet Union economically has a certain ring of irreality to it: the Soviet Union has the world's second largest economy, a large and skilled work force and an enormous resource base. Administration claims notwithstanding, there is no evidence that the Soviet Union is in any danger of economic collapse or even of being paupered by the sacrifices required in its pursuit of strategic defense. The Central Intelligence Agency and the Defense Intelligence Agency, in a recent report to the Joint Economic Committee, point out that since the mid-1970s, major investments in defense industrial facilities have resulted in a substantial expansion and upgrading of the Soviet defense industry. "As a consequence, most Soviet weapons expected to be delivered to the Soviet forces through 1990 will be manufactured in plants already built and operating."[13] Moreover, the Defense Department claims that the Soviet Union already has made, and continues to make, substantial investments in the strategic defense sector—the Soviet laser weapons program alone would cost roughly $1 billion per year in the US according to *SMP*.[14] If this is true, then marginal shifts in resource allocation—and not large-scale new investment—are all that will be required for the Soviet Union to undertake short- and mid-term responses to SDI.

Even so, as a recent Council on Economic Priorities Newsletter notes, "US efforts to force the Soviet Union into an accelerated SDI race . . . may, in the long run, backfire." At issue "is the question of how a strategic

11. Andropov, *Pravda*, March 27, 1983.
12. *Soviet Military Power*, p. 3
13. CIA-DIA, "The Soviet Economy Under a New Leader," March 19, 1986.
14. *Soviet Military Power*, p. 46.

defense effort might influence and change the Soviet scientific community." This is admittedly speculative, but "should SDI become a high priority," the newsletter continues, "such a program might lead to a greater emphasis on the building of the Soviet scientific infra-structure. Strategic defense relies heavily on computers and sensing technologies. The Soviets may, in turn, pursue these programs more aggressively and choose to accord higher priority to the coordination and efficient control of its research and development efforts. Thus, pressuring the Soviet Union toward this policy change may not be in the best long-term interests of the US" and, rather than leading to an "exhaustion" of the Soviet economy, may very well serve to reinvigorate it.[15]

There is no question but that SDI represents a technological challenge to the Soviet Union which it would rather not undertake. But the Soviet leadership will meet that challenge if forced to; this is the record of the last fifty years of Soviet behavior, and there is no sign on the horizon that Soviet determination to "keep up with the Joneses" will change.

The Fallacy of the Last Move

Hope springs eternal, it seems, especially when dealing with technology or the Soviet Union. The latest manifestation of this is the president's call for science to create defensive systems to render nuclear weapons "impotent and obsolete." This belief in the ability of science to solve what are essentially political problems is often linked with the fourth fallacy, that of the "last move"—the belief that somehow the next great technological breakthrough will knock out our adversary, put an end to the nuclear threat, and/or return strategic superiority to the United States.

In the US, at least, it is quite difficult to be rid of this extraordinary belief both in the ability of science to save us from our political follies or in the irrepressible fallacy of the last technological move. But as one of the scientists (David Parnas) engaged to review the computing demands of SDI put it, "it is our duty, as scientists and engineers, to reply that we have no technological magic that will accomplish that [i.e., making nuclear weapons impotent and obsolete]. The President and the public should know that."[16]

As for the fallacy of the last move, Senator James Exon (D-Neb.) observed: "I think it should be clear to all if there is one thing we have learned, although we may not have learned it well, [it] is that for every action there is a counteraction, and certainly if we proceed in this area, the

15. Council on Economic Priorities Newsletter, December 1985. John Kiser (in *Foreign Policy,* No. 60) says that "in the long run the threat posed by the SDI may actually strengthen the Soviets, pushing them more quickly into a broad-based computer culture and leading them to develop their own defenses ahead of the United States."

16. *Los Angeles Times,* September 22, 1985.

Soviet Union is not going to sit idly by. So, I guess while we must build up our defenses from a deterrent standpoint, I suggest that we never will find the ultimate weapon because whatever we do, the other side is going to take countermeasures."[17]

History gives us good reason to believe that, as in the past, the USSR will respond to new strategic threats. "Their reaction inescapably will be to match us in a defensive systems race, to increase their nuclear missiles and nuclear warheads and to develop decoys, chaff and other techniques to make sure that they can overwhelm any US defense."[18]

More fundamentally, the Soviet Union sees SDI as a direct threat to the US-Soviet strategic relationship, which is uneasy, but currently not critically fragile. As such, this threat must be met and mastered and the previous terms of the relationship reestablished. Most likely, the Soviet Union will attempt the latter: reestablishing the strategic relationship—that is, offense-dominated deterrence—by instituting a wide range of counter-threats and counter-measures to SDI. This is a predictable and endless cycle in which both sides have engaged since the beginning of the nuclear competition. A partial listing of the SDI counter-options available to the Soviets includes:

anti-satellite weapons	ground-based lasers
electronic countermeasures	X-ray lasers
space mines	pellets in orbit
paramilitary forces	weapons proliferation
depressed trajectories	clustering ICBM launches
booster hardening, spinning	fast-burn boosters
quick PBV release	maneuvering
salvage fusing	penetration aids
decoys	anti-simulation
masked warheads	saturation attack[19]

These and other changes in the Soviet strategic force, such as greater Soviet reliance on air- and sea-launched long-range cruise missiles, could be initiated and put in place in a comparatively short time—that is, compared to any substantial deployment of SDI hardware. Ironically, as in the economic sector, these changes in Soviet hardware and strategy could lead to a situation where the Soviet Union actually improves its relative standing vis-à-vis the US as a result of the stimulus of an SDI program that the US might not actually deploy. Thus, rather than being the "last move," SDI is likely to trigger a massive series of intermediate moves and counter-moves on both sides with unpredictable and—with one exception—uncertain out-

17. Testimony before the House Foreign Affairs Committee, May 1, 1985.

18. Clark Clifford, testimony before the House Foreign Affairs Committee, May 1, 1985.

19. *SDI: Progress and Challenges,* Staff Report Submitted to Senators Proxmire, Johnston, and Chiles, March 17, 1986.

comes. The one certain outcome is that neither side will permit the other to improve its relative strategic position over the long term.

The Strategic Fallacy

For a number of socio-psycho-political reasons—including an attachment to an offense-dominated strategy, traditional adversarial distrust, and an ineradicable knowledge of weapons technology—it is not likely that the present strategic relationship based on offense-dominated nuclear deterrence will be abandoned. Deterrence based on the capability to "punish" is not a strategy of choice; it is a "condition" rooted in man's socio-political behavior. It is, as Hans Bethe has said, "not a policy or a doctrine but rather a fact of life."[20]

If that is true, and I firmly believe that it is, then the US cannot simply "delegislate" mutual deterrence and transition from an offense-dominated to a defense-dominated world. It is partly this realization which led the administration to propose a discussion of a "cooperative transition" touched upon earlier. But even if we wished to declare mutual deterrence outmoded, no one knows how to proceed to the next stage. The recent comprehensive OTA study of BMD had this to say about an arms control agreement intended to phase in BMD:

> It would have to establish acceptable levels and types of offensive and defensive capabilities for each side and means for verifying them adequately. It would have to specify offensive system limitations that prevented either side from obtaining a superior capability to penetrate the other's defenses. It would have to specify the BMD system designs for each side that would not exceed the BMD capabilities agreed to. *It is important to note, however, that no one has yet specified in any detail just how such an arms control agreement could be formulated.*[21]

And no one can, from either within or without the administration!

I think it worthwhile to continue to quote from the OTA study on this matter:

> OTA was unable to find anyone who could propose a plausible agreement for offensive arms reductions and a cooperative transition. . . . Without such agreement on the nature and timing of a build-up of defensive forces, it would be a radical departure from previous policies for either side to make massive reductions in its offensive forces in the face of the risk that the other side's defenses might become highly effective against the reduced offenses before one's own defenses were ready.

20. *Scientific American*, October 1984.

21. Office of Technology Assessment, *Ballistic Missile Defense Technologies*, September 1985. Emphasis added.

Although this is the last of the "fallacies" I have chosen to discuss, it is the most basic one. The US, much as it may wish to, cannot single-handedly move out from under the umbrella—or threat—of mutual deterrence. It may not be a comfortable position, but it is a realistic one and one which can be better managed but not overthrown. This is where SDI comes up against its most demanding challenge: no one can satisfactorily explain how SDI will accomplish anything except provoke a spiral of offense-based deterrence to higher levels.

Implications for the Future of US-Soviet Relations

What are the implications of these five fallacies for the future of United States–Soviet relations?

In regard to the first or "cooperative" fallacy, despite the call—even the reliance—on cooperation to effect a move to a defense-dominated environment, there is no chance the Soviet Union will see the US strategic defense program as a cooperative opportunity to improve US-Soviet relations. Rather, the Soviet Union will continue to view SDI as a threat to the very basis of the existing strategic relationship, and as a challenge to the technological "manhood" of the Soviet industrial establishment. So instead of cooperation, we will have increased tension between the superpowers and an intensified—not diminished—adversarial relationship.

As for the second or "technological" fallacy, we are likely to find that rather than pushing the Soviet Union to the technological edge, we will have provided it with a mobilizing factor for the next decade or two. Also, quite unintentionally, we might force it to organize and forge ahead in just those areas where it is weakest and where it would be to US advantage to leave it so. Without question the Soviet Union is having difficulty in adjusting to the computer-driven, information-processing revolution. As Loren Graham observed in an article two years ago in the *Washington Post,* "if we can gain time by controlling the military technology that can so easily destroy us all, the civilian computer technology that is now penetrating to the lowest level of society—the individual—will give a real advantage to societies that do not try to control information."[22] A vigorous US SDI program risks doing exactly the opposite.

As for the third or "exhaustion" fallacy, even Secretary Weinberger admits that Moscow shows no indication of reducing the percentage of resources dedicated to the Soviet armed forces. In brief, it is very unlikely that SDI will cause the "evil empire" to throw in the strategic towel. Marshal Akhromeyev has made it very clear that "if this process goes on we will have nothing to do but to take up retaliatory measures in the field of both

22. March 11, 1984, Outlook Section.

offensive and defensive weapons."[23] There is no reason not to believe the Soviets on this issue, and there is no indication they will not make every sacrifice to assure the resources are available for whatever effort they may undertake to counter SDI.

If the Soviet Union is forced to mobilize in order to respond to SDI, then it is likely that the US will become more useful as a whipping boy than as a partner. This will make it almost axiomatic that the hard-liners in foreign and domestic policy will dominate deliberations in Moscow and that diversionary challenges to the US in the Third World and elsewhere will become even more attractive to the Soviet leadership than they are at present.

As for the fourth fallacy, that of the "last move," rather than bringing about the end to nuclear weapons, we are more likely to see SDI stimulate a very vigorous and broad-gauge offensive response. The most direct Soviet response to SDI will be to proliferate, harden, and decoy offensive systems. Secondly, the response will focus on defensive counter-measures to SDI, including ASAT systems and ground-based laser threats. Thirdly, the Soviets will work on their own version of SDI. It may not be an elegant system, but as Stephen Meyer has noted, "the innovation and development philosophy that allowed the USSR to deploy the first ICBM, ABM system, and SSB ('don't make it fancy, don't worry if it doesn't quite work initially, and let people run it') may triumph again."[24]

Another result of striving for the ultimate technological "fix" of SDI is likely to be the destruction of arms control as we have understood it for the past fifteen years. This is not meant as a value judgment, but it is unclear what, if anything, will substitute for the loss of this process. For good reason, the US military has favored retaining what little arms control is still in place: it places predictable constraints on the Soviets; it simplifies US force planning; and it conserves resources. All of these advantages would be lost without arms control: SDI would provoke an unpredictable series of reactions from the Soviets; it would enormously complicate US force planning, both offensively and defensively; and it would represent an intolerable drain on the budget.

The last of our fallacies, the strategic one, has the most serious implications for US-Soviet relations. An effort by the US to move away from offense-dominated deterrence is doomed to failure, and the development and deployment of strategic defenses will be viewed by the Soviet Union as both provocative and destabilizing. At a minimum it increases fears that the actual goal of an imperfect strategic defense is to protect against a ragged retaliation after a first-strike by the side with the defensive systems. We can expect a vigorous quantitative and qualitative response from Moscow, one

---

23. *New York Times,* December 17, 1985.
24. *Survival,* November-December 1985.

that—together with the demise of traditional arms control—will revise the nature, if not the terms, of the strategic relationship. Specifically, we will stimulate the growth of strategic arsenals on both sides, waste untold billions on deploying useless defensive systems, increase tension, and decrease stability. And after all this, mutual deterrence will still be with us as the primary means of keeping the uneasy peace.

In short, our five fallacies lead to the following five implications for the future of our relations with the Soviet Union:

—rather than cooperation we will have tension;
—rather than decreasing our technological edge we may well dissipate it;
—rather than exhausting the Soviet Union we will mobilize it;
—rather than making the "last move" we will trigger another arms cycle; and
—rather than abandoning an offense-dominated strategy we will reinforce it at higher and less stable levels.

It is difficult to conceive of a more serious, threatening, and negative set of implications for the future of US-Soviet relations.

8

The Soviet Union's View of SDI

Sergei Kislyak

WE in the Soviet Union regard SDI from a different perspective than you Americans do. You have schools of thought about SDI, pros and cons. I am afraid that we have more unanimity about it, and we regard it as a real threat for several reasons.

One, we do not judge American intentions, American military programs, just by the arguments presented in favor of them. We do not believe that the SDI, which is promoted as a peace-keeping technical fix, would be so. We rather judge such a program by the military capabilities it would give the United States. From our standpoint, SDI is just another attempt, on a more sophisticated technological level, to gain military superiority; we could not see it otherwise. In our view, SDI is an attempt to create a shield in order to obtain a first-strike capability. If the United States deploys such a shield, it would gain the capability of striking first and knocking out our reprisal weapons. I remind you of the quote by Mr. Weinberger that if the Russians ever created such a shield, it would be a great threat to the United States. I can assure you that we regard American SDI in this way, and our military regards it just the same.

The second thing which is dangerous about SDI is that it is not conceived as a means for stabilizing peace, which, proponents argue, would come as a consequence of defensive programs—never! SDI comes in a package: a further offensive build-up and so-called defensive weapons. If you take a look at the budgetary estimates of this administration for the next fiscal year, you will see that strategic modernization programs will be continued along with SDI. Thus, in our view, what we face in this country is the attempt to continue an offensive build-up, plus SDI research and development.

This is an edited transcript of Mr. Kislyak's remarks delivered during a conference on SDI at the Virginia Military Institute, April 7, 1986.

We have analysed SDI quite thoroughly. Our Academy of Sciences' public report, which analysed the technical aspects of it, also came up with twenty-five or thirty countermeasures which we can take in order to make SDI obsolete. They vary from building up weapons to saturate ABM systems to other weapons to neutralize the systems. There are many technical options to apply against it, and our scientists believe that they will be taken at a fraction of the cost of SDI.

I do not know at this moment what would be our response, what would be the countermeasures, whether they would include an increase in appropriate offensive capabilities or something else. I am not a military man, and I am not a scientist; I do not know what it would be. But I am sure that we would allow no military superiority over our country. We have never allowed it in the past; we have faced many prior threats from this nation in the form of nuclear weapons, MIRVs, cruise missiles or whatever. We have always been successful in restoring the balance. I will take up what this balance will be afterwards. It will not be of our choosing. I can assure you that if the need comes, we will make our best resources even better.

As a net result, we will face greater arms competition. We will face a less stable world. The SDI is, in our view, a destabilizing system. First, taking the technical characteristics of SDI, the mere participation of a human being in the decision-making for these weapons would be impossible. Any decision taken with regard to an SDI or ABM system would have to be taken in a fraction of a second. So everything must be delegated to the things themselves. Second, since computers fail, they may cause the strike systems to fire on their own. A computer failure might lead to a nuclear exchange and a complete catastrophe.

Third, just imagine that our response includes some systems of an SDI nature; your scientists would be trying to create systems to overcome them. And, in fact, as far as I know from reading newspapers here, your scientists are already doing research in countermeasures against this type of thing. So in principle, the arms race would be on an even higher level. Then let us imagine the situation in which a battle station that is a part of an American system is placed into outer space. Suppose that the battle station disappeared, and we have nothing to do with it. It is a mistake or a malfunction of the system. A military person in the United States would have to make an immediate decision, so there would be a major reaction, major decisions, and the world as a whole will be destabilized.

What does SDI mean in terms of the arms race or the international system? As I have already said, we would never allow any superiority over us. We have made sacrifices to secure our existence and our freedom in the past, and if the need comes, we will exert efforts to defend ourselves in the future. SDI will mean a new level of arms competition. The arms race will continue and will take new shapes and destinations; I do not know any other possible result but the accelerated arms race and its spread to outer space, which would be quite dangerous. Secondly, it might cause us to increase our long-range missiles, our ICBMs, or the construction of dummy missiles. So SDI is pushing both us and the United States into new dimensions of the arms race.

At the same time, the existence of SDI, for the reasons I have already outlined, makes it almost impossible to get agreements on nuclear disarmament because we will not allow the balance to be upset. With SDI, we cannot proceed toward reductions in offensive weapons. I say at the onset that we do not take seriously any notions about sharing technology, of sharing cooperatively in this program. There are restrictions for us even to buy computers here; I just will not buy the notion that we could have SDI technology.

Every indication points toward deployment, despite pronouncements that the administration is not resolved on deployment. But I heard General Abrahamson say on a TV show, the 23d of March, which was the third anniversary of SDI, that the technical feasibility of SDI is no longer an issue. We take such statements as a serious sign of intentions. Secondly, we would not believe that after spending twenty-five or thirty billion dollars for pure research, such a program would easily be abandoned by those who want its deployment. So everything indicates that this country is going to work further and further in order to deploy such a system. And we will have to respond.

SDI makes the possibility of agreement on offensive weapons almost impossible. That is why we insist that we have to deal first of all with SDI at the Geneva negotiations. I would like to remind you that the subject matter of the negotiations at Geneva is to prevent an arms race in outer space and to stop it on the earth. And no talk about cooperative measures, about shifts of strategy and so forth should mislead us. We have said in the negotiations on January 8, 1985, and we have confirmed that our aim is to prevent an arms race in outer space and to stop it on earth. The SDI program is a major obstacle, and I agree that it is the major divisive issue between our two nations now.

Another dangerous aspect of SDI is that if this program is pursued further, it would lead to the abrogation of the ABM Treaty. And there must be no illusion about that. There are already pronouncements being made in this country about taking a broader view of the ABM Treaty. But I am sorry to say that for us any wider interpretation is an abolition of this treaty. A few days ago, if my memory serves me well, Mr. Nixon said that within a couple of years there should be some adjustments to SDI in order to proceed further; appropriately, he continued, there must be negotiations with the Soviet Union to make amendments to the ABM Treaty. In fact, what seems to be proposed is to get rid of the limits established by the treaty. That means only one thing: the abrogation of the treaty.

If the ABM treaty is abrogated, it means automatically the end of the entire SALT network of agreements. It would make arms competition unlimited, and the only thing we would face would be an even greater arms race. There are other treaties which would also be automatically undermined if SDI continues: for instance, the treaty dealing with outer space, where there is clearly need for cooperation. What we face is an X-ray laser to be used from outer space, the utmost importance of which has already been confirmed by Dr. Teller. Another treaty which would be automatically damaged by the X-ray laser is the partial test ban treaty of 1963.

In fact, the whole network of agreements between our two nations concerning the arms limitations would be jeopardized. The only thing we would face would be a competition in arms without any solution. It would not be in our interest and I think not in yours. Like many times in the past, it would be imposed on us. And it is not the way we would like to proceed. We do not believe that mutually assured destruction is the best way to deal with our security problems. Our view is that you have to get rid of nuclear weapons. We have made our proposals, and we have proposed that we get rid of nuclear weapons by the year 2000.

There are critics who say that our proposals are only Communist propaganda. Mr. Reagan has welcomed the initiative. But whatever you call it, it is the concept of dealing with nuclear weapons. And it is all very simple: in order to get rid of mutually assured destruction, we have to get rid of nuclear weapons.

9

The Technical Feasibility of SDI

Louis Marquet

ONE of the most frequently asked questions concerning the Strategic Defense Initiative is "will it work?" or perhaps more unfortunately, all too often, the absolute assertion that it cannot work. We should be very clear on definitions.

First of all, we do not yet know what "it" is, since we have not yet completed the process of defining a systems construct. We have funded a number of architectural studies over the last couple of years, and they have indeed given us some excellent insights as to what systems performance requirements must be in order to meet certain offensive threats. Our concepts, appropriately, are still very much in the preliminary stage.

Next, we have some difficulty with the word "work." Work against what, when, under what circumstances? The true meaning of this word is that our system, whatever it is or can be and whenever it may be deployed, will result in a strategic posture vis-à-vis the Soviet Union with a specific effect: they would be convinced that an offensive nuclear ballistic missile strike against the United States and/or our allies has such a low probability of success in achieving its military and political goals that they are effectively deterred from the launch in the first place. Some persons have said that perhaps a 90 percent kill rate is sufficient for this purpose. But I must emphasize that this is a purely subjective estimate. In fact, the subjectivity is in the minds of the Soviets.

Allow me to differentiate between a couple of other terms which give us trouble. They include "scientific feasibility," "technical feasibility," and "practicality." The Manhattan Project is perhaps an excellent example of this. Until Albert Einstein came to the conclusion that $E=mc^2$, physicists could be excused for postulating that the destructive potential of the nucleus was, in

This is an edited transcript of Dr. Marquet's remarks delivered during a conference on SDI at the Virginia Military Institute, April 8, 1986.

fact, scientifically impossible according to the laws of nature as we knew them then. The laws of nature may be immutable, but our perception of them changes. Over the years, it became very clear that on a technical basis there was energy to be derived from the nucleus. But it was not until after the Manhattan Project that the technical feasibility was established with the first explosion at the Trinity site on the top of a tower. But again, there is a great difference between detonating a device on the top of a tower and converting it into a practical device that could be delivered and used in an operational sense. Thus, we have the entire spectrum of scientific, technical, and finally practical feasibility. Therefore, we have to be really cautious as we discuss total feasibility in the context of the SDI.

Where are we now in our understanding of the SDI problem? First of all, I would like briefly to describe the nature of the problem. What are we trying to accomplish? Let us imagine that the Soviet Union, for whatever reason, decides to unleash a massive offensive ICBM strike against the United States accompanied by a complementary SLBM strike against shore installations and other targets and, perhaps, an IRBM strike—a tactical missile strike—against our allies. Depending on one's assumptions, we could postulate upwards of two or three thousand missiles simultaneously launched from around the globe, mostly from the Soviet Union itself. Each missile could be MIRVed, that is carrying some ten to twenty independently targetable reentry vehicles (RVs), bearing nuclear warheads. As if that is not enough, for each of these RVs there could be from ten to a hundred decoys or penetration aids designed to confound any defenses we may deploy. This entire attack might last only twenty-five to thirty minutes, during which time some thirty thousand nuclear warheads could rain down upon us.

I should emphasize that this scenario, although somewhat extreme, must be taken very seriously as it is entirely feasible from the standpoint of the current state of Soviet capabilities. Is it a practical scenario in an economic and military operational sense? That is difficult to say, and I hope we will never find out. However, the scenario does at least provide us with a challenging model against which we can postulate and evaluate potential defensive systems.

I might add that the problem that confounds us in solving the ballistic missile defense system has been likened to both the Manhattan Project and the Apollo Project. I do not believe that this is a particularly apt analogy in either case. In fact, I would say that the challenges are considerably more stressing than either of those examples. For instance, in the Apollo Project, when President John F. Kennedy set us the goal of going to the moon, we did not have to worry about the moon moving out of its orbit to dodge us, hiding by some stealth technology, or shooting back at us as we approached. Obviously, in this case, we have to be concerned about all possible countermeasures and countermoves by a very clever and determined adversary. The Soviet Union certainly will not sit back passively and allow us to develop capabilities that would nullify an enormous investment that they have made in offensive strike capabilities.

Shortly after President Reagan raised his historic March 23, 1983, challenge that our scientific and technical communities render nuclear weapons impotent and obsolete, the Defensive Technology Study Team, headed by Dr. James Fletcher, both former and now newly appointed administrator of NASA, was commissioned. Their charge was to determine whether a technical solution to the problem was possible and to develop a research and technology program designed to outline the technical basis for an informed decision in the early 1990s, whether to proceed to the development and deployment phase. In other words, the goal of the research phase was to determine and to demonstrate the technical feasibility of an antiballistic missile defense system.

Over fifty government, academic, and industrial specialists involved in the technologies, representing all extremes of viewpoints, served on the Fletcher Study Team. We—and I include myself, because I was a member of the Systems Panel—worked continuously for three solid months in the heat of the Washington summer pouring over all the work which had been hypothesized about the problem. Believe me, we were our own worst enemies in the sense that we created a very competent and aggressive Red Team which invoked or invented every conceivable—and some inconceivable—countermeasure, including proliferation and hardening of current boosters, invention of fast-burn and/or rotating boosters, exotic decoys, and counter-defense threats. You name it, we had it.

The principal conclusion of the study team was that certain emerging technologies indeed do offer us the possibility that an effective defense system might be developed. What is new about the emerging technologies as opposed to what I will call the old BMD, that is the BMD of the 1960s and the 1970s based on terminal interceptors guided by radar systems and carrying nuclear warheads, is the opportunity to engage the ballistic missiles throughout the entire length of their trajectories all the way from cradle to grave, as they say, or from silo to impact. This includes the boost phase, the post-boost phase, the mid-course phase, and finally the terminal phase. A number of so-called "long-poles," which represent the *sine qua non,* were identified by the study. These include the practicality of a boost-phase intercept, mid-course discrimination, low cost kinetic energy interceptors, a battle management system, survivability—particularly for space assets. More recently Ambassador Paul Nitze has explicitly added cost-effectiveness at the margin.

These "long-poles" represent the primary technical challenges, but I believe that it is fair to say that all of these issues have already passed the test of basic scientific feasibility. For example, this last year, we were able to demonstrate at White Sands, by using a very large chemical laser, that indeed a laser beam held fixed on to a stationary booster strapped down in front of the laser device could easily poke through that booster. It blew it up in a very dramatic test. In a sense, that kind of scientific feasibility is a far cry from being able to prove that under operational conditions, under dynamic leads, in the face of potential hardening and countermeasures, we can accomplish the same objective. In fact, it is the proof of this technical feasibility that is the objective of our current program.

At this point I would like to run through some of these emerging technologies that were identified by the Fletcher Panel. I am not going to try to convince you of technical feasibility because, as I say, that is the goal of our program. But I would like to point out that we do have, as the Fletcher committee concluded, very promising technologies which certainly offer us the potential of solving the problems.

The first one I want to talk about is directed energy, the one that I perhaps know the most about, because it is the one for which I am responsible in Washington. Directed energy is perhaps the most exotic of the new technologies in the strategic defense program; of course, I am referring to high-energy laser beams and particle beams. The good news about directed energy is that it travels near or at the speed of light. It allows us to contemplate applying destructive forces essentially around the world in less than a second. And this, of course, is consistent with the time scales in the boost phase which could be perhaps as short as one hundred seconds or even less. By engaging the boosters over the Soviet Union as they are rising up, we have the option of destroying in a single kill, so to speak, perhaps as many as ten to twenty warheads. Furthermore, kills in the boost phase are particularly advantageous because they allow us to break up structured attacks and to avoid the deployment of decoys and penetration aids. It therefore makes the job of the subsequent defenses that much easier.

That is the good news about directed energy. The bad news, in a sense, is that in order to engage the boosters over the Soviet Union, some of our assets must be deployed in space. That means that they must be survivable. Allow me to say one word about survivability. We find ourselves arguing this point all the time; it is a very complex issue. Survivability should be viewed in the context of system survivability, not one-on-one combatant survivability. Nothing is survivable in the sense that if the opposition is willing to pay enough of a price, it cannot take it out. The question is: can we exact a high enough penalty for that process? Furthermore, there is an asymmetry: although this is counter to the current wisdom, it may be easier to protect the shields and structures of space assets from, for example, ground-based lasers than to protect boosters which have to operate under much more dynamic and stressing conditions. It is unfair to say that there is a perfectly symmetrical situation.

Recent advances in high-energy lasers, in particular in free-electron lasers, plus the demonstration which showed that we can compensate for the turbulence introduced by the atmosphere, allow us to contemplate putting the high-energy lasers on the ground and bouncing the beams around the earth into the engagement battle. This would clearly make the problem much simpler; our most complex systems would now be on the ground, and we would not have to go about maintaining and supplying these laser systems in orbit.

One additional asset that directed energy brings to the table is the possibility of solving the mid-course discrimination problem. The good news about the mid-course phase is that it is much longer than the boost phase: almost twenty minutes. The bad news is that it has been very difficult for us to handle this discrimination problem. It is relatively straight forward to design decoys that act very much like reentry vehicles. The emerging technology which is new, as opposed to the game twenty years ago when we looked at the problem, is that by using directed energy we may be able to reach out and touch these boosters, so to speak. We can perturb them in their orbits and trajectories; we can heat them up; we can zap them and cause them to recoil; we can probe them with particle beams and measure the radiation that comes out of them, thereby making a direct measurement of the weight of the object in orbit, which in fact is the test for the merit of the decoy. Thus, directed energy offers us another opportunity now of solving this very vexing mid-course problem, where the actual intercepts will be carried out by more conventional kinetic energy systems, but in a much more cost-effective fashion since we will not have to waste all of our assets on decoys.

Allow me finally to deal with battle management. Battle management has taken a severe rap in the technical community. We are being told that this will be the most complex system ever invented by mankind, that it cannot possibly work. It would be very difficult for someone to argue what the implications of computer technology are going to be because we have certainly seen a dramatic increase in capability, both of hardware crunching, of number crunching, and of a number of systems over the last several decades. For example, the Japanese are now marketing a gigaflop machine; that is a machine capable of one billion floating-point operations per second. One can actually put in an order for one of these things now. This indicates a tremendous number crunching capability, and the curve of growth is still ever upward; in fact, it is exponentially expanding. Thus, how can one argue what will work in number crunching capability tomorrow?

Perhaps more to the point is that our whole approach to using computers is undergoing a dramatic change. We are moving away from what is called the von Neumann machine, which is a serial operation, sequential, one after the next, and has certain intrinsic limitations, towards parallel processors and perhaps even beyond this to what are called locative processors, which emulate the human brain. We are approaching expert systems, artificial intelligence, if you like, which, we believe, offer us enormous potential.

The entire system does not have to be perfect; it has to be good enough to serve the primary purpose. That does not require perfection; it requires sufficient effectiveness so that the attack is deterred in the first place.

10

The Technical (In)feasibility of SDI

G. W. Rathjens

IT may be possible to develop many—perhaps most—of the technologies that have been seriously discussed for SDI to the point where they will "work," after a fashion. To answer the questions of whether a defense system is feasible and desirable requires more, however, than affirming the feasibility of individual technical developments. We must be concerned especially—indeed, as matters of first order—about what the defense is supposed to do, i.e., what it is to defend, and how well, and about plausible and likely adversary reactions to what we might do. With a necessity for brevity, I shall concentrate on these last questions and on a few general reasons why defense, especially of cities and population, is likely to be difficult—much more difficult than defense of selected military targets, such as missile silos.

Missile silos are typically hardened to withstand overpressures of one thousand pounds per square inch or more. This means that the *keep-out volume* for defenses, the volume within which nuclear explosions cannot be permitted, is small. This has two important implications for defense: adversary warheads that appear destined to impact more than about a quarter of a mile away from any silo need not be intercepted at all; and for those that might strike within this distance, intercept can be delayed until they approach the keep-out volume. This means that any decoys or other penetration aids must be *sophisticated*—read *heavy*—thereby imposing a substantial penalty on the offense; otherwise, it would be relatively easy for the defense to discriminate between warheads and decoys on the basis of their interaction with the atmosphere.

In the case of defense of cities, the radii of keep-out volumes must be at least an order of magnitude larger. This implies longer-range interceptors, earlier commitment to intercept, greater difficulty in discrimination between warheads and decoys, and greater likelihood of disruption of defenses by nuclear explosions at high and intermediate altitudes. Moreover, a population must be protected from the radiological effects of fallout by providing shelters, by defending against the impact of adversary warheads up to several hundred miles upwind of cities, or by a combination of the two.

The level of effectiveness required of a defense is at least as important a factor as any judgment about defense feasibility. In the case of defense of intercontinental ballistic missiles—and some other military targets—a relatively low degree of effectiveness may be quite acceptable. An expectation that, say, a quarter of an ICBM force would survive ought to be a powerful deterrent to attacking it. Indeed, in referring to defense of retaliatory capabilities, it is often suggested that just introducing uncertainty into an adversary's calculus may be a worthy ballistic missile defense objective.

If low levels of survival are acceptable, the defender can, in principle, use the same interceptor missiles (or other defense assets) to defend any of several targets. He could thereby realize the enormous advantage of "preferential defense," while the offensive would have to attack the undefended targets as well as the defended ones with a weight of attack appropriate to the latter if it is to have any hope of overwhelming the defense. In contrast, if the defender's objective is to limit damage to very low levels, as would generally be the case for defense of population, the offense could realize the advantage of "preferential offense" targetting. It could concentrate its attack on a small subset of targets; the defender, not knowing the offense's targetting plans, would have to be prepared to defend the whole set, including those not to be attacked.

The problem arises in extreme form with President Ronald Reagan's proposal of March 23, 1983, for a defense that will render nuclear weapons "impotent and obsolete."[1] It is clear from that speech and from subsequent remarks by Mr. Reagan and by Secretary of Defense Caspar Weinberger that the president had in mind a defense so effective that the survival of Americans, and of others, would not depend upon the Soviet Union's being deterred from executing a nuclear attack; rather it would depend only upon our capacity to prevent damage in the event that it were to launch such an attack.[2] The implication is a perfect defense, i.e. a defense that is 100

---

1. Address to the Nation, "Peace and National Security," March 23, 1983, *Department of State Bulletin*, vol. 83, no. 2073. Secretary Weinberger has made it clear that he envisaged defense not only against ballistic missiles but against other means of delivery as well. (NBC Television, "Meet the Press," April 27, 1983.)

2. Other remarks by President Reagan and Secretary Weinberger have been ambiguous. On some occasions they have reiterated their objections to basing security on the threat of nuclear retaliation and have stressed their commitment to "astrodome-type defense." At other times they have argued for strengthening nuclear deterrence by increasing uncertainty in an adversary's calculus. And sometimes they have argued both cases in the same document! (President Ronald Reagan, interview with representatives of Soviet news organizations, October 31, 1985; President Ronald Reagan, remarks before the National Space Club, March 29, 1985; and Report of Secretary of Defense Caspar W. Weinberger to the Congress on the FY 1986 Budget, FY 1987 Authorization Request and FY 1986–90 Defense Programs, February 4, 1985, p. 54.)

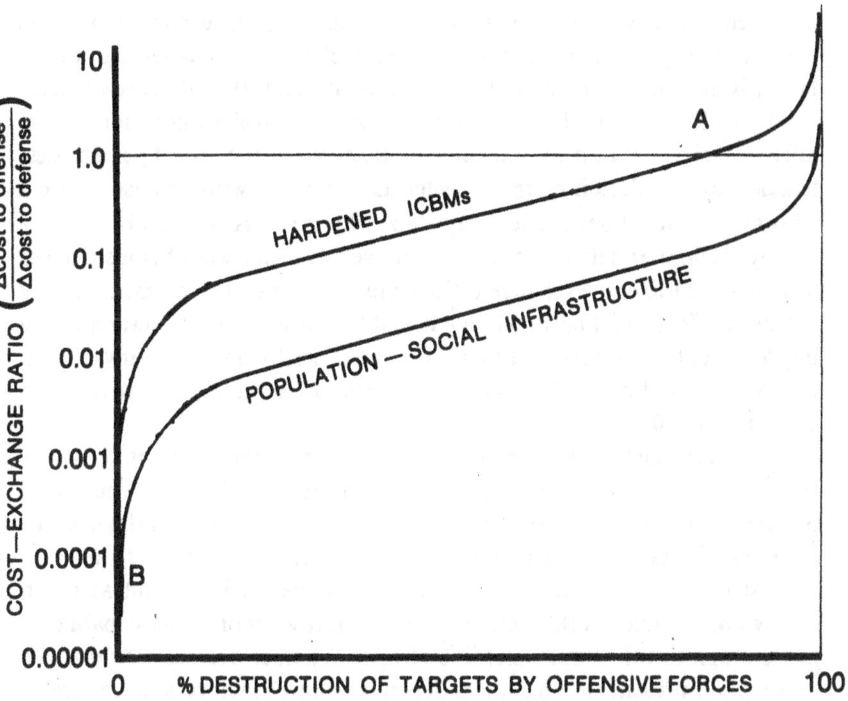

Figure 1. The cost-exchange ratio (i.e., the cost to the offense of negating an incremental expenditure on defense) as a function of level of destruction. The top curve might be appropriate for defense of missile silos and the bottom curve for defense of population and social infrastructure. The scale on the cost-exchange ratio scale is very approximate. The spacing between the two curves is meant to suggest that at any given level of effectiveness, defense of population is an order of magnitude or so more difficult than defense of hardened missile sites. The shape of the curves is meant to show how much more difficult defense is an near complete effectiveness than for more substantial levels of destruction are acceptable.

---

percent effective, not just, say, 99.9 percent effective. The reason is that if 0.1 percent of Soviet offensive warheads, about ten in number, could be expected to penetrate American defenses, nuclear weapons could hardly be considered "impotent and obsolete." Political leaders could not be expected to behave in crises as if such weapons had never been invented—at least, one would hope they would not; nor would the general public be completely relieved of its concerns about them.

The extraordinary differences in the demands on defense, depending on objectives, are illustrated in Figure 1. In this figure, the cost-exchange ratio, i.e., the cost to the offense of negating an incremental expenditure on defense, is shown as a function of level of destruction. Two curves are shown: one that might be appropriate for defense of population and social infrastructure and another appropriate to defense of missile silos. The scale on the ordinate, the spacing between the curves, and their shapes should be regarded as only very approximate. But the spacing between the two curves is meant to suggest that at any given level of effectiveness, defense of population is an order of magnitude or so more difficult than defense of hardened missile sites. The shape of the curves is meant to show how much more difficult defense is at near 100 percent effectiveness (i.e., for 0 percent destruction of population or missile sites as compared with defense where a substantial level of destruction, say 80 or 90 percent, is acceptable.

Any attempt at defense against a reactive adversary will obviously make little sense if the cost-exchange ratio is much less than 1.0 inasmuch as any defensive effort could be, and likely would be, offset, at much lower cost, by improvements in adversary offensive capabilities.[3] Conversely, investment in offense would make little sense if the ratio of marginal costs were much greater than 1.0.

The circle marked A in the figure is meant to reflect present reality as regards our ICBMs. The Soviet Union could presumably destroy nearly all of them in a disarming strike, assuming we did not launch on the basis of warning of attack. But it is quite conceivable that if we were to make a modest investment in defenses, it would cost the Soviets about as much (say, within a factor of three either way) to improve its offensive capabilities just enough to offset our defensive efforts. Whether defense would make sense in our present circumstances depends on three factors: a more careful estimate of the cost-exchange ratio; whether we believe that there is a real threat to our ICBM forces, considering both possible Soviet motivations to attack us and the retaliatory capabilities inherent in the other legs of our strategic triad; and whether defense is to be preferred to other alternatives, such as greater emphasis on mobile strategic forces, as a way of dealing with any threat to our ICBMs.

It is likely that if the Soviet Union were to attack the United States with the objective of maximizing destruction of our population, it could, at present, inflict about 80 to 95 percent fatalities in a preemptive strike. Let us, however, hypothesize a level of 0 percent. This assumes a capability on

3. This point has been made forcefully by Paul Nitze in what has probably been the most thoughtful discussion of SDI by an administration spokesman. Ambassador Nitze also made the point that defense would make sense only if the components of defense were not vulnerable to destruction by an adversary. ("On the Road to a More Stable Peace," Speech to the Philadelphia World Affairs Council, February 20, 1985.) These criteria have been generally accepted by analysts concerned with the problems of defense. A different view has been expressed, however, by some advocates of ballistic missile defense efforts, including notably General James Abrahamson. He has suggested that "when they [Soviet leaders] see that we have embarked on a long-term effort to achieve an extremely effective defense, supported by a strong national will, they will give up on deployment of more offensive missiles." (*Science,* August 10, 1984, p. 598.) More recently he has said that the criterion for an affirmative decision on an SDI deployment should be simply whether it is "affordable."

our part to intercept all the nuclear warheads that might be thrown our way. Such a balance might be indicated on the figure by the heavy line marked B on the figure. As drawn, it suggests that the cost-exchange ratio is favorable to the offense by about four orders of magnitude. This would mean that an incremental investment by the US of, say, a hundred billion dollars on defense might be offset by a Soviet investment of ten million for offensive improvements. I believe the cost-exchange ratio would, indeed, be at least this unfavorable to the defense at the *perfect* defense (i.e., zero destruction) level, although it would probably rise very rapidly (as shown in this figure) as the level of destruction that might be accepted goes from zero up to a few percent.

The figure is meant to suggest that we need several orders of magnitude in improvement in the ratio of ballistic missile defense effectiveness to offensive missile effectiveness if we are to achieve President Reagan's SDI objective; and similar improvements would be required in other kinds of defenses to cope with all other means of nuclear weapons delivery. However, if the objective of defense is simply to "enhance" deterrence, the requirement for improvements in defense is very much less demanding. Indeed, cost-effective defense to enhance deterrence may be possible even without exploitation of exotic technologies, the most promising of which are, at best, remote possibilities. In other words, such defense may even now be within the state of the art.

I should reiterate that the relative ease of defending ICBMs derives in large measure from the possibility of preferential defense. Such defense is possible only if the decision to attempt intercept can be delayed until the defender can ascertain which particular targets are destined for destruction by particular adversary warheads. From a practical standpoint, with present offensive technology, preferential defense would be best accomplished with terminal defenses. It would be less feasible with mid-course defenses, still less so with a post-boost-phase defense, and not feasible at all in the boost phase. In the event of adversary deployment of maneuvering reentry vehicles, preferential defense would likely be infeasible except in the terminal phase. Whatever the nature of the offense, such terminal defense is likely to be much easier for other technical reasons, and certainly a nearer-term prospect, than defense based on intercept earlier in a missile's trajectory. One can probably go further and argue (although I will not claim a "proof") that if the objective is a modestly effective defense of ICBMs or other hardened military targets (one that would assure survival of 10 to 50 percent of the targets), all, or nearly all, resources available for defense should be allocated to the terminal phase. That is not, however, where the emphasis presently is in SDI. It is rather on boost-phase, post-boost-phase, and mid-course defense. Thus, the research and development program for SDI is ill-designed to the extent that the objective is to "enhance deterrence."

To the extent that the objective is to deny damage to population and social infrastructure—President Reagan's original objective—the research and development program seems more rational. Indeed, if the objective is "perfect defense," there can be no case at all for terminal defense, and a highly effective boost-phase defense would seem to be a sine qua non.

That terminal defense would be irrelevant is clear from the following syllogism:

(1) A 100 percent effective defense of US interests would require preventing the detonation of nuclear weapons virtually anywhere—not just at all points in the United States but on territory of allies and other friendly states as well. Because of the possibility of fallout, detonations in the seas and even on the territory of adversaries would also have to be prevented.
(2) It would be infeasible to defend large areas with terminal defenses; indeed, there are many parts of the world that can not be defended with them at all.
(3) Defense of such areas must, then, depend entirely on other-than-terminal defenses, probably on boost-phase defense, at least in large measure.
(4) If this other-than-terminal defense could provide "perfect" protection for points anywhere, it would presumably provide such perfect protection everywhere.
(5) Any terminal defense anywhere would, therefore, be redundant.

One is driven to the acceptance of the essentiality of boost-phase intercept for a perfect or near-perfect defense of population. This is not only because of the infeasibility of world coverage with terminal defenses but also because of the extraordinary number of threatening objects that are likely to confront any post-boost defense system: many thousands of warheads and at least tens, if not hundreds, of thousands of decoys, not easily distinguishable from warheads prior to reentry. Thus, there is a broad consensus that realization of President Reagan's SDI objective, or anything close to it, would be critically dependent on development of an effective boost-phase defense, however much the superposition of post-boost and/or mid-course defense might help mitigate the effects of boost-phase leakage.

But for a variety of reasons, boost-phase defense is likely to be extraordinarily difficult. I touch here on only two reasons that seem fundamental.

First, there would be very little time in which to conduct the engagement. This is because the boost phase does not last long: three or four minutes for current ICBMs and less for submarine-launched ballistic missiles, with the possibility of reducing the time with more advanced boosters. Moreover, the presence of any atmosphere between boosters and defenses may interfere, making sensing and tracking of the booster difficult, and making the kill of it even more so. Thus, only a small fraction of total boost time could be utilized by the defense. Indeed, with "short-burn" boosters, some much-touted kill mechanisms—notably the nuclear-driven X-ray laser—may be totally ineffectual.

Second, because the time available for intercept is short, the destruction of ICBMs would have to be accomplished over the interior of the Soviet Union. Therefore, boost-phase defense would have to depend on space-based sensors; also, in some schemes it would have to rely on mirrors in space to reflect laser beams of terrestial origin; in others, on satellite-based instruments of kill—kinetic energy projectiles, lasers, and/or particle beams. Such defense dependence on space-based components is troublesome in two respects: (1) Any system will require placing a great deal of material in orbit. Many satellites would be needed if they are in orbits of two or three hundred kilometers. The reason for the large number of satellites is the so-called *absentee effect:* at crucial times many of the orbiting satellites would not be located over enemy territory, but would instead be over other parts of the earth. Alternatively, very large individual payloads (large mirrors and/or much fuel for beam weapons) would be required if the satellites are in more remote orbits of tens of thousands of kilometers. (2) Satellites will almost certainly be highly vulnerable. Dealing with this vulnerability—the problem in meeting Ambassador Paul Nitze's second criterion for SDI—poses truly difficult requirements for the designers of a defense. They must confront the possibility of space mines: of the adversary's ability to place destructive satellites near the defensive satellites during peacetime, and then to detonate them on command in order to destroy the defenses. There is also the likelihood that any defense system with significant boost-phase intercept capability would be even more effective as a means of destroying adversary space-based defense components. Thus, assuming boost-phase defenses of some effectiveness deployed by both the US and the USSR, each would have the capability—and in a crisis, perhaps the incentive—to destroy the adversary's defenses in a first strike. It is hard to see how one can get around the vulnerabilty problem, notwithstanding much talk about self-defense of space-based defense components.

Permit me now to give a summation.

President Reagan's objective of a defense that can make nuclear weapons "impotent and obsolete" really requires a 100 percent effective defense, not just of the United States but of all the world, and not just against ballistic missiles but against all means of delivery of nuclear weapons. Such an objective would seem impossible to attain. The two criteria spelled out by Paul Nitze are just too demanding. Yet, it is this version of SDI (now sometimes referred to as SDI-I) that is the basis for public support of the program. What is so troublesome about it is that the vision is cruelly illusory and diversionary. The president has offered the country—and the world—a quintessential example of a technical fix to deal with the dual problem that inheres in having a troublesome adversary in a world of nuclear weapons. He thereby diverts attention from political solutions, which can be the only real alternative to nuclear deterrence, which he so rightly denounces.

Enhancing deterrence by providing a limited defense of military assets—ICBM silos and perhaps others—(SDI-II) is a much easier task and the one that seems to provide the rationale, if not the motivation, for those who work in, and defend, the program. It is not at all obvious, however, that such enhancement is necessary or desirable, given the cost and other demands on resources. Of fundamental importance is the fact that the technical requirements for SDI-I and SDI-II are very different: so much so that a research and development program optimized with one objective in mind would look radically different from one optimized for the other.

We have had far too much ambiguity—indeed, doublespeak—about the program. The public and the Congress should demand that the administration make clear its objective. Is it the president's version (SDI-I), which is politically attractive, or is it enhancing deterrence (SDI-II), which may be technically feasible? If we could have a clarification, there might be a positive national response. If so, we would at least then know what kind of a research and development program makes sense. More likely, the result of clarification would be a negative response by the public and the Congress: negative in the case of SDI-I because there would be no technical basis for defending the program; and negative in the case of SDI-II because there would no longer be even a pretense of escape from the world of Mutual Assured Destruction and because most people would question whether a defense designed to "enhance deterrence" would actually make nuclear war less likely at all or enough less likely to justify investing in it.

I suggest that the administration understands that a clarification about objectives would sound the death knell for the SDI program. That is why we have had so much confusion about it, and why the administration will continue to resist speaking about it with disciplined clarity.

11

Why SDI Can and Must Be Done

Louis Marquet

I think that the question of whether it can be done is a fair question. It is the question I encounter almost inevitably. But it is a trick question, in a sense. You have to be very careful to define the terms. It is only a four-word question. Frequently what we will find is people who have already made up their mind for political or other reasons that they hate the idea of the SDI or they hate even the idea of investing research funds on anything military. They have their own agendas, which they are entitled to in a free country. But when they argue or attempt to argue in a technical symposium or in a technical community that "it cannot be done," their general technique is first of all to define "it" in such a way that is patently absurd. And then they, in fact, prove what they defined was absurd.

I will just give you a few examples of the pitfalls of this question. First of all, the "it" that we are talking about does not necessarily mean eliminating one hundred percent of the penetrating warheads on any target anywhere in the world. That is not necessarily what "it" is. "It" does not mean having the capability of intercepting simultaneously three to five thousand ballistic missiles launched from silos in the Soviet Union within a few seconds of their launching without any warning whatsoever. Again, that is not what "it" is all about. And "it" does not necessarily mean having our space assets survive during a so-called peacetime attrition period wherein the Soviets are allowed to take any measures they like, including chipping away at our assets, blinding our sensors and indeed installing space mines, and from time to time blow them up, etc. That is hardly what "it" involves.

I think it is very clear that what "it" really does involve is bringing about a new strategy, a new posture for our strategic survival, or strategic balance, based on defense, not on offense. That is to say, "it" really means making

---

This is an edited transcript of Dr. Marquet's remarks delivered during a conference on space at the Virginia Military Institute, October 16, 1986.

mutual assured destruction obsolete, and along the path, making ballistic missiles impotent and obsolete, as the president so eloquently stated.

Having been a little more careful about defining what the technical aspects of the problem are, let us talk about some of the things which are giving us hope that we can, in fact, solve the problem. This is to say that we can bring about a world based on deterrence by defense. I would like to go back to a quote from 1960 by General Medaris, of which I will give you the gist. He was lamenting the fact that at that time our only strategic options seemed to be based on weapons of mass destruction. Of course, the atomic and hydrogen bombs were what he was referring to. He regretted that we did not have weapons of surgical precision to use in a very discriminating way to provide a kind of balance.

What has happened in the interim as a consequence of the continual progress of technology is that we have now identified those surgical weapons, and there are a number of them. They fall into two classes, and I will give you a couple of examples. There are both kinetic type of weapons, and there are weapons based on directed energy.

In the area of kinetic energy weapons, the army carried out an experiment in the summer of 1984 called HOE, which stands for Homing Overlay Experiment. For the first time, we demonstrated with that experiment that it was possible to launch an interceptor from the ground, have that interceptor rise up, look out into space, detect and home in on a ballistic missile warhead, the reentry vehicle of a ballistic missile. In fact, it intercepted and utterly destroyed that warhead.

Since then there are other notable examples of kinetic energy surgical precision which have received a lot of publicity. In the fall of 1985 the Air Force carried out a similar test against a satellite, in this particular case, using their miniature homing vehicle. The vehicle had an infrared sensor on board. It was launched from an F-15 aircraft, rose up out of the atmosphere, looked out into space, detected the target and made a direct impact, steering itself into the target. And finally, very recently in an experiment sponsored and carried out by SDI called Delta 180, we conducted a similar experiment and demonstrated that we were able to satisfy all of the functional requirements of making an intercept in space.

These types of devices, these types of weapons are surgical weapons. They are very precise, and they only destroy the target which they are intended to destroy. Marvelous as they seem, we have even greater wonders in the realm of directed energy. You might ask why we do not just rely on kinetic energy alone. The difficulty with kinetic energy when talking about intercepts in the boost phase, which of course is the area of major leverage for SDI, is that it simply takes a fair amount of time to cover the enormous distances of space. And one of the things the Soviets might turn to could be fast-burn boosters, or they might maneuver out of the way, et cetera. One could conceivably argue that it would be possible to outmaneuver a kinetic-energy interceptor.

But it is not at this point conceivable to outmaneuver the speed of light. And that is where directed energy offers such great potential.

In the area of directed energy, what we are discovering, first of all, is that we have been able to build very large and powerful devices, lasers, particle beams, and so forth. That is not all of the job, of course. One of the major uncertainties in the area of directed energy is what it takes to kill these targets, to kill these boosters. We made a very important first step along those lines last fall when we carried out an experiment with a Titan booster, which was simulating operational conditions: it was pressurized, it was loaded, it simulated the stresses of launch, and we directed a laser beam—in fact our largest laser beam in the country, located at White Sands Missile Range—and within a very short period of time, we were able to demonstrate catastrophic destruction of that particular booster.

We did not do that just to get coverage on television, although we have been accused of that. It was a very important test from the standpoint of understanding what it really takes to kill a booster. But that is not all. We have still got to go considerably further in understanding this because, first of all, that particular booster was a current booster, essentially taken right out of the inventory. There were no measures taken to protect the booster from directed energy weapons. Of course, if we have an effective directed-energy weapon, we should expect that the Soviets would take every measure they could possibly adopt, harden up their boosters to protect them against such weapons.

The next part of our program is to understand what are the limitations of such hardening. I might add that hardening does not come for free. Every ounce of hardening, every ounce of protection you add to the booster means that much less warhead that you can put into space. If you could even conceivably make the booster invulnerable, but have to take all the warheads out to do so, we will in effect have accomplished our objective.

I want also to say a few words about the laser devices themselves. Last year, a team of researchers at the Lawrence Livermore National Laboratory demonstrated with a device called the Experimental Test Facility a 35-gigahertz free-electron laser which achieved better than 40 percent electrical efficiency. Those who understand laser physics know that electrical efficiency is an extremely important commodity in such devices. And typically speaking, lasers operate in the area of 1, 2, or 5 percent at best. To achieve efficiencies in the order of 40 percent is truly phenomenal. In fact, we are so encouraged by the results of these tests that we are currently planning to do some experiments to address the other side of the coin.

Building a laser is not enough in itself. You have to be able to demonstrate that you can get the laser beam from the device, through the atmosphere, and onto the target. We will be carrying out experiments for the next few years to address that issue under realistic, high-power conditions. We are already encouraged about the problem of compensating for the atmospheric turbulence because we have demonstrated, again last fall, in a very exciting series of tests carried out in Hawaii, that it is possible to put a low-power laser beam through the atmosphere and essentially take the twinkle out of the stars, as Dr. Teller puts it. Up to now, atmospheric turbulence has limited astronomers to effective apertures, from the standpoint of resolution, of only on the order of several inches or so. When you go larger than that, you collect more light, but you do not get any better resolution. The opposite occurs when you try to transmit the beam from the ground into space. The experiments at Maui demonstrated that in fact we can now postulate very, very large apertures which are meters, perhaps even tens of meters in diameter, and use those effective apertures to transmit over the enormous distances of space.

Let me say a few words about survivability, and then I will get back to the fundamental question. Survivability is again a tricky question. There are people who are willing to grant us the opportunity to actually technically build all of these things, but they ask, "can they survive against a determined attack?" Of course, if we cannot develop a convincing argument that they can survive, then we are only contributing to or precipitating a further arms race, this time in space. That is not our objective, and so it is very important for us to understand the issues of survivability and to make a compelling case that such systems are survivable.

Basically, if you put yourself in the position of a potential attacker who is interested in taking out your space systems, what must he do? First, he must find you, he has to put you into track, he has to intercept you, he has to destroy you, and after all that is through, he has to confirm that you have been destroyed. Every step along the way in this sequence of functions involved in attacking us gives us an opportunity to thwart him. We can hide; we can provide decoys of our own to provide false aiming points; we can jam his sensors; we can spoof him and give him false targets; we can maneuver out of the way as he comes close; we could harden up our assets so that if he does explode or put a laser beam on us, that does not allow him a cheap shot; we can shoot back at him as he is coming up to get us; and we could even play possum and pretend that we are dead when we are really still functionally alive. In other words, there are a great number of tactics that are involved, as well as technologies, in this whole issue of survivability. It is not at all an easy issue to convince yourself one way or the other that either you are a dead duck or that you are invulnerable. We are looking at these aspects. We believe that a compelling case for survivability can indeed be made.

In conclusion, let me return to the original question. Can we do "it"? My answer to that question is not only can we do it, but we must do it.

12

Strategic Defenses and Future US Defense Strategy

Steven A. Maaranen

STRATEGIC defenses have been discussed and investigated intensively in the United States since 1983. But exactly what the Strategic Defense Initiative means, or aims at, in terms of US defense strategy remains a matter of considerable uncertainty and controversy. The most widespread notion is that SDI would allow the United States and its allies to defend themselves against ballistic missile attack, rather than try to deter attacks by threats of nuclear retaliation. This would allow us to reduce (or better yet, eliminate) our reliance on nuclear weapons generally and greatly to reduce the chances and dangers of nuclear war.

Many people accept the desirability of a strategy like this. However, it is widely argued that defenses good enough to make such a strategy possible could never be built. Accordingly, most arguments about SDI focus on the technical feasibility of extremely effective defenses. This is still an open question, and it skirts the most important and pertinent issues of strategy.

To examine these issues, it is necessary to approach the strategic possibilities associated with development and possible deployment of strategic defenses in a different way. We should acknowledge that there are reasons in addition to our desire to reduce reliance on retaliation that impel the United States in the direction of strategic defenses. Principal among these reasons are the difficulties the US is encountering in retaining a satisfactory strategic policy on the basis of offensive nuclear forces alone. We also need to acknowledge that, barring a breakthrough on a defensive superweapon, the practical option ahead with regard to defenses is progressively to develop and deploy more effective defensive weapons, allowing them initially to

---

The views expressed herein are the author's alone, and do not necessarily represent those of the Los Alamos National Laboratory or the University of California.

supplement and gradually to substitute for some of our offensive nuclear forces. But to make this approach possible, the United States would have to update its strategic thinking and strategic policy to accommodate a changing mix of defenses and offenses. This way of looking at strategic defenses addresses the practical world and the choices which lie ahead of us. This approach to SDI strategy also engages the most serious strategic objections to SDI.

The greatest concern with a gradual adoption of strategic defenses is the belief that any move in that direction would undermine the stability we have known in the nuclear age, and institute a dangerous and perhaps indefinite period of stratetgic instability. On the other hand, critics of SDI usually assume that if the US does not adopt strategic defenses, a stable strategic policy will be assured. Both of these assertions are open to challenge. First, strategic stability as it has been known since the 1960s, and the strategic policy which the United States has pursued to enforce and exploit that form of stability, are rapidly passing into history. Ample evidence of this fact was provided by the heated arguments in the years immediately before the Strategic Defense Initiative concerning nuclear strategy and which new nuclear forces to buy—and why. One need only to consider the debate over the role and configuration for the MX intercontinental ballistic missile. The debate since 1983 has amplified those controversies.

If it is true that the conditions that underpinned US strategic policy for the past twenty-five years are crumbling, then everyone—whether supporters of strategic defenses or proponents of one or another form of an offense-only policy—needs to rethink how stability will be assured in the future. To do that, we need to break away from the narrow definitions of stability that were appropriate for the environment of the 1960s and 1970s and reexamine what it really is that US strategic policy is trying to achieve. Stability is an objective; there are many different ways to achieve it. But even then it is only one of several essential elements of a satisfactory US strategic policy. The question here is: can a combination of offenses and defenses, evolving over time, support a new form of strategic stability, while satisfying the complex of national security objectives which the United States has traditionally pursued?

Current Strategic Policy: Stability With Strategic Competence

Over the years, the United States has achieved stability with very different strategic policies. In the 1950s the United States had the ability to attack and destroy Soviet strategic forces. The world was stable because, while we could defeat the Soviet Union, we only wanted to defend ourselves. In the 1960s, realizing that we would soon not be able to destroy new generations of Soviet offensive forces with the weapons then foreseeable, we developed a new strategic policy which attempted to assure stability in a different way.

In essence, the United States accepted and tried to make a virtue of the fact that no matter what it did, the Soviets would be able to attack the US with a considerable number of nuclear weapons. As long as the United States maintained a similar minimum capability, which it could easily do, Mutual Assured Destruction (MAD) would be a technical fact of life. The United States believed that both it and the Soviet Union should accept the notion that the MAD relationship existed and should exist. Further, the United States believed that a stable relationship between the two sides could be constructed on this basis. It believed that such a relationship would allow nuclear offensive forces to be greatly reduced.

In pursuit of this objective, the US avoided building a strategic offensive force posture that could threaten to destroy all of the Soviet land-based offensive force. It also signed and perpetuated the Anti-ballistic Missile Treaty, which it believed would ensure that an aggressor's society would remain vulnerable to a retaliatory strike by his victim. Accordingly, the US has pursued an arms control policy designed to make sure that our mutual destructive capabilities are unquestioned, while reducing the total number of nuclear weapons in the world and avoiding serious imbalances between US and Soviet military power.

Even while pursuing stability, the United States has recognized that a force posture and policy aiming at stability alone cannot satisfy US security needs. That is the case partly because the bare threat of destroying the Soviet Union as a modern society, except in response to an unlimited and indiscriminate attack on the United States, is not a particularly desirable, useful, or believable action. In addition to this ultimate threat, the United States must be able to pose limited responses to limited attacks on itself and its allies. And the United States together with its allies must be able to pose credible threats to use nuclear weapons first, in response to serious aggression.

Accordingly, the second US nuclear policy objective has been to maintain what I will call strategic nuclear competence. A competent force is one which can respond credibly to a full range of Soviet attacks upon the United States and which can effectively threaten nuclear escalation in the event of aggression against US allies. Maintaining strategic competence has required the US to construct survivable, flexible nuclear forces capable of destroying a wide spectrum of targets which the Soviets might value, and to plan many options for selective, discriminatory, and controlled employment of nuclear forces in response to aggression. A competent nuclear offensive force is kept partly in order that US forces could be employed wisely and with the best hope of limiting and terminating a war if deterrence fails. But a strategically competent force has also long been seen as an essential underpinning of a credible, stable strategic deterrent.

Strategic Policy As We Have Known It May No Longer Be Feasible

What is wrong with continuing to rely on offensive forces? There is obviously some tension between the demands of stability and those of strategic competence. But if those tensions have been resolved in the past, why cannot they be in the future?

The problem is that the United States probably cannot maintain the technical capability to rely on this strategy indefinitely. The US had hoped that the ABM Treaty and the Strategic Arms Limitation Talks (SALT) offensive force limits would permanently establish the mutual vulnerability of both the US and the USSR, and that the Soviets would accept the US version of a stable strategic policy and force structure. Instead of this the Soviets, following their own objectives and strategy, developed and deployed offensive nuclear forces capable of destroying US ground-based strategic forces. Certainly this does not eliminate the capability of the United States to retaliate *in extremis* with our surviving sea-based weapons. However, these Soviet actions have undermined our strategic competence. This is the case because it is the ground-based nuclear forces alone that possess the technical capabilities to destroy small groupings of valuable targets whose destruction could cause the Soviets to pause, limit a war, and seek a negotiated peace. Consequently, the Soviets have seriously undermined our confidence that we can deter or if necessary respond effectively to aggression.

The second blow to US offensive nuclear policy has been the recognition that the Soviet Union has developed active and passive defenses of those assets which the United States now believes the Soviet leaders value most highly. These assets include, preeminently, Soviet leadership and its ability to control both the Soviet Union and Soviet-occupied states, Soviet residual nuclear forces, and also Soviet ability to prosecute a non-nuclear war. It is the destruction of these highly valued assets which, it is believed, constitutes assured destruction of the USSR in Sovet eyes and which, at a minimum, the US seeks to threaten in order to achieve deterrence. Our ability to threaten these assets is diminishing. The USSR is protecting its leadership with strategic defenses, by means of hardened, deeply-buried bunkers, and perhaps in the future by mobile command posts. The Soviets are protecting their nuclear forces by further hardening missile silos, deploying mobile missiles, employing deception and concealment measures, and possibly deploying ballistic missile defenses. As our ability to hold these most highly valued assets at risk declines, our ability to assure deterrence by effective retaliation diminishes. This undermines stability.

It is possible that major additions and improvements to US offensive nuclear forces and to the military planning process could allow the US to restore the effectiveness of an offense-only policy. These could include weapons that could destroy hardened, deeply-buried command bunkers, systems that could find and destroy mobile ballistic missiles, or approaches that would allow us to re-plan attacks on the basis of real-time intelligence. But such improvements would be difficult to develop technically and expensive to build. Also, building a force posture designed to restore our strategic competence could appear to be destabilizing.

Another solution to the competing demands of stability and competence in an offense-only regime would be to attempt, through arms control, to negotiate a nuclear force balance allowing for stability alone and to abandon the requirement for competent strategic offensive forces. However, a successful outcome from negotiations aiming at this goal is unlikely. Also, the US has traditionally viewed a force designed to be stable, but with little operational competence, to be inadequate to the nation's needs.

For A New Definition of Stability, We Must Return To The Basics

If perfect defenses will not be available for a very long time, and if an offense-only nuclear policy faces serious risks within the next decade or so, the use of imperfect defenses, in combination with offenses, seems to be a potentially valuable option for US strategy. But, we are told, defenses would create instability and therefore are unacceptable. The charge that strategic defenses would be destabilizing is based largely on narrow, abstract mathematical calculations which were developed in the United States more than twenty years ago. According to such calculations, a country that believed it could improve the outcome of a war by striking first would be tempted to do so in a crisis. This, presumably, would give rise to crisis instability. Anyone with a simple computer model, sharing the assumptions of the US strategic analyst, could easily calculate the degree of crisis instability for a given set of forces.

It is obvious, however, that real stability is much more complicated than such calculations and that many other aspects of the situation would have as much, or more, influence over national decisions in time of crisis. The value of potential losses versus gains in a war, the absolute damage a nation might suffer, in its own terms of value, the situation a country would be in vis-à-vis other potential hostile states after a war, the potential for and consequences of escalation, the postwar recovery potential of the states involved all are factors which must weigh in any comprehensive calculation of stability or instability. Perhaps most important of all are the character and objectives of the states involved in a confrontation. France, as Winston Churchill noted in the 1930s, was "armed to the teeth but pacifist to the core." The world, he believed, had nothing to fear from French military superiority over Germany, but a good deal to fear from efforts to bring French and German military power into balance.

Certainly Soviet writings indicate that Soviet leaders understand stability in a very complex way and do not place great weight on simple mathematical calculations of crisis stability. In the United States, the greatly increased interest in crisis control centers, such as those advocated by Senators Nunn and Warner, and the extensive examination of many other confidence-building measures indicate a growing recognition that ensuring stability and preventing war pose complex requirements which have multifaceted solutions. Technical crisis stability is certainly a desirable attribute of a satisfactory US defense strategy. But it should not be allowed to place a straight-jacket on our thinking about alternative strategies or to predetermine the outcome of that thinking.

Stability And Strategic Competence In An Offense-Defense Strategic Policy

How might less-than-perfect strategic defenses be deployed, and how could they contribute to a satisfactory and stable strategic policy? One early objective for strategic defenses could be to support and rehabilitate US offensive forces so that they can continue to carry out their roles effectively in the face of Soviet offensive and defensive challenges to our strategy. In this role, defenses could help restore the competence and maintain the deterrent certitude of US nuclear policy. Such limited defenses could help insure stability.

One challenge to the competence of our forces is a so-called "decapitating" or "leading-edge" attack. A small number of ballistic or cruise missiles launched from submarines close to US shores, or of aircraft and cruise missiles flying under US radars, could be used in a surprise attack against US political and military leadership and its attack-warning, attack-assessment, and communications facilities. Such a strike could be followed by a major attack against US nuclear forces. The objective of the small surprise attack would be to destroy or paralyze our ability to detect and evaluate the subsequent attack on US nuclear forces, reach a political decision to respond to that attack, and communicate that decision to our nuclear forces before they were destroyed. An attack of this kind, which would require stealth and surprise, might be severely compromised by limited US defenses. Defenses could force the Soviets to increase the number of missiles dedicated to the decapitating strike, making their intentions more detectable and less ambiguous. That would either deter such an attack altogether or allow the US to make a better informed and appropriate response.

Strategic defenses, even if quite limited in capabilities, could also contribute to the deterrence of a massive Soviet attack. A limited defense which intercepts missiles or warheads before they begin to reenter the earth's atmosphere would tend to subtract warheads randomly from the Soviet attack. The advantage of such a defense is that it would help ensure the survival of an unknowable number of randomly distributed US assets. That would be a problem for Soviet planners, who are believed to be required to assure very extensive and very certain destruction of the groups of targets that they would attack in the United States. A defense that subtracts missiles and warheads randomly from the Soviet attack would wreak havoc on the Soviet planners' operational plans and calculations of certainty. By decreasing Soviet certainty that even a massive attack would achieve its objectives, strategic defenses would contribute to deterrence generally. Finally, a limited defense would allow the United States to identify and defeat an accidental launch of a ballistic missile, unlikely as this might be.

At the same time, while limited US defenses would disrupt Soviet confidence in the success of highly structured attacks with stringent goals, similar Soviet defenses would probably be leaky enough that the US could still impose massive damage on the Soviet Union in retaliation. Therefore, much of the basic deterrent effect of offensive nuclear weapons would remain in effect. Perhaps most importantly, by deploying limited strategic defenses of this kind, the United States would serve notice that it intended to retain the effectiveness of its strategic nuclear policy and was determined to restore an acceptable balance in US-Soviet nuclear capabilities.

More effective and ambitious strategic defenses, which might be achieved by adding more interceptors to first-generation defenses and by deploying subsequent generations of defenses, could underwrite a new, significantly different US strategic policy. Such defenses could reduce or eliminate the need for limited offensive nuclear options and establish a new basis for strategic deterrence and stability.

If the United States developed an increasingly defense-heavy mix of defenses and offenses, US and allied forces should become competent to defeat limited ballistic missile attacks against their leadership, military forces, and supporting infrastructure. Limited US nuclear offensive options to deter or respond to limited threats against the US homeland would no longer be essential. Furthermore, effective layered strategic defenses might protect the bases, forces, and industry which allow the US to project military power overseas and to conduct an effective military campaign in Eurasia. The Soviets appear to place considerable importance in being able to destroy these capabilities. It is the inability of the US to count on a successful land campaign in Europe, along with the threat of European devastation by Soviet theater nuclear weapons, that create the need for US nuclear first-use options for Europe. Very effective strategic defenses could greatly increase our confidence that essential US power-projection assets would indeed survive a Soviet attack and that we could succeed in a war in Eurasia without resorting to limited first use of nuclear weapons.

Defenses effective enough to reduce the need for limited nuclear options would also be able to impose a heavy, randomly distributed toll on any large-scale Soviet attack. While the defenses would not be perfect, they might be good enough to convince the Soviets that they could not count on effectively destroying the ability of the United States to carry on as a viable society, recover from damage inflicted, and prosecute a land war against the Soviet Union in Eurasia. If strategic defenses could help deny the Soviets the possibility of a swift victory against either the United States or its allies, Soviet leaders would face the prospect of a prolonged military campaign, facing the vast industrial potential of the West.

Equally competent Soviet defenses would reduce the ability of the US to attack targets in the Soviet Union. Thus, as our ability to deter and defend against Soviet attacks improved, our ability to retaliate would decline. But even without defenses, our ability to attack will probably decline some time in the future. And if the kind of policy described should come to pass, the United States might be satisfied with a minimum ability to retaliate against the USSR with offensive nuclear forces. Overall, this should be a deterring and stable strategic balance.

Conclusion

In summary, the United States is facing serious problems in maintaining the adequacy of its nuclear policy on the basis of offensive nuclear forces alone. Solutions to these problems generally either undermine the competence of our forces or create instabilities. Strategic defenses of limited effectiveness, properly combined with nuclear offensive forces, offer solutions to the competency problem; but such defenses are usually thought to be very destabilizing. However, if we set aside the narrow definitions of stability prevalent in recent years—which are based on terrifying and unbelievable final outcomes from massive nuclear exchanges—we can find ways in which defenses contribute to a more relevant form of stability. In particular, they could reinforce the effectiveness of current strategic policy by helping ward off the limited, militarily-oriented uses of nuclear weapons which are the most credible avenues toward actual nuclear use today. Beyond this, increasingly effective strategic defenses might also contribute to stability and US security. Rather than just reinforcing current policy, the deployment of limited strategic defenses could begin a transition toward a stable, fundamentally new strategic policy for the United States and the West in which defenses play a major and growing role, and in which offensive nuclear forces play a diminishing one.

13

SDI and US Nuclear Doctrine

Barry M. Blechman

I would like to bring us back to fundamentals in order to understand what the purposes of our nuclear forces are, and what doctrines, policies, and plans we utilize to achieve those objectives, both politically and militarily. I think it important to step back here before launching into the debate and becoming enmeshed in nuances and variations. It is hard to establish such a list of objectives—why the US maintains nuclear forces. What I have done is to review the posture statements of the secretary of defense and statements of the president and secretary of state, trying to reduce what is really a very complicated set of goals and objectives to four basic purposes. Let me list these four objectives and describe them. I will then go back and talk about the doctrines that have been established to achieve these purposes and how the Strategic Defensive Initiative might impact on these doctrines.

First of all and obviously most importantly, we maintain nuclear forces to prevent any military attack on the United States or any attempt to coerce the United States by threatening such an attack. This is a self-explanatory goal and clearly the most important purpose of the government, the reason we have a government: to protect ourselves from external threats.

Secondly, if we cannot prevent such attacks, we maintain nuclear forces to defeat any military attack on the United States. This objective is not often stated openly because it does not sound very good politically. It sounds as if we are almost planning to fight a nuclear war. But, naturally, it is only prudent and responsible for our military officials and defense officials to plan how to fight such a war should one occur. That is their responsibility. And, as Secretary of Defense Caspar Weinberger said earlier in the administration, US nuclear requirements are determined by analyses of needs to "prevail" (that is the word that is used) in a nuclear war. In fact,

---

This is an edited transcript of Dr. Blechman's remarks delivered during a conference on SDI at the Virginia Military Institute, April 8, 1986.

one major reason why our nuclear force posture is so large is that we have established requirements as to how many nuclear weapons would be needed in such a conflict. This is not new to this administration; it has gone back to the earliest atomic days in the late 1940s and early 1950s. Quietly the US has always planned, if necessary, to prevail in a nuclear war.

Thirdly, we maintain a nuclear force posture to prevent, or if necessary to defeat, a nuclear or conventional attack on our allies. We have explicitly made such a commitment to our allies in NATO and ANZUS and to Japan. This is a very important part of our policy. An explicit tenet of NATO doctrine is that, should conventional defense fail, NATO would initiate a nuclear war. First, tactical nuclear weapons would be used on the battlefield. Then, if necessary, that conflict would escalate up to and include a nuclear exchange between the United States and the Soviet Union.

This policy is essential for the security of NATO. There is a great asymmetry in the basic geography of world politics. Throughout the postwar period, we have attempted to contain the Soviet Union in its own territory, or at least in the territory that it occupied at the end of the Second World War. To do that, we created a system of alliances and committed ourselves to the defense of those countries along the margins of the Soviet Union, particularly in Western Europe. But the Soviet Union is a huge country and a great military power. It is an intrinsic part of Europe, and as such its military power is factored implicitly in any decision taken by any European country. The United States, on the other hand, is situated across an ocean, three thousand miles away. It always has the option, at least in the minds of our allies, of withdrawing or otherwise not fulfilling that commitment.

The US attempts to offset this asymmetry and the political weakness that it causes in the alliance by stating that we would share our nuclear forces and that we would risk the devastation of our own homeland, if necessary, in order to defend Western Europe. To make that commitment credible, we not only deploy large numbers of American forces in Europe; we also maintain a "first-use" policy, with the battlefield intermediate-range weapons which make that policy credible.

Implicitly, we also provide nuclear guarantees to other states beyond NATO, Japan, and other treaty states. We have, in fact, put our nuclear forces on alert or threatened to use them at various times in the postwar period: in the defense of Korea before the close of the Korean peace negotiations in 1953, and in the defense of Taiwan in 1958, for example. More recently, we implicitly made a nuclear threat in support of Israel in 1973, when we increased the alert level of our forces, including the alert of nuclear forces. In short, nuclear guarantees are the ultimate foundation of US foreign commitments, even beyond those states with whom we maintain treaty commitments.

The fourth and final objective of maintaining nuclear forces is to limit nuclear proliferation. Many industrialized countries which are allied with us and such newly industrializing countries as Korea, have a potential to develop nuclear weapons, but have chosen not to. One reason they feel able to avoid such a step is because we have made the guarantees that I have just described. This is very important in the case of Korea and Taiwan, for example. Both had nuclear weapons programs in the 1970s, but they terminated them, in part, because of our security guarantees. US nuclear guarantees are also important in the case of other industrial states, such as Japan. If its people chose, it could certainly develop nuclear capabilities in a very short time.

Let us now look at US doctrines, at how we have sought to achieve these four objectives. We attempt to prevent attack on, or the coercion of, the United States through the doctrine of mutual assured destruction (MAD). At least historically, we have maintained nuclear forces capable of surviving any attack and retaliating with great devastation on any enemy which would initiate the conflict, whether against its economic or military targets. By maintaining that retaliatory capability, we have tried to persuade any adversary, meaning the Soviet Union, not to strike this country with nuclear weapons.

I would maintain that the "mutual" part was never a conceptual part of US policy. With only one exception the United States has, in fact, never deliberately refrained from gaining the capability to strike Soviet forces. The one exception is the brief time in the early 1970s when Congress stopped a program to improve the accuracy of our land-based missiles. More commonly, the United States has pursued programs, for example, to be able to destroy Soviet strategic submarines. As Admiral Watkins and Secretary of the Navy Lehman have stated, it is US policy that, in the event of a conventional war, we should be in a position to identify and destroy Soviet strategic submarines. It will be remembered that strategic submarines are that part of our own strategic posture which we consider the most secure, the most survivable. The US also has pursued continued improvements in its land-based forces, particularly with the MX missile program. If that program were extended to the degree which the administration would like to see it, the MX force would have quite a significant capability against Soviet land-based ballistic missiles. It is thus the "assured destruction" which is the important part of the doctrine not the "mutual" part.

What would the deployment of strategic defenses do in terms of this situation? First of all, we must note that neither country would have perfect defenses. As President Reagan said in a press conference last year: "I've never asked for 100 percent. That would be a fine goal. But if you have the most effective defensive weaponry, even if it isn't 100 percent, it would strengthen deterrence because the other fellow would have the knowledge that if they launched a first strike it might be such that not enough of their missiles would get through and that in turn we would launch the retaliatory strike. If SDI is, say 80 percent effective, then it would make any Soviet attack folly." In the president's words, even partial success of SDI would strengthen deterrence and keep the peace. In other words, SDI would continue our past policy of seeking to maintain our ability to deter attack on this country.

One should note here the president's figure of 80 percent is quite optimistic. I do not know of any complex system which works, or that can work, with that degree of reliability in the sort of environment which one would be imagining for strategic defenses. There is no air defense system which has ever worked with anything approximating those figures. I believe that the most effective air defense system the world has ever seen occurred early in the Second World War; it was the German defenses against British bombers carrying out daylight raids in central Germany. Those systems worked something on the order of 25 percent to 33 percent effectively. Typically, the most sophisticated, complicated air defense systems, if they work at all, work on the order of 10 percent or 5 percent. The very sophisticated air defense used by Syria against Israel in 1982 worked at 0 percent effectiveness. Thus, I do not know on what basis, what scientific, empirical basis, figures like 80 percent can be founded.

We must also assume, when thinking of the effects of SDI, that the Soviet Union will have its own ballistic missile defense system. It may not have the same technologies as ours; it may work somewhat worse or somewhat better, but the USSR will certainly have its own system. We cannot make the fallacy of assuming that we get the last move, that we put in our system and nobody reacts.

In terms of the effects of SDI on deterrence, the question, therefore, is "what is the effect on our ability to prevent, to deter, an attack on this country in a world in which *both nations* have *somewhat effective* defenses?" I would say that in most cases the consequences are zero. That is, we would have spent a lot of money. Strategic defenses would have major effects on arms control and on relations with our allies and so forth. But in terms of deterrence, SDI would have no effect. The present military balance is extremely stable, as the president has said. In peacetime or situations of modest tensions, no one is going to risk a nuclear war, whether there are defenses or not.

The urgent question, though, is the effect of SDI in crises, especially in that ultimate crisis, when the two sides' military forces have already exchanged fire, when the two nations have already clashed, and when both sides begin to ask themselves: "Well, maybe war really is coming. How will I protect my country? What am I to do about this ultimate catastrophe? Maybe, if I go first, I can take out enough of the other side's strategic forces and disrupt its command and control systems such that, *with my defenses*—which I know do not work perfectly, but which might work to some degree—I can withstand his retaliation. Does my defense nullify the effect of his retaliatory forces?"

This is the most dangerous aspect of SDI, from my perspective. It is the effect on crisis stability. Uncertainty in this type of situation is not desirable. The adversary must be certain that if he strikes us, he is going to be destroyed. There should be no uncertainty in his mind as to whether, perhaps, if he goes first, if he gets in the first shot, he will be able to survive our retaliation. That is when war becomes thinkable. And that is when both sides begin to have incentives to initiate the first exchange.

I believe that this would be the major impact of defenses. An exception, of course, would be the deployment of a very limited system designed to defend only US retaliatory forces. If one could define such a system so that it did not lead, meld, into a broader system and thus complicate the whole situation, and if it made sense on a cost-effectiveness basis (that is, if compared to other alternatives to protect and modernize land-based forces, it was the most efficient) then one should consider it. But we are talking here of quite another matter which should be considered on its own merits, not on the basis of images of rainbows protecting people and so forth. When contemplating comprehensive defenses, one has to be most concerned about crisis stability.

Let me quickly run through the other objectives of US nuclear forces. In terms of our ability to defeat enemy attacks and prevail in a nuclear war, should it occur, the main effect of a strategic defense system is to increase greatly the requirements placed on our offensive forces. Again, one cannot make the fallacy of the last move. If the USSR put in defenses first (as in fact it would if we broke the ABM Treaty, because it is in a much better position to do that), we would increase offensive capabilities, multiply warheads, perhaps increase the number of penetration aides and so forth. If we put in defenses, he does the same thing. It gets us into a very destabilizing offense-defense race. Both sides would continue to try to meet their military requirements. Both sides would continue to offset the other's defenses. And it would be a very dangerous situation if we got into a bidding war with nuclear forces. It would distinctly harm the political relationship. It would create the kind of world conditions in which crises became more likely.

Let us look at the impact of SDI on our ability to extend deterrence to the allies. This is the most controversial aspect of US nuclear doctrine, the cutting edge where our doctrine is most sharply challenged. It has been challenged because of the changing military balance. It was easy to maintain a nuclear commitment in the 1950s, because we had superior strategic nuclear forces. The Soviets closed that gap in the 1960s, and in the 1970s they closed the gap in medium-range forces. They are now closing the gap in tactical nuclear weapons. In this situation, our threat to initiate nuclear war has more and more become a threat to commit national suicide. We are saying, "if you attack, our policies are such, our forces in place are such that we will be committed to launch nuclear strikes. We know what the consequences will be, but the stakes are so important that we will do it anyway."

We remain committed to the policy, as the best available option.

What would be the effect of a US strategic defense, or both US and Soviet strategic defenses, on the credibility of that doctrine? I think what SDI does is to confirm the Europeans' worst fear: what we call "decoupling." It separates American military power from the defense of Europe. It creates a situation in which the Europeans believe that when push comes to shove, if the United States and the Soviet Union fight a nuclear war, they will fight it in Europe and spare one another's homelands. The political implications of that perception are extremely negative from the American perspective. It means that over time the Europeans will attempt to go their own way. They will attempt to create an independent entity, perhaps centered on the French nuclear force. Perhaps the Germans would move to get nuclear weapons for themselves. It means the destruction of what has been a very peaceful situation, a very positive situation for the United States in Europe for more than forty years.

Acknowledging this problem, the proponents of SDI say, "No problem, we'll extend SDI to Europe." Nevertheless, "decoupling" remains a problem. First, Europe faces a different kind of missile threat; short-range tactical missiles cannot be halted with the same kinds of systems that we use to defend the US homeland. Secondly, SDI has a tremendous impact on NATO strategy. If these defenses existed, then one would have to move away from the first-use policy, greatly strengthen West European conventional capability, and try to maintain deterrence that way. It is an extremely expensive proposition added to the expense we would already have with SDI.

Thirdly, it would force us to draw lines. Which allies are under this rainbow? Is it just the treaty allies, Western Europe and Japan? How about South Korea? How about Israel? Do we want to set out demarcations like that? Do we want to put everyone in, commit the United States to defend everyone in the world under every situation? I do not think so.

Finally, transition to the strategic defense regime would have a tremendous impact on our relations with Europe. When we move into this situation where we are changing our strategic doctrine, where we are scrapping the ABM Treaty and any possibility of arms control, where we are nullifying or seeking to nullify the British and French nuclear forces, it is going to be an extremely difficult period. It will be rife with chances for misunderstandings and tensions between the US and its closest friends.

14

SDI: A European View

Pierre M. Gallois

ANY discussion of the Strategic Defense Initiative and European perspectives requires a transition, in history and in military scenario. With the exception of the Soviet Union, the European nations are relatively small and densely populated. Their longest period of peace in modern history has been that since the end of World War II.

During that period of peace, however, Europe has known containment. It has known brinksmanship. It has known the Berlin Airlift. It has known Korea, Indo-China, Vietnam, Algeria, Afghanistan, Iran, Lebanon, and many other skirmishes that could have had devastating effects on it and on other global regions.

The conditions for peace since 1945 have often been strained, but peace has nonetheless been maintained. Much of the credit—or at least the impetus—for maintaining the peace is, in my opinion, due to the creation and maintenance of nuclear arsenals by the United States, the Soviet Union, Great Britain, and France. The nuclear arsenals have made us less venturesome. But will they continue to make us safer?

As a point of departure, we must remember that the most devastating war in the history of mankind ended just forty-one years ago. And it was also forty-one years ago that the devastation that could be wrought by nuclear weapons became apparent.

At that time, however, only the United States possessed nuclear weapons. That situation changed dramatically in 1949 when the Soviet Union detonated its first atom bomb. The events since then are well known. The atom bomb has been succeeded by the hydrogen bomb. Our vocabulary has since been expanded to include a plethora of nuclear weapon systems with acronyms such as ICBM, SLBM, GLCM, ALCM, ABM, ATBM, and a number of others.

As the nuclear arsenals of both East and West grew, reasonable people sought to control their proliferation. Throughout the last two decades the Western Allies have many times called for talks on a balanced and verifiable arms reduction. Repeatedly, the Soviets have failed to respond to these calls for talks. Their decisions taken in the early and mid-1970s to produce the SS-20 and SS-22 missiles and to deploy a new generation of ICBMs were typical of their response relative to new arms-control initiatives.

Thus, after numerous disappointments, the East and West evolved a series of nuclear strategies that, by the 1960s, had culminated in the policy of Mutual Assured Destruction or MAD. MAD is predicated on an offense or retaliatory capability being superior to any known system of defense. It is a strategy of retaliation, based primarily on a threat to destroy centers of population and static military installations, thus inflicting crippling devastation on the aggressor. For twenty years, it has been the West's operative policy in spite of the fact that, in 1985, Soviet Marshal N. V. Ogarkov acknowledged that in the post–World War II period, "Means of defense were developed at an accelerated rate . . . whose skillful use at a certain stage balanced the means of offense and defense."

Those of us in Europe have been perhaps more conscious of these developments than our American counterparts—more conscious because the accuracy of recently developed Soviet short- and medium-range missiles acquire staggering significance when placed in a European setting. For example, a surprise attack consisting of Soviet SS-20 missiles could paralyze or even destroy allied conventional forces with only minor collateral damage. Thus, the Soviets could accomplish much with little. I believe that the ensuing discussion will convince you of the dilemma that has occurred due to advances in missile technology.

Conventional forces are the strength of the Federal Republic of Germany and that country's main contribution to the common defense of Europe. Consequently, former Chancellor Helmut Schmidt, noting how vulnerable his nation's forces had become vis-à-vis newly deployed Warsaw Pact missiles, requested in 1977 that matching NATO armament be arrayed to counterbalance the Eastern bloc. Two years later, the Western Allies agreed to the deployment of 572 Pershing II and ground-launched cruise missiles in Europe.

This sense of urgency was in contrast to the previous allied attitude. Prior to the SS-20 deployments, the allies had not focused a great deal on the potential threat represented by the Soviet SS-4 and SS-5 missiles. The reason was in the missiles themselves. Though more than five hundred SS-4s and SS-5s were aimed at European military and industrial targets, the weapons were known to have relatively poor guidance systems and hence to have a significantly high circular error probable (CEP) or miss distance.

To compensate for their high CEP, the SS-4s and SS-5s were in the one-megaton range. Their employment would have inflicted tremendous collateral damage. Moreover, their fireball would have made ground contact, creating dust storms and significant nuclear fallout. Because of the earth's rotation and prevailing winds, the Soviets consequently ran the risk of inflicting as much damage to their territory as to the West.

The potential benefit of the SS-4 and SS-5 missiles to the Soviets was thus questionable. NATO planners largely discounted their threat. Indeed, when President John F. Kennedy decided to withdraw the Thor and Jupiter IRBMs that had been deployed in Europe during the Eisenhower administration, only minor concern was voiced by the US's European partners. In general, the resulting asymmetry was accepted by the Allied Powers. The latter were of the opinion that a missile threat with the potential of inflicting as much damage on the aggressor as on the intended victim was not very likely to materialize.

The deployment of the SS-20 missiles removed European complacency. This new generation of nuclear-tipped missile enjoyed a CEP that was 1,000 percent improved over its predecessors. Its yield was reduced accordingly and, for the first time, the Warsaw Pact had the capability to strike at distant targets with pin-point accuracy, little collateral damage, and only limited fallout.

As a result, the SS-20 was viewed not only as an instrument of destruction but also as an instrument of policy that reinforced the theory and doctrine of Soviet military specialists, especially that dealing with the control of crisis situations through the threat of selective destruction. A corollary was their newly acquired ability to conduct nuclear operations with only limited collateral damage and minimal fallout.

Those of us who had assumed a specialized status as nuclear planners postulated that, as a general rule, the accuracy of ballistic missiles could be increased two-fold approximately every six years. With the increasing accuracy, steps had to be taken to guard against the new feasible "surgical" strikes against our conventional forces. We had to anticipate a surprise nuclear strike, which would give us only four or five minutes warning.

Just as progress does not stand still, neither does related weapon systems development. The Warsaw Pact has continuously upgraded its family of SRBMs and IRBMs. The new family of Soviet missiles, the SS-21, SS-22, and SS-23 have (according to Dr. Richard DeLauer, the former US undersecretary of defense for research and engineering) achieved accuracies making possible CEPs of thirty to fifty meters or less.

I am convinced that the Soviets are not making exaggerated claims about the proficiency of their weapon systems. Convergence between official Soviet statements and the weapons necessary to implement the corresponding doctrine seems now to be fact. As Soviet Colonel M. Shirokov has stated: "The objective is not to turn the large economic and industrial regions into a heap of ruins . . . but to deliver strikes which will destroy strategic combat means, paralyse enemy military production, reduce the enemy capability to conduct strikes." Marshal A. A. Grechko had, in fact, preceeded Shirokov in the following fashion when he explained the basis of adapting weapons in development to specific strategic concepts: "The tactical missiles units are the basis of the firepower of the ground troops. . . . The missile units will be capable of hitting targets located at the range of several scores to many hundreds of kilometers with great accuracy . . . using nuclear ammunition. Chief attention is given to a further rise in the range and aiming precision of these missiles."

Interestingly enough, concern has been voiced, most critically and most recently, by Israel. Former Prime Minister Shimon Peres has stressed that the state of Israel is now under the threat of twenty-four SS-21 launchers (albeit with conventional warheads) and that no existing Israeli defensive system is adequate to neutralize this threat. As one consequence of this threat, Israel has become a participant in the Strategic Defense Initiative program.

We in Europe are thinking along the same lines and for similar reasons. Let us examine the scenario:

(1) Our conventional forces, especially those of nonnuclear armed nations, can be crippled by a surprise first strike. Organized, structured resistance would probably be ineffective in a matter of hours, maybe even minutes.
(2) Increased accuracy of short- and medium-range missiles has led to a concomitant reduction in yield.
(3) Almost all critical West European targets are at surface level; hence they are virtually unprotected. They could be severely damaged or destroyed by a low-yield nuclear strike that produced only limited fallout and caused very little collateral damage.

Because of its closed society, the USSR will probably enjoy the benefit of initiative and surprise. If it chooses to employ a disarming first strike, the accuracy of its short- and intermediate-range missiles would give the USSR a big plus. In this vein, we must remember that when democratic nations face an expansionist adversary, the strategic buzzwords are usually the following:

(1) Initiative (a political option more easily taken by an autocracy);
(2) Surprise (a military capability);
(3) Accuracy (a technological achievement).

The Free World has accurate weaponry, but the Eastern bloc can be reasonably expected to have a monopoly on initiative and surprise. Hence, stationary elements of NATO forces deployed in Western Europe could well serve as targets for selective, accurate, preemptive nuclear strikes. Examples of potential targets abound: antennae of long-range detection systems; combat aircraft; ammunition dumps; fuel storage depots; communication centers; pipe lines; transportation areas; major headquarters, and many similar emplacements.

In Europe, and certainly in France, much thought has been given to the exposure of our air forces. They are very vulnerable, as are the majority of

Western European conventional forces, to a potential aggressor assuming the initiative and making a first strike with highly accurate weapons.

At a typical French military airfield, aircraft are hangered in individual concrete shelters. Though the shelters are blast resistant up to 21.3 pounds per square inch, four (low yield) one-kiloton warheads would destroy the aircraft. Yet, even with a CEP of 150 meters, the damage inflicted by the warheads to the areas adjacent to the airfield would be minimal.

We have made such evaluations for several airfields, including those hosting bombardment and interceptor aircraft. The configuration of each airfield being somewhat different, we have calculated the differing minimum yields to achieve a 0.8 probability of destruction and minimal collateral damage. The results of our calculations are startling. A total of thirty-one warheads, yielding fifty-nine kilotons, would effectively cripple operations at France's eleven main military airfields. Aircraft and facilities would, in the main, be largely destroyed. Due to the precise delivery systems involved, however, collateral damage would be slight. The surgical removal of French combat aviation capability by accurate low-yield nuclear-tipped missiles would leave intact and unscathed more than 99.5 percent of French territory!

It is difficult to conceive how the combat air forces of other Western European nations, such as West Germany, would not be similarly affected. As stated earlier, the Soviets now have the capability to deliver low-yield nuclear-tipped weapons discretely to achieve a desired effect on military targets with minimum damage to surrounding areas. No site is immune, and many Europeans perceive that the SS-20, when coupled with other improvements in Soviet theater nuclear forces, has shifted the balance in Europe in favor of the Soviet Union.

The threat is credible. A surprise nuclear strike against critical military targets is feasible because corresponding weaponry could initially be launched without alerting the defense. The concurrent launch of a few hundred short- and intermediate-range missiles could be accomplished quickly and by a relatively small number of people. Further, it would not allow for the same defense-alert posture (or defcon) that would be triggered by the potential movement westward of some fifty to sixty army divisions supported by artillery, armor, and aviation.

Given their range and numbers, the SS-21s, SS-22s, and the SS-23s have the capability effectively to neutralize NATO ground and air forces with one orchestrated salvo. The destruction of some two hundred to three hundred of the most critical Western European targets could be achieved with very little collateral damage.

The Soviets could also adopt the tactic of crippling only small, but vital, portions of Western Europe's armed forces such as radars, combat aviation squadrons, and nuclear depots. By such a tactic, they could hope to impact Western European public opinion and then seek to negotiate Western

Europe's (partial or total) demilitarization. This scenario can be viewed as unrealistic or even ludicrous, but it should not be ignored.

Conventional forces in Central Europe are not generally intermingled. National units are generally separate from NATO allies. Accurate nuclear strikes could cripple, even destroy, specific allied targets but avoid the collateral destruction that would more likely trigger a dangerous escalation in the conflict. German, Dutch, and Belgian units, for example, could be hit without damage being inflicted upon the US forces' zones of deployment. This postulated attack, if successful, could jeopardize a united NATO response. The US could be pressured by European public opinion to withdraw while damage is slight. The US could be placed in a "no-win" situation. On one hand, it would be at peril if it launched a massive retaliatory attack, and on the other hand it would be at peril if it did nothing. Not having suffered losses and with its NATO allies in complete disarray, however, its reaction to such an attack could conceivably be mild and conciliatory.

Thus, no matter how unrealistic it may be viewed, such a scenario should not be ignored. To ignore it gives strength to the threat of a disarming Soviet first strike—and its effect as a political instrument. In such a scenario, NATO would have neither the time nor the opportunity to galvanize its response. This would abet the potential decoupling between the American and the European allies, especially if the aggressor's demands were moderate, and the European populace were in panic. Such indeed could well be the sad end product for a Europe which has based its defense strategy largely on a US nuclear umbrella and has not concurrently generated or maintained a sufficiently credible conventional deterrent.

Moreover, it is my considered opinion that national armed forces solely dependent upon conventional deterrents and conventional defenses are at a disadvantage vis-à-vis forces which are similarly equipped, but which also possess a nuclear capability. It is axiomatic that the more powerful the adversary counterstrike forces appear to each side, the more stable will be the peace—because of each side's awareness of the risk they take in resorting to the use of force.

Consequently, a surprise first strike by a nuclear-armed aggressor could yield disproportionate results. Active defense must be examined as an appropriate, albeit partial, response to the dilemma created by the relative vulnerability of the Western European conventional forces. The West must clearly and resolutely deny a potential aggressor the opportunity to neutralize or destroy NATO deployed forces and to blackmail the Western European allies by a threat which, at this time, cannot be countered effectively.

Part of the process of denial could be accomplished by the creation of a ballistic defense system for Europe which, in part, would serve to deny the potential aggressor the element of a disarming first strike by low-yield nuclear-tipped missiles. The special conditions relative to Europe are different than those unique to the American continent. These special conditions include:

(1) Enemy launchers at close proximity to their targets. Warhead trajectories would be lower, and total warning time may be limited to a maximum of four or five minutes.
(2) The (tactical) launchers are sited and compressed into much smaller areas than ICBM silos.
(3) Space-based defenses would probably be less practical in a European scenario than a US one because of the shorter reaction times and the geographical constraints. Hence, ground-based lasers, supported by "pop-up" or space-based mirrors and interceptors (i.e., small interceptor missiles) may be more suited for Western Europe's special conditions.
(4) Western Europe's critical military targets—two hundred to three hundred in number—are far fewer than those in the US. The targets are generally on the surface, and they are not nuclear hardened. Their destruction could be achieved relatively easily by low-yield nuclear warheads, even those of limited accuracy.
(5) Soviet warheads released against Western European targets would have more depressed trajectories.
(6) Soviet use of Multiple Independently Targeted Reentry Vehicles (MIRVs) in Western Europe is unlikely, as is the extensive use of penetration aids or decoys. The ratio between Soviet launchers having a single warhead and the critical NATO targets renders unnecessary the use of MIRVed vehicles, with their corresponding additional complexity.

The different European conditions call for a slightly different prescription for a defense against ballistic missiles. In the European theater, the interception of the single-warhead missiles during the boost phase may prove somewhat less imperative than it would be for the multi-warhead ICBMs. Mid-course and terminal-phase intercepts remain vitally important. Saying this, however, is not tantamount to making it happen: much research and much work must be accomplished before even the rudiments of a strategic defensive system against ballistic missile warheads travelling along depressed trajectories and arriving on target only minutes later, as in a European scenario, can be introduced.

The problems (but conversely, the opportunities) facing the US and European scientific, engineering, and military communities are enormous. Nonetheless, the SDI program has acquired considerable definition and momentum since President Ronald Reagan's speech of March 23, 1983. The progress to date reinforces my belief that SDI will create new opportunities for innovative operational concepts and new tactics. The technologies under examination and development are not so narrow that their use must be restricted to merely the potential construction of a space shield. They could, in fact, serve as the beginnings of new families of defensive weapon systems, characterized by high velocity, selectivity, and precise destructive power.

If Western Europe is not to fall prey to a potential aggressor, it must act to strengthen and protect its conventional forces. One manner of so doing would be through the adaptation of SDI technologies and their utilization within a European scenario.

Some vital NATO targets could be better protected through the progressive modernization and transformation of our existing air defense systems. An add-on system to improve detection, acquisition, tracking, and guidance is badly needed. Some of the improvements could be obtained from technologies that the US is developing for the terminal phase. The efficiency of any European air defense system could be significantly enhanced by utilizing the concepts and the technologies being developed in support of the SDI Airborne Optical Adjunct and the Kinetic Interceptor programs.

Clearly Western Europe must also look ahead and identify the parallel developments and linkages between its needs and those of the overall SDI program. It could limit itself to building up only a rudimentary defense against ballistic missiles, as above, or it could proceed much further, in building-block fashion, to improve its ability to deter Soviet aggression. I prefer the latter, and I can envision an evolutionary defense network that appears to be within reach of Europe's combined capabilities and efforts.

I have previously mentioned a combination of ground-based lasers, supported by "pop-up" or space-based mirrors and interceptors as appearing well suited for Western Europe's special conditions. They could be employed to counter incoming SS-22s and SS-23s which would be launched a maximum distance of five hundred kilometers east of the Iron Curtain borders. Hypothetically, sensors could lock-on to the missiles and the mirrors would have line-of-sight at an altitude of approximately twenty kilometers. Relative to missiles deployed approximately one thousand kilometers from the nearest NATO installation, a launch apogee of eighty kilometers would be adequate to permit the reflection and focusing of laser rays adequate to destroy the incoming missile.

In this vein, batteries of high-energy lasers could be positioned and deployed from elevated sites, such as those in close proximity to the Alps. Admittedly, considerably energy would be required to operate the lasers, but they would be sited so as to be near a major energy source.

Positioned on level terrain forward of the lasers would be launchers capable of sending reflective, focusing mirrors, and infrared telescopes up to altitudes of two hundred to three hundred kilometers. Before the mirrors complete their arc and disintegrate in the atmosphere, they could reflect the ground-based laser beams against incoming enemy missiles and warheads.

Such a system would function automatically, with split-second accuracy and without human intervention. Geosynchronous satellites would activate the sequence to launch the mirrors and to energize the lasers. Eventually, it may be possible to position space-based, relatively fixed platforms to substitute for the "pop-up" mirrors.

Relative to Soviet missiles of greater range such as the SS-20, as well as the strategic SS-11 and the SS-19 launchers, Western European defense needs parallel those of the US. Thus, its needs, apart from terminal defense, could be satisfied by a global strategic defense system.

To date, Europeans have greeted SDI with a curious mixture of interest and opposition. The interest in SDI has been spawned naturally by the program's promise of significant, bold research that serves to advance the frontiers of knowledge in space-related technologies. Clearly, Western Europe, so much and so often a partner in technology sharing, wants to maintain this relationship. On the other hand, opposition to an initiative that could unbalance the combination of the US nuclear umbrella, French and British nuclear arms, and conventional forces that have helped maintain peace in Europe for four decades should not come as a surprise. France and Great Britain, for example, have evolved a capability for proportionate nuclear deterrence. This capability has become a cornerstone of their national security policies.

This puts a special burden on our great American ally. It alone has the requisite resources to develop the technologies necessary for an effective strategic defense system. Europe, by pooling its resources, can greatly help, and it can be a positive catalyst in this process.

Europe recognizes that it can no longer remain complacent in the face of the Soviet nuclear threat. Otherwise, it may not long exist in its present form. The Strategic Defense Initiative, adapted for and adopted by Europe, may provide the most honorable and pragmatic solution to its dilemma.

15

SDI and the European Defense Debate

Jacquelyn K. Davis

IN the course of his March 23, 1983, speech on strategic force modernization issues, President Ronald Reagan raised publicly his interest in reexamining the strategic defense concept. For the United States's NATO allies this came as a complete, and in many ways, an unpleasant surprise, since most European and Canadian defense analysts had long since discarded the strategic defense concept in favor of a deterrence posture based on the putative threat of an offensive nuclear weapons employment policy. No less an important factor influencing European reactions to the president's speech was their immediate concerns relating to the "prevalent imbalance of NATO–Warsaw Pact conventional forces" and the perceived importance of offensive nuclear forces—US central strategic systems as well as intermediate-range forward-deployed systems—to the deterrence equation in Europe. Especially in the aftermath of the extremely divisive European debate over the stationing of new generation intermediate nuclear forces, ground-launched cruise missiles, and Pershing II ballistic missiles, the nascent American interest in the strategic defense concept was widely greeted in Europe with apprehension. It stimulated echoes of the NATO defense debate of thc 1960s when flexible response was put forward as the basis of alliance strategic planning and operational concepts.

Indeed, with the exception of a very few West European defense analysts, virtually no one among the United States's NATO allies had appreciated the extent to which the strategic-military environment had changed since the signing of the ABM Treaty. This agreement had limited drastically the superpowers' deployment of strategic defenses and, in effect, rejected the concept as not "cost-effective." That is, most European and Canadian analysts had conformed to the prevailing US assessment of that time which held that in military-technological terms the strategic defense concept was not viable. Defenses could be overwhelmed relatively easily by the deployment of additional offensive weapons systems. Together, the advent of multiple independently targetable reentry vehicle (MIRV) technology and cold launch-refire techniques appeared to render obsolescent the then "state of the art" strategic defenses. Those were based on what are now called "conventional" terminal phase, ground based, nuclear-tipped interceptors and associated radar systems.

Since, it was widely held, offensive technologies were cheaper to produce than defensive systems, and because defense effectiveness at nearly perfect levels could not be guaranteed technologically, a majority of European defense analysts and their North American counterparts concluded that strategic stability could be most effectively maintained on the basis of offensive deterrence. Moreover, in the context of NATO strategy, it was widely assumed that the coupling afforded by US central strategic systems to the direct defense of Western Europe could only be maintained in the absence of defensive weapons deployments. In fact, it was argued that a protected US deterrent posture would enhance the prospects for a nuclear war being fought in, and limited to, Western Europe. This argument was somewhat paradoxical given the European assumption that strategic parity created the condition in which it could not be presumed that the United States would be willing to risk the destruction of its own territory to defend Western Europe. Contentious even though they were, these two lines of argumentation supported the prevailing reliance on offensive nuclear weapons in NATO strategy and opposition to rational discussion of the strategic defense concept. Europeans generally attributed to the Soviet Union a similar conception of deterrence theory. These views bolstered their opposition to reconsidering the defensive concept, as proposed by President Reagan in his enunciation of the Strategic Defense Initiative.

Initially, Western European concerns relating to the president's proposal of March 23, 1983, focused on political-strategic arguments against the strategic defense concept. This was largely due to their concern over the potential implications of SDI for the extended deterrence concept, and more specifically, to the legacy of their protracted INF debates during which supporters of the NATO modernization program had vested much of their credibility in supporting the NATO program. From the perspectives of those who supported the INF deployments, the strategic defense concept appeared to undermine the rationale for INF deployment in Europe and, worse, threatened to promote the strategic decoupling of the United States from its NATO allies. According to this view, the "decoupling" or fear of America's NATO allies that the US, if protected by some type of strategic defense deployment, would be less willing to participate in the direct defense of Western Europe, at least less inclined to use nuclear weapons. Such unwillingness would undermine the escalatory link that is inherent in NATO strategy.

Intrinsic to this perspective was the assumption that the United States sought a situation in which conflict could be limited to the European theater, leaving the continental United States virtually unscathed. Militarily this would place a greater premium on the conventional force balance in the European theater. However, European NATO allies were solidly against a greater reliance on conventional military power, given the adverse trends in NATO–Warsaw Pact nonnuclear forces. Beyond this, few NATO allies had met even the alliance's goal of increasing military spending by 3 percent in real terms. This objective had been established in 1977 as part of the Carter administration's defense improvement program. For the allies to include in their national budget allocations to NATO the sizable increases that would be necessary to support an enhanced conventional defense posture was, and remains, unthinkable. Indeed, in this respect, it is important to note that most of the NATO allies, including the United States with its Gramm-Rudman legislation, will be hard-pressed even to meet the spending needs for the range of force improvement programs required by the Follow-On Forces Attack concept, which had already been adopted by the Atlantic Alliance.

Technological arguments against the American research program were aired in Europe early in the SDI debate. Based largely upon the arguments of American detractors of the program, few Europeans conceded the possibility of a technologically feasible strategic defense. In early 1983 there was a limited European pool of knowledge relating to the evolution of defensive weapons technologies and component capabilities that had taken place over the last decade in the United States. New technical possibilities for strategic defense had emerged in the computer, electronics, components, and microchip industries. These provided increased substance to the arguments of those who supported a rethinking of US defense-deterrence concepts. These technological developments were largely ignored outside of the United States because the European allies were slower in coming to the realization that the Soviet Union did not share a symmetrical approach to defense and deterrence issues; indeed, it emphasized in its own strategic force posture the synergism of both offensive and defensive techniques. Even now, amidst the debate over Soviet arms control compliance, less concern is generally evinced in Europe over alleged Soviet violations of the SALT I and SALT II treaties than there is about the potential for US abrogation of the ABM Treaty if and when a decision is taken to pursue vigorously the defensive option.

In terms of such things as trade and human exchanges, the European allies have benefitted from the policy of détente to a far greater degree than has the United States. For fear of upsetting détente, West European defense elites and policy officials have, by and large, been unwilling to support politically the concept of strategic defense. Whereas there is growing support among West Europeans over various enhanced air defense and theater defense proposals that have been advanced in NATO, the West European Union, and the United States, there remains a fundamental unwillingness to embrace wholeheartedly the strategic defense concept. Especially in France, but also in other NATO countries, including Canada, there is a persistent emphasis on the alleged dangers inherent in undermining the East-West cooperative frameworks that evolved in the aftermath of SALT I.

Parenthetically, it should be noted that NATO allies, among them France and Britain, perceive that they have much to lose if the ABM Treaty limitations are revoked. This is only presumed and not demonstrated. In fact, one could argue that under certain conditions the deterrence credibility of countervalue forces would be enhanced, not diminished, by deployment of ballistic missile defenses. Nevertheless, Britain and France are concerned about the possibility of a Soviet "ABM breakout" which would undermine the deterrence potential of their respective national nuclear forces. Both countries not only deploy small, national nuclear forces, but in recent years each has alloted scarce defense resources to modernizing their respective deterrent capabilities.

In both countries, especially France, nuclear modernization programs have undercut conventional force improvements. French critics of the former Socialist government's five-year defense bill charged that, as a result, the conventional First Army is poorly equipped and undermanned. At the heart of French and European concerns is not just the future credibility of their nuclear forces but their inability to compete in the global arena where national and regional issues are at stake. For the French, the national deterrent force continues to rest on a domestic consensus in support of its existence, in part because of the prestige it is believed to confer upon France in the global arena. As in other West European nations, the French are suspicious of what they perceive to be superpower efforts to make decisions without reference to European interests, which often diverge from those of the United States. Thus, even though the new Conservative prime minister of France has publicly expressed an interest in the possibilities afforded by strategic defense technologies, there remains great skepticism in French strategic policy circles over the desirability of emphasizing SDI to the detriment of exploring more fully East-West cooperative solutions to the nuclear arms problem.

Moreover, Europeans fear that another round in the arms race would ensue, this time in space. Such a race would provide yet another arena in which they would be subservient to the designs of one or the other superpower. To those in Europe and elsewhere who emphasized this particular aspect of the anti-SDI campaign, which had evolved in America, a derivative argument was put forward. Critics argued that the United States sought strategic superiority over the Soviet Union by means of SDI. They often argued that because the United States was pursuing SDI at the same time that it was modernizing its offensive nuclear forces, it would attain a first-strike capability over the Soviet Union. Despite efforts to educate the allies on the motives and limitations of US strategic modernization programs, critics remained predisposed to pursue this line of argumentation. It was only when Congress limited the deployment numbers of MX missiles that some European opponents began to relent on this line of argumentation.

In Western Europe today there is growing recognition that the military-technological environment has evolved dramatically from the time when President Reagan established the US Strategic Defense Project Office. Political endorsement of the strategic defense concept is not yet forthcoming in most West European countries, in large measure due to its allegedly profound implications for the arms control process. Nevertheless, West European defense elites and policy officials are increasingly prone to consider the technological and economic benefits that likely will accrue from European participation in the US research program. Thus, while it remains politically expedient for allied governments to separate the SDI project from attempts to augment NATO's defense-deterrence posture in Western Europe, it is more directly acknowledged that the technological potential of SDI may have relevance to theater defense issues.

The perceived importance of augmenting alliance air defense assets as a priority NATO concern has awakened West European interest in the potential theater applications of SDI-derived technologies. More than this, however, the intricate relationship between conventional, theater-oriented, and strategic-nuclear air defense issues has legitimized European participation in the US research endeavor. At the same time, it has somewhat dampened criticism among a wide range of European elites of the concept of strategic defense and, more specifically, of antitactical missile defense. In fact, there has emerged a broad European consensus in support of the "enhanced air defense" concept, which is widely interpreted to include defenses designed to counter conventionally armed Soviet short-range ballistic missiles of the SS-21, SS-22, and SS-23 classes. Thus far, Britain, Italy, and the Federal Republic of Germany have signed Memorandums of Understanding (MOUs) with the United States governing their respective participation in the SDI program. In each of these three countries, the MOUs focus on industrial participation and offer no political endorsement of the strategic defense concept. Not only is the notion of relying on US research funds attractive to European and Canadian industries, but the perception of possible research spinoffs to the civilian sector and, more importantly, to conventional defense vulnerabilities, has provided an important stimulus for allied participation in the SDI program.

For the most part, Europeans focus on the air-defense-related potential for SDI spinoffs. Increasingly, though, there is discussion of other possible applications of SDI technologies for the European theater, including, for example, their likely contribution to the NATO-approved Follow-On Forces Attack and Counter-Air 90 concepts. Israel has also signed an MOU with the United States, stimulated in part by its expectation that railgun technologies may have an important antitank application. Almost exclusively, the potential contributions of SDI-derived technologies are discussed in NATO within the context of the current NATO strategy of Flexible Response, and they focus on deployment timeframes extending beyond the year 2000.

In the interim, the glaring vulnerabilities of the alliance's defense-deterrence posture, especially in the counter-air arena, are viewed widely in Europe as having been exacerbated by the relative lack of attention in the recent past to modernizing NATO's air defense assets. Prior to the president's speech in 1983, NATO had initiated a review of this aspect of alliance force posture. In its Counter-Air 90 study of 1982, it addressed the subject of the tactical missile threat to Europe, including possible defense deployment options. Thus, even before US revitalization of the strategic defense concept in 1983, West European defense specialists were increasingly disposed to examine a NATO antimissile defense option as an element of alliance air defense assets, not as a radically new strategic option that might alter the nature of alliance strategy. The immediate stimulus underlying a rekindling of the NATO air defense debate was the modernization of Soviet tactical surface-to-surface missiles, some of which (like the SS-22 and, marginally, the SS-23) display an endoatmospheric trajectory. Of greater perceived importance was President Reagan's SDI speech which provided a basis for the more precise delineation both in Europe and the United States of the spectrum of missile vector threats facing Western Europe and continental United States.

In Western Europe the perception that SDI does not explicitly pertain to theater defense issues gave rise to a proliferation of regional concepts—the European Defense Initiative being perhaps the most famous. These aim more specifically at defining the nature of the evolving threats facing Western Europe and the defense options available for consideration on a regional (NATO) or a transnational (Franco-German/IEPG collaboration) basis. To date, most European defense analysts are reluctant to include nuclear-capable and longer-range systems (e.g., the SS-20 IRBM) in their threat analyses. They concentrate instead on the shorter-range conventional threats posed by the SS-12/SS-21, SS-1/SS-23, and SS-22 tactical ballistic missiles, and by the evolution of Soviet ground-based cruise missile capabilities. Of greatest concern to a growing number of European, especially West German, defense analysts is the continued modernization of Soviet tactical missiles (SS-21 to SS-23 class) and indications that by the late 1990s such systems could deploy nonnuclear, precision-guided munitions with a high degree of accuracy.

This concern contributed to the articulation of the European Defense Initiative, which has attracted widespread support in Western Europe, even though the desired framework for collaboration remains less than clear. For some, EDI is viewed entirely as an entity separate from the American SDI program, although linkages would exist due to European industrial participation in the US research effort, as well as to certain compatibilities in the technologies involved, such as sensor and early warning systems. This is the position of the West German Free Democratic party, the Italian Socialist party, and even some members of the Christian Democratic parties in both countries as well as in Belgium and the Netherlands. Sparked by the initiative of West German Defense Minister Manfred Wörner, other Europeans emphasize the differences between the objectives of the SDI program and the requirements of NATO strategy. They envisage development of a European antimissile defense capability within the alliance framework. By articulating the antimissile concept within the context of NATO's broader conventional force improvement programs, Wörner was able to attract the support of both his Italian (Giovanni Spadolini) and Dutch (Jacob de Ruiter) counterparts.

Essentially promoted as an "extended air defense" concept, Wörner's thesis has found increased support within NATO defense circles, in part as a result of the conclusions of a new alliance air defense study. This study, which was prepared at the request of NATO's Military Committee by the "Advisory Group for Aerospace Research and Development" (AGARD), is a follow-on to a March 1980 AGARD analysis entitled "Defense Against Missiles." Its focus is on the possibilities for developing active, nonnuclear defense capabilities against Soviet tactical ballistic missiles (having a range of less than one thousand kilometers) up to the year 2000. The AGARD panel investigated a spectrum of weapons and sensor technologies, including space-based, airborne, and ground-based high-energy laser weapons, directed-energy systems, guided missiles, surface-to-air missile systems, anti-aircraft artillery, and electromagnetic gun technologies.

Other studies have been undertaken simultaneously in Western Europe, including in France where former Socialist Defense Minister Charles Hernu established a Space Studies Group within the President's Defense Council to examine the possibilities and opportunities of space-oriented technologies for military purposes. The French created EUREKA, a civilian-oriented research project, and called for European participation. However, the funding for EUREKA is allocated from existing programs and is based largely on collaborative endeavors already in progress. Europeans do not view it as an alternative to their participation in the SDI, although that may be precisely what the French president had in mind when he articulated the concept in the first place. Nevertheless, there is an expressed desire on the part of some Europeans to draw France into the SDI and EDI projects. The West Germans are particularly desirous of drawing the French into closer collaboration on regional defense issues. Simultaneously, there is an increased desire to draw France more directly into the NATO air defense debate and to consider collaboration on antimissile technologies under a framework that would allow for French participation. France already has a candidate system for European consideration: the SA-90 which is scheduled to replace French Hawk batteries in the mid-1990s. The French have also broached the subject of developing a high altitude surface-to-air missile with the West Germans. As a result, in the Netherlands, Belgium, and Italy, there has emerged a greater interest in a "European Defense Initiative."

European participation in the American research program is likely to increase, given the growing perception of the conventional defense spinoffs from SDI. However, this will not come without increased debate in allied countries, primarily because of concerns over a US deployment decision. With the announcement in the summer of 1986 that, failing a change in Soviet policy of noncompliance with existing arms control arrangements, the United States would exceed the SALT II limits on MIRVed launcher deployments, which the United States never ratified in the first instance, European interest in perpetuating the ABM Treaty regime appears to have been enhanced. As a result, there is likely to be renewed criticism against the strategic defense concept, especially as it becomes enmeshed in the domestic electorial debates of Britain, Italy, the Netherlands, and West Germany. Even so, support for the research program will remain staunch among allied industry, defense analysts, and policy elites who recognize its potentially important contribution to the defense of Western Europe.

Allied support for the program, including the defensive concept itself, may be broadened if it can be demonstrated that the logic of the emerging strategic environment points toward the need for strategic defense technologies not only to counter the instability that would result from increasing accuracy of conventional offensive systems targeted against NATO's most important military assets, but also to maximize the potential afforded by technology to enhance conventional deterrence at the battlefield level. In short, it may be that strategic defense and extended deterrence not only are likely to be mutually compatible. Strategic defense may be necessary for extended deterrence and conventional defense into the twenty-first century.

16

West Germany and SDI

Wayne C. Thompson

IN an important speech before parliament on April 18, 1985, Chancellor Helmut Kohl announced that the Strategic Defense Initiative (SDI) "will be the dominent defense policy issue in the years to come. It will decisively affect the East-West relationship and, in a particular way, influence the relationship of the United States of America to Europe." Professor Dieter Senghaas correctly noted in December 1985 that the German debate over SDI would go to "the core of deterrence and will lead into a fundamental and protracted security policy controversy. In it all the aspects of past conflicts over the premises and shaping of security policy will be revived." In the same parliamentary hearing, Professor Michael Stürmer added that "however far in the future the new reality might lay, the present has already been overshadowed by SDI."[1]

SDI has unleashed a public controversy in the Federal Republic of Germany (FRG), much as it has in the United States. This sometimes emotional debate came at a bad time for the government; it was trying to catch its breath after the furious public dispute over deploying NATO intermediate-range nuclear forces (INF), which lasted right up until their installation in December 1983. Also, President Ronald Reagan's March 23, 1983, speech caught West German leaders by surprise; they had not been told in advance about this initiative which portends a dramatic shift in Western defense.

1. I would like to thank the VMI Foundation for enabling me to conduct the interviews and research in the FRG which were necessary for this work. I am also very grateful to Dr. Raban von Westphalen, of the Federal Parliament's Research Commission on Assessing the Consequences of Technology; Dr. von Westphalen's help in assembling the key published documents on the SDI debate in Germany was indispensable for my study. Finally, I wish to thank Maureen Ward, whose patience and computer know-how were very important to me at a critical time during this project. *Stenographisches Protokoll der Sitzungen des Auswärtigen Auschusses und des Verteidigungsausschusses* (Bonn: Bundeshaus, December 9–10, 1985) (hereafter referred to as *Protokoll*), pp. 160, 173.

This oversight elicited the familiar complaint about not being adequately consulted by NATO's senior power, which in important defense matters seemed to prefer to act unilaterally and to expect European allies to fall into line.

The US government later bent over backwards to try to ensure that allied interests and viewpoints were respected. But major strategy changes in NATO have always originated in Washington, and what appears to be American unilateralism is a familiar problem for Europeans and Canadians. After protracted and careful deliberations, the West German government decided to give official support to SDI, hoping thereby to influence any changes in strategy which might be in the air. Nevertheless, because of widely differing opinions about SDI inside and outside the government, Bonn's position can best be described as "yes but."

In this chapter I will briefly describe the West German defense policy dilemma, which always influences the Federal Government's decisions regarding security policy. I will then explain why West German opponents of SDI think that this program would endanger security in Central Europe. This will be followed by the arguments which proponents of West German participation in SDI make. Finally, I will describe the government's actual decisions and the conditions which it placed on the FRG's participation.

The Setting for the West German Defense Policy Debate

Germany is a divided nation. The line, which runs 858 miles (1,381 kilometers) and which separates the most heavily armed peace-time alliances in history, cuts right down the middle of it. Its largest city and historic capital, Berlin, is located 110 miles (176 kilometers) inside the German Democratic Republic (GDR), a small country crowded with four hundred thousand well-equipped Soviet troops. One-third of the FRG's population and one-fourth of its industry are located within 60 miles (100 kilometers) of that line between the two German states. West Germany is only 140 miles (225 kilometers) wide at its narrowest point. Thus, the FRG is very sensitive to any situation or change in policy which might exacerbate tensions between the superpowers or endanger the links which have been carefully established between the two Germanies and with Berlin. West Germans tend to believe that they have a much greater need for détente, which they see as having benefitted them greatly. They regard not only bilateral agreements between the FRG and the GDR and Soviet Union, but also the arms control process itself, as an indispensable pillar of détente. If SDI appears to endanger the arms control process in any way, West Germans are bound to be leery.

The vivid memory of the Second World War creates strong feelings among many Germans against any form of armament. Slogans such as "arms race" and "militarization of space" are very potent in public debates.

This remains true despite the fact that most informed Germans have long known that offensive nuclear weapons, such as intercontinental ballistic missiles (ICBMs) and INF weapons, are designed to fly through space and that many satellites in space serve military purposes. According to a Defense Ministry publication, more than two thousand satellites have been launched since 1957 which perform military tasks.[2]

In order to soothe uneasiness among its future allies and potential adversaries, the FRG renounced the possession of atomic, biological, and chemical (ABC) weapons upon entering NATO in 1956. No major political party or leader in the FRG contemplates reversing this earlier decision. The price which West Germans have had to pay for this renunciation of atomic arms is heavy dependence on the only country in the world which has offered to extend its nuclear umbrella over West Germany: the United States. France and Britain have never been willing to provide nuclear protection to the FRG. Therefore, despite all the talk of a "European security pillar" built upon a German-French foundation, the FRG has never been able seriously to consider weakening its security ties with the United States. Every federal government has firmly believed that only American nuclear weapons can ultimately dissuade the Soviet Union from launching any form of attack against Western Europe. Since the United States provides that nuclear guarantee, the FRG must preserve its close security links with the US almost at any cost. It cannot easily reject an important American proposal such as SDI.

German politicians have frequently reminded their constituents of this constant need for demonstrating alliance loyalty. For instance, at a meeting of NATO's Nuclear Planning Group (NPG) in Würzburg, March 19, 1986, Franz-Josef Strauss, the governor of Bavaria and head of the Christian Social Union (CSU), warned that only participation in SDI could prevent "creeping alienation" between the US and Europe. The West German press has frequently given the same reminders. For example, the *Westfalenpost,* published in Hagen, wrote on May 28, 1985: "We should not forget that for a long time to come there will be no substitute for America's protective umbrella. That is the real departure point for a political decision for or against the SDI." The *General-Anzeiger* of Bonn argued on April 19, 1985, that "rejecting SDI, as called for by the opposition, underestimates America's will and determination to search for new approaches in any case, without Europe if necessary. What kind of security would such a Europe, without a self-confident strategic partnership with America, choose as the end of the 20th century approaches?"

2. *SDI: Fakten und Bewertungen, Fragen und Antworten, Dokumentation* (Bonn: Der Bundesminister der Verteidigung, Informations- und Pressestab, Referat Öffentlichkeitsarbeit, June 1986) (hereafter *Fakten*), p. 28.

West Germans must be concerned about the vulnerability of US nuclear forces, which are still the ultimate deterrent against Soviet aggression in Western Europe. Many have long feared that the steady build-up of the Soviet nuclear arsenal has so increased America's vulnerability to Soviet nuclear attack that the credibility of the American nuclear guarantee has been weakened. In Europe itself the Soviet Union has achieved numerical superiority in INF (through SS-20s) and short-range ballistic missiles (through SS-21, SS-22, and SS-23 missiles). Also Soviet troops are reportedly better trained to fight on battlefields where tactical nuclear weapons have been used. These facts would indicate that the Soviet Union has already gained superiority at every nuclear level below the strategic one. This unpleasant situation, combined with impressive improvement in Soviet strategic nuclear power, could possibly dissuade the United States during a war in Europe from ever crossing the nuclear threshold in the first place, as it has obligated itself to do under the terms of flexible response, the prevailing NATO strategy. NATO would be faced with the Soviet Union's overwhelming conventional power in Europe without realistic recourse to nuclear weapons. Thus the deterrence of war or Soviet blackmail in Europe could be seriously weakened. This would be especially true if the Soviet Union were to build up its ballistic missile defense (BMD) capabilities, which could render any American nuclear threat less harmful. German proponents and even most opponents of SDI are aware that the Soviet Union has not remained inactive in the field of strategic defense.

From the beginning of the INF debate in December 1979, the West German government has had to deal with Hiroshima scenarios in the minds of critics gripped by fear of nuclear war. Petra Kelly, a Green member of parliament, spoke for hundreds of thousands of West Germans when she told a *Newsweek* correspondent that "we will not accept mass-destruction weapons for defense, whether in the East of the West." Many West Germans can no longer follow the logic of threatening to destroy the world with nuclear weapons in order to save it. One town after the other has proclaimed itself to be a "nuclear-free zone," and most Greens and many Social Democrats, along with spokesmen from the churches, have openly asserted that nuclear weapons were "immoral." For example, the German Protestant Church Council (EKD) proclaimed in the fall of 1983 that "the system of nuclear deterrence and increase in armaments potential must not dominate us forever. The task is to reduce step-by-step the importance of the role of nuclear weapons."[3]

Ironically, these same critics of offensive nuclear weapons were to become strident critics of SDI. They have become strange converts to flexible

3. *Newsweek* (International Edition), July 12, 1982, p. 56. EKD quote in *Dokumente: Strategische Verteidigungsinitiative der Vereinigten Staaten von Amerika* (Bonn: CDU/CSU, April 18, 1985) (hereafter *Dokumente*), p. 15.

response and nuclear deterrence. What appeared to them yesterday as a crime against humanity is now seen to be indispensable for the maintenance of peace. Hubertus Hoffmann bluntly asserts that "viewed morally this way of thinking is perverse illogic." Chancellor Helmut Kohl undoubtedly enjoyed reminding his critics in an April 18, 1985, speech that "every possibility of moving away from the gloomy threat of a nuclear holocaust as the ultimate means for preventing war deserves conscientious examination." An SPD disarmament specialist, Egon Bahr, counters by asserting that "one need not go into space and begin a new destabilizing arms race in order to overcome deterrence."[4] Despite growing opposition to nuclear weapons in the FRG and elsewhere, West German governments continue to cling to the American nuclear guarantee as the only way to undergird the policy of deterrence.

West German leaders have also had to take seriously American complaints about the insufficiency of NATO conventional military strength in Europe, which, in some respects, remains dangerously inferior to that of the Warsaw Pact. The FRG has tried to respond to American calls for better conventional defense. It had done better than most NATO countries in achieving the target of a 3 percent real annual increase in defense spending. With a conscript army, it has been able to put an impressive amount of its defense spending into improving the battle capability of its forces. It has especially decided to improve air defense systems, a decision which could ultimately be benefitted by SDI research. But the government faces severe financial limitations. It has no mandate to increase defense spending so dramatically that the conventional weakness on the central front would disappear. There is no doubt also a latent suspicion in Bonn that if conventional defense capabilities were improved too impressively, then the United States might be tempted to refuse ever to release nuclear weapons in Europe and simply allow a NATO defense of Europe to be conducted exclusively by conventional means. This might be an attractive strategy for a country more than three thousand miles away. However, as Foreign Minister Hans-Dietrich Genscher wrote in the midst of the SDI debate: "Nothing must be allowed to happen which enables a war in Europe actually to be conducted because even an exclusively conventional war, in view of the present state of technology, would be a far greater catastrophe for the peoples of Europe than the Second World War."[5]

The SDI debate in the FRG takes place in a context in which many Germans look at American military policy with greater skepticism than they did a quarter of a century ago. Many fear that American conflicts with the Soviet Union outside the NATO area might spill over into Europe, and

---

4. *Dokumente*, p. 5. *Aktuell/Strategische Verteidigung. Kontrovers 1/86* (Hannover: Landeszentrale für politische Bildung, 1986) (hereafter *Aktuell*), pp. 67, 73.

5. *Neue Zürcher Zeitung* (hereafter *NZZ*), March 21, 1985.

there is general recognition that US and West German interests are not always the same. Most West Germans still assume that the USSR presents the most dangerous threat to their country. However, many respond to growing Soviet military power by bending over backward not to provoke Kremlin leaders, who always influence public opinion by threatening a worsening of relations with West Germany. A minority of Germans are even inclined now to view the United States, not the Soviet Union, as the primary threat to peace in Central Europe. The assumption that American military power would always be used wisely was badly shaken by US policy in Vietnam and by what some Germans see as an overly bellicose American reaction to the Soviet invasion in Afghanistan or to Libyan support for international terrorism. This distrust of American policy has grown since Ronald Reagan took office. The president is disliked by many Europeans, and some Germans criticize SDI simply because they are suspicious of any of his initiatives. Finally, the debate over SDI is taking place in the open, not only in the confines of think-tanks and parliamentary committees. Churches, trade unions, universities, ad hoc groups of scientists, as well as political party organizations have entered the fray. As with the INF debate, the controversy is emotionally charged and highly politicized.

West German Reactions to SDI: Opposition

Walter Stützle, one of Germany's most informed journalists on defense subjects and now director of the Stockholm International Peace Research Institute (SIPRI) wrote in the *Stuttgarter Zeitung* on February 12, 1985: "America's allies are torn between fascination and fear. President Reagan's vision of harnessing outer space possibly to provide total protection from missile attack hits the pact hard." It is hardly any wonder, as the *Süddeutscher Rundfunk* commented on April 18, 1985, that "SDI . . . has become a preeminent theme in the domestic political debate, with a volatility resembling that of the NATO two-track decision." West Germans are as divided as Americans on this subject. Critics in the FRG also face the difficulty of knowing exactly what it is that they are criticizing. Hans Günter Brauch told a parliamentary hearing in December 1985 that the US administration says many different and sometimes contradictory things about SDI; it also seems to change its positions. Therefore, "there is no presentation of the goals of SDI which are non-contradictory and are capable of consensus." Thus, it is hard to say how SDI would affect NATO strategy, the economy, or scientific research.[6] As we look at the German criticisms of SDI, it must be noted that even proponents of SDI share some of the opponents' anxieties and doubts about the program. The principal German criticisms of SDI are as follows:

6. *Protokoll*, pp. 116–17.

1. The earliest fear was that a defensive shield over the US would lead to the decoupling of the United States from Western Europe. Defense Minister Manfred Wörner stated several times in 1984 that "unequal security zones" might emerge within NATO. What this meant is that if the US were invulnerable to Soviet nuclear strikes, it would either be less inclined to use its own nuclear weapons to defend Europe, since its own security would no longer depend on the defense of Europe; or it would become more "trigger-happy," since it could safely fire nuclear weapons at the Soviet Union without fearing retaliation. In the first case, deterrence might be undermined; in the second case, American military restraint might be lessened. The result in both cases might be a less secure Europe.

The moderately leftist *Frankfurter Rundschau* wrote on February 12, 1985: "Should the Strategic Defense Initiative, against all expectations, turn out to be significant for America, Western Europe would stand completely exposed." Theo Sommer, an editor of the weekly *Die Zeit* and a persistent opponent of SDI, also wrote of "precisely those zones of unequal security, which we must avoid like the plague. That would lead psychologically and strategically to the feared decoupling. If a war broke out, it would not be a war between client states, but rather a war against the clients. Europe could in fact become the shooting gallery of the superpowers." The moderately conservative *Frankfurter Allgemeine Zeitung* feared that all the talk about decoupling could become a self-fulfilling prophesy, "if one distances himself [from SDI] for too long a time."[7] The fear of decoupling has diminished somewhat as the confidence in America's ability to build a perfect system has declined. Wörner later became a proponent of SDI. However, the Kohl government never dropped one of its conditions for supporting SDI—that there be no unequal zones of security and that deterrence not be weakened until there existed something better to replace it.

2. A more persistent criticism is that the purpose of SDI is not really defense at all, but instead military superiority over the Soviet Union by means of a "first-strike capability." The idea is that with a perfect defense, the United States would be free to launch a nuclear attack against the USSR. Under such conditions, the US would be in a position to dictate to the Kremlin leaders. Critics see this alleged move toward an aggressive US military doctrine as a logical consequence of a series of aggressive steps: new offensive strategic systems (such as MX missiles and B-1B bombers), Pershing II and cruise missile deployment in Western Europe which could "decapitate" the Soviet command structure, as well as Follow-on Forces Attack (FOFA) and Air-Land Battle tactics, which call for NATO military actions beyond the front line of NATO. Hans Günter Brauch warned in a parlia-

7. *Die Zeit*, April 4, 1986. *Frankfurter Allgemeine Zeitung* (hereafter *FAZ*), March 29, 1985. See also articles by Kurt Kister in *Süddeutsche Zeitung* (hereafter *SZ*), April 5, 1984, and February 14, 1985.

mentary hearing of a "cult of the offensive" à la 1914. Retired General Gert Bastian, a leading member of the peace movement and a former member of the Green party in the Federal Parliament, asserted that the US wanted armaments in space "to strengthen its shield in order to be able to make the sword even sharper." Even Wörner was initially worried about the appearance of an offensive doctrine. The government has been very careful to say that it would support an SDI program only if that were not a part of an attempt to gain military superiority.[8]

3. A third criticism of SDI is that it could siphon funds away from conventional defense, which most NATO allies agree must be strengthened. If the US assigns too high a priority to SDI, then it might not be in a position to contribute to its conventional might in Europe. Critics can refer to the supreme NATO commander, General Bernard Rodgers, who worried outloud that "I am just concerned that if from now on we turn all our attention to SDI, we then fail to undergird our efforts in the more mundane areas such as sustainability of conventional forces and modernization of nuclear weapons." If the FRG ever had to bear some of the costs for SDI, which it has not been asked to do, then critics fear that it could indeed be at the expense of more pressing conventional requirements. General Wolfgang Altenburg, the Inspector General of the Federal Army and therefore chief military and security policy advisor to the defense minister and Federal Security Council, wrote in a report in late 1985 that a particularly distressing factor is that "the more effective SDI will be, the more important the conventional threat to which the Federal Republic is exposed will be." SPD critic Egon Bahr has similar worries: "If both sides were sure that their strategic weapons were unusable, then the danger for Europe grows."[9]

4. A fourth criticism is that SDI would escalate an arms race, not only in space, but on earth as well. Most German critics of SDI are aware that the Soviet Union has been steadily investing in strategic defense. For example, Christoph Bertram, an editor of *Die Zeit* who was director of the International Institute of Strategic Studies in London, admitted that "the Soviet Union has been especially busy. In the years since the ABM [Anti-Ballistic Missile] Accord, it has outspent the U.S. in missile defense." But opponents fear that the Soviet Union would step up its efforts and move quickly both to deploy strategic defenses and to build more offensive missiles to overwhelm an American counterpart system. They fear a "Sarajevo effect," in Brauch's words.[10] They often point out that the large sums involved could

8. *Protokoll*, p. 117. *Frankfurter Rundschau* (hereafter *FR*), November 2, 1985.

9. Jacquelyn K. Davis and Robert L. Pfaltzgraff, Jr., *Strategic Defense and Extended Deterrence: A New Transatlantic Debate* (Cambridge, Mass., and Washington, D.C.: Institute for Foreign Policy Analysis, 1986), p. 23. *FAZ*, November 29, 1985.

10. *Zeit*, January 4, 1985. Alfred Mechtersheimer, *Zeitbombe NATO. Auswirkungen der neuen Strategien*, 2d ed. (Cologne: Eugen Diederichs Verlag, 1984), pp. 134–35. *Protokoll*, p. 117.

feed millions of people in the world and help Third World countries develop. For some unknown reason, they seldom argue that they could make even larger contributions to Third World development if they were willing to make a small sacrifice in their own high standard of living. In any case, critics are inclined to see the basic danger of an arms race as lying in SDI, rather than in a Soviet arms build-up, which has never been significantly impeded by earlier arms control agreements.

5. One of the most serious charges against SDI is that it would wreck the arms control process. Christoph Bertram wrote before the resumption of the Geneva talks that "if President Reagan holds fast to his vision of Star Wars, arms negotiations must fail. . . . This vision—not the arms buildup on both sides—is what blocks the way to negotiations. . . . As things are now, the question of space weapons stands in the way of arms talks like a huge rock."[11]

Many critics insist that stability can only be achieved by reaching agreements with the adversary, not by technological leaps. For instance, in lashing out against Edward Teller, who allegedly "shares most of the obsessions of the ultraconservatives whose anticommunism can only be psychologically comprehended," the opinionated and unbalanced weekly magazine *Der Spiegel* wrote: "Teller . . . believes that the human mind can accomplish anything, but does not believe in the slightest progress toward rationality, not in the realization . . . that weapons, even defensive weapons, cannot create security in the atomic age." A more respected critic, Professor Hans-Peter Dürr, director of the Werner Heisenberg Institute of the Max Planck Institute for Physics and Astrophysics in Munich, argued in a parliamentary hearing that "long-term security of peace cannot be achieved through technical tricks, but only through political measures." Eckhard Lübkemeier added: "SDI could become a classical proof of the thesis that the best can be the enemy of the good. . . . Behind the inspired vision of the American President hides the illusion that a transformation of the international situation could be brought about by technology. . . . However, no good, even one so honorable and well-intended [as SDI], is worth leaving untried here and today the possibility of securing peace through détente and arms control."[12]

Opposition politicians regard this potential obstacle to arms control as being of utmost importance. All West German politicians talk about the goal of creating stability on a lower level of armaments, and all see the arms control process as the primary means of achieving this. Within the government the strongest voice warning of the danger of SDI disrupting arms control is Foreign Minister Genscher, backed by the Foreign Office. He has declared the ABM Treaty of 1972 to be the "Magna Carta" of arms control

11. *Zeit*, January 4, 1985.

12. *Der Spiegel*, Nr. 13, 1986, pp. 119, 148. Note Teller's contribution to this book, in which he argues the opposite.

and has insisted that the government take no stance on SDI which is not based on a restrictive interpretation of that treaty. Theo Sommer spoke for many of his countrymen when he wrote in *Die Zeit*, April 4, 1986: "Mutual disarmament must be the goal—atomic and conventional. It is already difficult enough to achieve without SDI. If Reagan's space plans are carried forward, then we must drop all hope for disarmament."[13]

6. Critics point to further diplomatic difficulties which could be created by the FRG's participation in SDI. Since the Soviet Union is adamently opposed and has told that to West German leaders in unmistakable terms, many West Germans fear that the FRG's political and trade contacts with Eastern Europe might be hampered. A further problem unique to Germany is Berlin, which is located inside the GDR and which is still under Allied rule. West Berlin has growing high-tech potential, but as long as SDI is viewed in primarily military terms, Soviet and East German leaders could object any time a Berlin firm becomes involved in an SDI project.

It is feared that the FRG's relations with Western Europe, especially with France, could also be endangered. Both Kohl and Genscher expressed their preferences for a "European" response to the United States, and Kohl conducted important meetings with French President François Mitterrand to try to work out a common approach. However, Mitterrand remained opposed to SDI, no doubt fearing that any determined move in the direction of strategic defenses might ultimately render France's own nuclear forces impotent. To use the words of the *Stuttgarter Zeitung* of April 14, 1985, the "declaration of an intention to seek a coordinated reaction by Western European Union governments to America's research program on space defense . . . belongs to a fair-weather vocabulary that hardly has any pertinence for European reality."

One project to emerge from this effort was the European Research Coordinating Program (EUREKA), elaborated by Mitterrand on April 17, 1985, and instituted within four months. SDI had again raised the specter of a widening of the technology gap between the Old and the New World and of a "brain drain" from Europe to America. Mitterrand explained that EUREKA would be designed to ensure "the technological independence of Europe in the vital fields of the future." It is neither a space nor a military program; for instance, one of its first projects involves finding a way of diagnosing AIDS. Its projects are backed by very little governmental funds and are geared to the development of high-technology products which can be sold successfully on the free market. EUREKA was not billed as a

13. Davis and Pfaltzgraff wrote in *Strategic Defense* that the ABM Treaty has an "aura of near sanctity" in Europe. See pp. 9, 33, 35–36. *White Paper 1985. The Situation and the Development of the Federal Armed Forces* (Bonn: Federal Minister of Defense, 1985), pp. 52–58. *Fakten*, p. 64.

response to SDI; therefore the US was able to wish it well, and the FRG and the UK were free to participate both in it and in SDI.[14]

7. As in the United States, a major problem on which critics focus is the technical feasibility of SDI. German critics are generally well-informed about American scientific criticism of SDI and do not use arguments which significantly differ from those heard in the US regarding costs, computer requirements, vulnerability of systems, and dangers posed by countermeasures. Resolutions by such groups as the Union of Concerned Scientists are widely reported. So are congressional reports, such as the one by the Office of Technology Assessment in September 1985 that the development of space-based nuclear systems would make a war more likely between the US and the USSR, and another by the Senate, dated March 30, 1986, disputing General James Abrahamson's testimony that SDI is technically feasible. Set-backs in US space research in 1986 were widely reported and have strengthened skepticism about sophisticated space-based weapons systems: the Challenger mishap on January 28, the explosion of a Titan rocket in April, and the destruction of a Delta rocket shortly after take-off on May 3.

Groups of scientists and university teachers have circulated petitions in the FRG and held conferences to attack SDI. The resolution of one group of left-wing scientists in Göttingen in 1984 became the text which the opposition Social Democratic Party (SPD) submitted for approval by the Federal Parliament; this call for a ban against all antisatellite and space weapons was soundly defeated. An anti-SDI resolution launched on the fortieth anniversary of the Hiroshima bombing was signed by more than fifteen thousand persons, including not only scientists, but also political figures such as Willy Brandt, athletes such as Olympic gold-medalist Michael Gross, and German rock star Udo Lindenberg. Even nineteen professors at the Federal Army University in Hamburg publicly condemned moves to "militarize space."[15]

West German journalists have coined catchy phrases to describe their lack of belief in the feasibility of any kind of strategic defense. Wilhelm Bittorf, in a series on SDI in *Der Spiegel,* spoke of SDI as a "mirage" and the "gospel of a protective shield" having "religious characteristics" and being like a Hans-Christian Andersen fairy tale: "welcome to the magic

---

14. Thierry De Montbrial, "The European Dimension," in William G. Hyland, ed., *Foreign Affairs. America and the World 1985* (New York: Pergamon, 1986), pp. 512–13. Davis and Pfaltzgraff, *Strategic Defense,* pp. 37, 41.

15. *FAZ,* October 16 and August 7, 1985, and July 8, 1986. *FR,* November 13, 1985. Perhaps the best presentations of opposing views on the technical feasibility of SDI were given by Hans-Peter Dürr, "Der Himmel wird zum Vorhof der Hölle" (against), and Hans Rühle, "An die Grenzen der Technologie" (for), in *Der Spiegel,* Nr. 29 and November 25, 1985. Both are republished in *Aktuell.*

empire of SDI where something is hardly ever as it seems." He alleged that never have leaders in Washington or Bonn discussed something "of which they had so little idea." In an article entitled "Sheer Madness" in *Die Zeit* on March 14, 1986, Grant Johnson, a computer and telecommunications consultant in Frankfurt, described the "bizarre visions of Ronald Reagan" as revealing "characteristics of political regression" and the desire to flee into fantasies in order to escape responsibility.[16] Most critics do not use such uncontrolled language. It is nevertheless true that almost no West German believes that the president's vision of a perfect defense can be realized in the foreseeable future.

8. Not only do many West Germans not believe in the technical feasibility of SDI, many do not think that there would be much economic benefit or technological spin-off of a civilian nature for German industries. Some critics rail against "the militarization of pure research" on ideological grounds. However, many thoughtful German experts have testified that the civilian spin-offs from military research are normally meagre and that if civilian technology is what is wanted, then purely civilian research should be done. Others argue also that American secrecy requirements would severely limit the possibility of German firms' using their newly discovered technology for civilian purposes. The US has in the past always demanded very tight definitions of strategic materials in COCOM, and this practice would no doubt continue. The result would be a limitation of the FRG's ability to trade with the East. Since the US government would be paying for the research, it would retain the right to determine how the results would be used.

Congressional actions could further limit German participation in SDI. The Export Administration Act of mid-1985, empowering Washington to control exports, gave no inferences on possible US attitudes toward technology transfer in the SDI context; later interpretation could work against European firms. Also, in August 1986 the US Senate voted to require that contracts for SDI research go to American companies unless the secretary of defense can certify that the work could not be carried out in the US. Even though such a measure may never be enacted, West Germans are leery that such restrictions could be levied at any time. In general, critics do not regard SDI as a cornucopia for German firms, and industrial leaders are generally not enthusiastic about the economic benefit for the FRG.[17]

9. A final criticism against SDI, of which most participants in the debate are aware, is that no one can really know how lasting will be the support for SDI within the United States and what priority future American govern-

16. *Der Spiegel*, Nr. 13–15, 1986. Johnson has a chapter in *Krieg und Frieden. Streit um SDI* (Berlin: Kursbuch Verlag, 1986).

17. *Die Welt*, January 16, 1986. *FAZ*, April 3, 1986. *Kölner Stadt-Anzeiger*, January 24, 1986. *Zeit*, November 10, 1985. *SZ*, April 4, 1986. See also *Aktuell*, pp. 68–73.

ments will give this project. Most informed Germans know that the issue is quite controversial in America. Congress is already whittling away at the funds alloted to SDI. Before any large-scale SDI systems could be developed and deployed, several subsequent presidents must continue to support it. Many German critics warn about jumping on the wrong horse, but many more sense that the horse which their country has already mounted may not even survive the decade.

Although opposition to SDI can be found in all five parliamentary parties, only the Greens and the SPD have taken firm positions against it. The official view of the largely anti-NATO, anti-defense Greens was predictable. In parliamentary debates on SDI, Green spokesmen argued that it is no defensive measure. Instead SDI is allegedly a part of America's over-all effort to force the Soviet Union to its knees by depriving the latter of its second-strike capability. The US would be able either to attack the USSR with impunity or successfully blackmail it. Also, by "supporting the war-mongering foreign policy course of the Reagan Administration," the Federal Government reflects German firms' greed for profits and "makes itself the grave-digger of the existing arms control agreements."[18]

The SPD approaches the subject much more rationally and less hysterically. Its position, which is backed by parallel resolutions by the Socialist International and Social Democratic parties in the NATO countries, is expressed in a policy statement published in May 1985, following bitter parliamentary debates. The party expresses the fundamental reservations of all responsible critics of SDI. It approves and sympathizes with President Reagan's desire to create greater strategic stability, to move away from deterrence based on weapons of mass destruction, and to replace fear by hope. However, "the solutions which President Reagan offers to achieve these objectives are . . . unsuited for the task. As it appears now, they will result in instabilities, rather than stabilities, insecurity rather than security, armament rather than disarmament, and possibly an increase in offensive nuclear weapons. Social Democrats wish to achieve stability and security, not only through adequate defense capability, but primarily through dialog, treaties, and confidence-building measures. In contrast, the American president assumes that a political problem is to be solved by technical means. . . . The goal must be to overcome deterrence between East and West in a long-term process . . . through a security partnership. . . . Reconciliation cannot be forced by technology; enmity cannot be overcome by technology." This policy was reaffirmed at the party's congress in August 1986. Some SPD spokesmen do not weigh their words quite as carefully as the authors of this official statement. For example, the party's disarmament

18. *FAZ*, April 19, 1985. *SZ*, March 29, 1986.

specialist, Egon Bahr, called SDI "fantastic and crazy," and he asserted that "not Soviet action, but American, creates the crisis in the alliance."[19]

Beginning several years before the president's 1983 speech, the party has been trying to work out a new approach to defense. The party's chancellor candidate, Johannes Rau, reportedly prefers to call this "common security" with the adversary, while others like the more ominous term "security partnership." The general goal is to transform the country's defense policy into one which cannot give the slightest appearance of being offensive or aggressive. Through arms control agreements, it hopes to induce the Soviet Union to assume the same posture so that both sides are no longer capable of an attack; the slogan is "structural non-aggression capability." It is not surprising that the party's parliamentary leader, Hans-Jochen Vogel, should announce that the SPD would terminate West German commitments regarding SDI "as soon as we are in a position to do so."[20]

Without question, many Social Democrats would ideally like to see West Germany's security policy change in the direction indicated. However, there are very important domestic political reasons which incline the major opposition party, which had governed the FRG very responsibly for fourteen years from 1969 until 1982, to take such a critical stance toward SDI. The SPD's bid to regain power in Bonn has been frustrated primarily by the ability of the Green protest party to attract voters whose support the SPD once received and now needs again. The SPD must respond to the anxieties on which the Greens feed. Thus, with its vision focused on the January 1987 and 1991 federal elections, the SPD calculated that it would be politically advantageous to oppose any form of West German cooperation in SDI, at least until after the elections. If it were again to come into power through a coalition with any party but the Greens, it could carefully reconsider its promise to terminate the country's SDI agreements with the US. Most responsible Social Democrats still consider the American pledge to defend the FRG to be indispensable.

West German Proponents of SDI

Many West Germans see some promise in SDI. They distinguish between President Reagan's "vision" of a perfect defense, which almost no one believes will be possible in the foreseeable future but which they realize provides the domestic support within America for the program, and a more limited version of SDI involving a mix between offensive and defensive

19. *Krieg der Sterne. SDI und die Interessen Europas.* Aktuelle Informationen der SPD, Nr. 3 (Bonn: Vorstand der SPD, May 1985). *Aktuell*, p. 11. *FAZ*, October 20, 1985. *Rheinischer Merkur*, June 14, 1986.

20. *Rheinischer Merkur*, June 14, 1986. *FAZ*, April 24, 1986. *This Week in Germany* (New York: German Information Center) (hereafter *Week*), April 25, 1985.

systems. All are concerned that European, particularly German, interests be considered by the US administration. In general terms, their support for SDI stems from the following factors:

1. Many thoughtful Germans realize that there is a problem in the credibility of NATO's present policy of flexible response. The powerful chairman of the CSU, Franz-Josef Strauss, stated in the Bavarian Radio on New Year's Day 1986: "It may be doubted that the principle of atomic deterrence, which guarantees us freedom and peace, will remain valid in times to come." They have noticed that the past several US administrations have tried to do something about this problem by means of changes designed to increase options in times of crisis. These involve a reorientation of nuclear targetting strategies culminating in Presidential Directive 59, an improvement in American INF capabilities in Europe, and all kinds of measures aimed at improving NATO's conventional capabilities in Europe. SDI is another step in the direction of trying to cope with the nuclear dilemma in which the US and its allies find themselves. Work related to SDI may not eliminate this problem entirely, but it could perhaps lead to some improvement in NATO's situation.

Hubertus Hoffmann has argued that "stability must be newly defined; it should no longer be a product of mutual vulnerability, but a process of equally phased reduction of atomic vulnerability on both sides." He believes that "a build-up of a real missile defense by the United States would enable one to expect not a decoupling but a strengthened coupling to the security of Europe." Gerd Schmückle also asserted that "defense must progressively be made into the strongest and more reasonably priced form of fighting." The most authoritative statement was made by Defense Minister Wörner: "In Europe the reduction of America's nuclear options would shift the defense burden further toward conventional weapons."[21] No one knows better than he does just how much sacrifice would have to be made to prepare NATO for an effective defense of West Germany.

Germans sense that there is a change of military doctrine in the air. The *Deutsche Tagespost* of Würzburg wrote on October 31, 1985, of "America's efforts to adapt security policy to galloping technical developments," and it chided Germans who hesitate to take part in this effort. General Altenburg noted that "SDI is for us first and foremost an across-the-board strategic challenge." In an article in *Die Welt* on May 31, 1986, Cay Graf Brockdorff wrote that "SDI is the logical consequence of a credible deterrence strategy" and that MAD, which Secretary of Defense Caspar Weinberger called "a mutual suicide pact," is outdated. Even some opponents of SDI, such as Hans Günter Brauch, note that "SDI requires a strategic revolution." The Defense Ministry sees it the other way around: "The SDI research program

21. *Aktuell,* pp. 9, 65. *Fakten,* p. 64.

is a response to negative trends in the strategic balance of power." Finally, Claus Richter noted that "the misleading slogans, 'star wars' or 'militarization of space,' are ultimately none other than a helpless reaction to a new dimension of military use of space which had begun long ago. There are many indications that the world is at the onset of a profound military-technical transformation. . . . One may be horrified by this prospect, but one should in the long term adjust himself to its realization."[22]

Three German experts, who testified before the parliamentary hearing preceding the final government decision to participate in SDI, discussed the need to draw the right conclusions from changes in the strategic situation. Those changes stem from an impressive Soviet military build-up across-the-board which could not be stopped by the arms control process and which makes the application of flexible response increasingly difficult. Retired General Franz-Joseph Schulze, former NATO commander for Central Europe, testified: "I believe . . . that SDI is the only possibility for stopping this trend [toward enlarging and improving nuclear potential] and even for reducing nuclear weapons. If that should succeed in lowering the danger of a first strike, . . . then the world is, in fact, many times safer. That is not a technical solution of the problem . . . but a decisive alteration of the strategic culture between both superpowers." He explained that this would not require a perfect defense system.[23]

Uwe Nerlich, of the Foundation for Science and Politics at Ebenhausen, argued that "the strategy of flexible response requires for its continuing effectiveness specific measures in the conventional, as well as the nuclear area." He suspected that SDI came prematurely and might have distracted attention from more immediate questions of strategy and defense. Nevertheless, "SDI could offer here a stimulous for a new rethinking of the strategic situation of Western Europe."[24]

His colleague at Ebenhausen, Dr. Karl-Peter Stratmann, does not share the view of many critics both in the conservative camp and the peace movement that the present nuclear strategic situation is dangerously unstable. However, he believes that West Germany should see whether it could derive some benefit from SDI research. Like most German proponents of SDI, and unlike many critics, Stratmann clearly sees the distinctions among the three SDI stages—research, development, and deployment—and does not view participation in the research phase as irrevocable support for further phases. Nor does research automatically have to lead to deployment. He recognizes that "SDI is not the cause, but the consequence, of

---

22. *FAZ*, November 19, 1985. *Protokoll*, p. 116. *Fakten*, p. 25. Claus Richter, "Strategische Verteidigungsinitiative (SDI)," in *Beilage*, *Das Parlament*, April 6, 1985, p. 14.

23. *Protokoll*, p. 71.

24. Ibid., pp. 149, 151.

profound difficulties and problems which for a long time have burdened East-West relations, arms control negotiations, and the development of defense policy in the Western Alliance. . . . The mere rejection of SDI would not mean the solution of these difficulties." In any case, "the controversy over SDI offers a further opportunity for a comprehensive security policy stocktaking."[25]

Proponents of SDI argue that it is in West Germany's interest to influence any changes which might take place in Western security policy and that the only way to gain any influence at all is to participate in the project. Nerlich argued that "the more German politics waits, the smaller become the possibilities of having a formative effect on the reorientation of strategic thinking." General Altenburg concluded that the FRG could be fully informed about the risks involved and actively try to minimize them "only insofar as we have influence, not by standing on the sideline." Wörner stated this point most clearly: "Whoever cooperates has the most influence."[26]

Stratmann called on his countrymen to pay attention to how Americans might react to German demands for being consulted about SDI, "but consultations please without the slightest appearance of an approval of the whole program. . . . If one puts himself in the shoes of the Americans, something which is no longer very usual here [in Germany], then based on our behavior up to now, one must be highly confused about what the political process in the Federal Republic really wishes to achieve." He concluded that "there remains only the way of a qualified participation, in order to be informed about the motives and options of the American policy and to be able to make our own influence felt in crucial decisions."[27]

General Schulze also called on Germans to think about American impressions: "A European policy of wait and see could awaken the impression among the American population that the Europeans see their salvation in the maintenance of assured vulnerability of the USA and shun all efforts toward building up strategic defense systems; this could expose the Alliance to a breaking test which it could not survive." Perhaps the *Rheinischer Merkur* said it best: "No one expects Bonn to stand at attention and salute the SDI flag, but the United States can be sure to attach less importance to European interests in general and German interests in particular, if all we do is shake our heads in dismay."[28]

2. A further argument which all proponents of SDI make (and few opponents need to be reminded) is that the Soviet Union has not only built up its conventional and nuclear offensive forces at all levels in such a way as to threaten any notion of offensive parity. It has also long engaged in

25. Ibid., pp. 102, 166, 170.
26. Ibid., p. 153. *FAZ*, November 19, 1985. *Fakten*, p. 59.
27. *Protokoll*, pp. 102, 168.
28. Ibid., p. 160. *Rheinischer Merkur*, October 26, 1985.

strategic defense work. It has modernized its ABM system around Moscow, which it is permitted to do under the ABM Treaty, and it has constructed a large radar at Krasnoyarsk which could be used for strategic defense. In view of the importance which the Soviet Union has attached to civil defense and to a mix of offensive and defensive weapons, many Germans are sensitive to the enormous danger which would arise if the Soviet Union were unilaterally to develop even a limited strategic defense capability. Nerlich noted at the parliamentary hearings that "in contrast to the Soviet side, a missile defense for the United States would not necessarily be a logical consequence of its military build-up heretofore: active and passive national defense measures (air defense, civil defense, etc.) have for twenty-five years played a subordinate role."[29]

3. Participation in the research and assessment phase of SDI could provide the FRG and other NATO countries with important military spin-offs. These could include such things as "fire and forget" weapons against tanks, battlefield lasers, sensors, and battle management (C^3I) capabilities. Most important, though, could be break-throughs in air-defense capabilities. In the past decade the Soviet Union had greatly increased its numbers of short- and medium-range ballistic missiles. Also, the accuracy of those missiles, which is assumed to be within fifty meters, has so greatly improved that the Soviet Union will eventually be able to arm them with conventional or chemical warheads, rather than nuclear warheads. Thus, it would be able to destroy many of NATO's most important military targets, which are soft targets, without crossing the nuclear threshold. According to Hans Rühle, head of the Defense Ministry's planning staff, there are two to three hundred targets, the destruction of which would make it impossible for NATO "to build up a cohesive forward defense, to mobilize reserves and to land reinforcements from overseas." The inability to defend those targets increases NATO's military need to use nuclear weapons first. On top of that land-based threat is the presence around Western Europe of Soviet ballistic-missile submarines which could attack Europe "from all azimuths," as the French say. Further, with the development of Soviet air- and ground-launched cruise missiles, as well as a quantitative and qualitative increase in Warsaw Pact combat aircraft, the European NATO allies have become confronted with a diversified threat which would have to be met whether SDI had been born or not. It is not surprising that the alliance had begun to study this air-borne threat to Western Europe almost a decade before SDI had been announced.[30]

In October 1983 an American team of experts led by Fred S. Hoffmann prepared a report advising the president to add to SDI a defensive system against tactical missiles. This anti-tactical missile (ATM) was intended to

29. *Protokoll*, p. 150.
30. Ibid., p. 159. *Die Welt*, June 23, 1986.

reduce allied worries that SDI might weaken American resolve to defend Europe. This could be done either by devising a new missile as a part of the SDI research program or by further developing the existing Patriot missile. The latter was designed to fight airborne targets at medium and high altitudes, but it is capable of upgrading to cope with a missile threat as well. The US Air Force launched a project parallel to SDI called Air Defense Initiative (ADI) to shoot down cruise missiles and bombers. Lieutenant General John F. Wall, head of the army's strategic defense command, indicated in 1986 that existing systems, such as the Patriot anti-air system, would be upgraded as an initial step to meet the Soviet missile threat.[31]

In Europe two related but distinct responses emerged. One was a European Defense Initiative (EDI), which, in the words of former West German Defense Minister Kai-Uwe von Hassel, could "parallel and supplement the Strategic Defense Initiative (SDI) of the United States." This effort, which was in part stimulated by the familiar fears of "decoupling" the defense of Europe from that of the US, has also been supported by the renown French nuclear strategist General Pierre Gallois. In a report produced by a group of experts supporting the plan, attention was drawn to Soviet missile defense systems, integrated with a vast air-defense system. "By what logic should Europe be denied the deployment of a missile defense system, when the Soviet Union already has deployed such a system in Europe?" The defense against nuclear or conventionally armed ballistic missiles is "a legitimate military requirement for NATO. It is just as logical to deploy defense systems against these offensive weapons as it is to defend against similarly armed manned aircraft or other forms of delivery systems."[32]

Both SDI and EDI are, at this stage, research and development initiatives, which could reinforce the basic goals of NATO: peace through deterrence. There could be some carry-over from the technology developed to protect against ICBMs and that to protect against shorter-range threats. A European system could certainly benefit from American tracking and target-acquisition capabilities carried out by space-borne assets. Of course, the vulnerabilities of Europe and the United States are different. Enemy missiles trained on Western European targets cover their shorter trajectories in five or six minutes rather than thirty. This shortened warning time presents the allies with a threat similar to the submarine threat against the US. A European system would also have to protect against a greater variety of threats. Further, it would paradoxically require more satellites. Finally, fewer military targets in Europe are hardened, compared with the US.[33]

31. Ibid. *SZ*, February 14, 1985. *Washington Post* (hereafter *WP*), May 4 and April 25, 1986.

32. *The European Defense Initiative—EDI: Some Implications and Consequences* (Rotterdam: High Frontier Europa, 1985), pp. 4–5.

33. Ibid., pp. 13–14, 28–31. See also Davis and Pfaltzgraff, *Strategic Defense*, pp. 29, 33, 35–36. *White Paper 1985*, pp. 52–58. *Fakten*, p. 64.

It is not surprising that the United States is extremely interested in Western European air defense; the US provides the second largest contingent to European defense, and its air bases, command centers, munitions depots, and equipment warehouses are inviting targets for Soviet air attacks. Therefore, when asked about EDI, General Abrahamson answered that it "is an exciting prospect. . . . The technologies, in many cases, overlap each other. . . . The research . . . should be an integrated activity. . . . We are as committed to theater defense against the short-range and intermediate-range ballistic missile threat as we are to the intercontinental-range ballistic missile threat to the United States. Frankly, the expertise and experience with short-range threats is centered in Europe. Therefore we want Allied and Japan's involvement in shaping that part of theater defense."[34]

On a separate question, Abrahamson stated that until now, "Soviet military planners have had a free ride" in their missile threat against Europe; "even a modest defense against tactical ballistic missiles would have a major effect on the situation." The potential political benefit of such American attention to European problems was underscored by the *Neue Osnabrücker Zeitung* on March 22, 1986: "The announcement that the problems created by short-range and intermediate-range weapons will be taken up within the framework of SDI could conceivably put a convincing argument for intensive participation in the hands of supporters of the project. This might have a decisive bearing on the Europeans, who will not be able to abstain with good reason if SDI will also deal with their own interests, namely the protection of their own territory."[35]

The West German defense minister is also interested in expanding air defenses against Soviet missiles. Already the FRG provides half the land-based air defenses in Central Europe, a third of the fighter aircraft, and all the NATO air forces in the Baltic, in addition to half the ground forces in Central Europe. At an International Defense Science Conference in Munich on March 1, 1986, Wörner said that the Soviets would soon have a kind of first-strike capability in Europe and would be able to paralyze NATO forces by far-reaching and precise attacks "without crossing the threshold of nuclear warfare."[36]

Three weeks later, in a meeting of the NATO Nuclear Planning Group in Würzburg, he stressed that his concept of air defense is to be understood as separate from SDI. He pointed out that even if there were no SDI, a greatly improved air defense system over Western Europe would be a long-term necessity. The NATO defense ministers' meeting on April 30, 1986, accepted a project known as "extended air defense" proposed by Wörner. This defense system, which would extend into outer space to protect against

34. *Armed Forces Journal International*, March 1986, p. 87.
35. *WP*, April 25, 1986.
36. *Week*, March 7, 1986.

bombers, cruise, and ballistic missiles, was publicly elaborated by Hans Rühle at a transatlantic conference on SDI in Kiel. Like his cabinet colleagues, Wörner favors efforts to reach arms control agreements. However, he does not try to deal with mortal military threats by parroting simplistic rhetoric about technical innovations not being able to help restore stability. He would agree with Gerd Schmückle, who pointed out that "weapons especially suited for an offensive in Europe could be blunted with the help of the most modern technology: the tank in the army, the fighter plane in the air force, the submarine in the navy—and a large part of the rockets. European and American SDI would merge at that point in research work where medium-range missiles could be intercepted." Unlike many Germans, Wörner accepts the notion that important military assets must be protected from potentially fatal airborne threats if stability and deterrence are to be achieved.[37]

Wörner does not use the term EDI to describe his objective, and his ministry has not given official approval to it. He is aware that SDI is controversial both inside and outside the government. He does not want to see a significantly upgraded air defense system in Europe stand or fall with SDI. Already the Federal Army has alloted funds to develop a tactical air defense system from 1990 on, but such a system is unlikely to materialize, in any case, during this century.

4. Before deciding to support SDI, the Bonn government consulted with West German industrial leaders about the potential economic benefits of SDI. In general, industrialists recommended that German firms be permitted to accept contracts. SDI could offer some economic benefit to the FRG. However, they have not predicted a cornucopia for the German economy. Few Germans think that SDI contracts, even in the best of circumstances, could significantly reduce unemployment. Certainly, some high-tech firms will profit from it. But there are neither illusions nor euphoria about the economic promise of SDI.

5. German proponents are more inclined than critics to stress the military and political goals of SDI. The official government position is that these criteria, not economic and technological ones, must override. Nevertheless, some of the FRG's support for SDI stems from a fear that has gripped many Europeans for several years: that Europe might be falling behind the US and Japan in the field of high-technology. SDI could be an important research boost for the FRG and good medicine against what some Europeans have called "Eurosclerosis." Professor Klaus Haeffner of the University of Bremen warned that "the Federal Republic must not forget in the discussion of SDI . . . that it is progressively falling behind in the area of

37. Ibid., March 27, 1986. Montbrial, "European Dimension," pp. 513–14. *Aktuell,* p. 77. *SZ,* May 2, 1986. *Die Welt,* June 23, 1986.

transforming high-technologies into products capable of competing on the world market. A renunciation of SDI . . . can hardly be a guarantee for changing this development." Eckard John von Freiyand, director of research for the Federation of German Industries (BDI), agreed that "an immense technological stimulus can be expected from SDI" and that it could help overcome the tarnished German image in high technology.[38]

This opinion has also been stressed at the highest political levels. Chancellor Helmut Kohl stated in February 1985 that "a highly industrialized country like the Federal Republic of Germany and the other European allies must not permit themselves to be left behind technologically." The minister-president of Baden-Würtemberg, Lothar Späth, warned that West German abstinence in SDI could do irreversible damage to the technological position of the FRG; it could so destabilize the economic structure that "political and social shocks" could be felt. The *White Paper 1985*, which is the official public statement of West German defense policy, also asserts that SDI "may have revolutionary implications for civilian purposes" and that "this economic-technological interest enhances the basically responsive attitude of the Federal Government towards participation in the research programme." The Defense Ministry compares SDI with the earlier Apollo Program of the 1960s and points out that "on the basis of the experience gained from that time, a broad technological impulse in the civilian sector could be expected from SDI." The *Rheinischer Merkur* concluded on October 26, 1985, that "SDI research can be sure to develop technologies that will take us beyond the turn of the century. It would be criminal negligence for Europeans to opt out of the process."[39]

Of course, the FRG would not only benefit from participation in SDI research, but it could also make valuable contributions. Almost half the SDI program is devoted to technologies that acquire, process, and communicate data. A study directed by the director of the planning staff in the Defense Ministry, Hans Rühle, determined that the FRG was weak in certain areas of SDI research, but is strong in five of eleven key areas of SDI research. It is on top in optic sensors and certain subsystems in space technology; it is strong in raw material technology, high-frequency technology, and signal analysis and in systems components for extremely high acceleration of fast rockets. Thus, the FRG has the basis for constructive cooperation in the program.[40]

6. West German proponents of SDI are committed to the arms control process, but they are less positive in their evaluation of past successes in this

---

38. *Aktuell*, p. 79. *Protokoll*, p. 132.
39. *Aktuell*, p. 82. *SZ*, July 25, 1985. *White Paper 1985*, p. 33. *Fakten*, p. 18.
40. *FAZ*, March 28, 1985. See also *NATO Review*, April 1985, pp. 8–16.

field than are opponents. Stratmann noted that "all the attempts in the SALT and INF negotiations to stop the continuous increase and modernization of Soviet intercontinental and regional-strategic nuclear potential have failed. . . . [SDI] could under certain future circumstances move the Soviet position in negotiations to a more flexible policy. In any case, in the military field the USA no longer has any other effective inducement." The Defense Ministry noted that "arms control until now has shown that it has not been possible to negotiate strategic stability by establishing a military balance at a lower level of armaments." It accepts the US government's assumption that SDI is more likely to bring drastic reductions in offensive weapons than would the prevailing methods. It recognizes the risk of further fueling the arms race, but "this arms competition would change in a qualitative way if its emphasis were shifted from offensive to defensive category of armaments."[41]

Most West German specialists consider the ABM Treaty to be very important. Foreign Minister Genscher has even called it the "Magna Carta of arms control." However, Hans Rühle pointed out an important weakness of this treaty: it halted progress toward an American defensive system without having the same effect on Soviet research and modernization programs. Gerd Schmückle also pointed out something which many persons seem to have forgotten: that the US government signed the ABM Treaty in 1972 under the assumption that significant cuts in offensive weapons would follow; this did not happen. Also he reminded readers that the ABM Treaty has a provision which allows it to be changed or terminated within six months if the highest national interests of one of the parties are endangered.[42] Proponents of SDI note that from the very beginning the US government offered to negotiate with the Soviet Union any transition from an offensive to a defensive doctrine. The Bonn government also favors "cooperative solutions." Those who favor SDI point out that this program is more likely to stimulate, rather than retard, the arms control process in the long run. The governing CDU's disarmament spokesman, Jürgen Todenhöfer, described SDI as "the most successful disarmament locomotive in recent years." Not only did it bring Soviet negotiators back to the table in Geneva, but all the important disarmament recommendations which Soviet party chief Mikhail Gorbachev made in 1986 were made primarily in response to SDI. Todenhöfer concluded that "we would have to have lost our wits if we, as the SPD demands, were to give away our best trump card, SDI, in international disarmament poker."[43]

---

41. *Protokoll*, p. 167. *Fakten*, pp. 29, 31.
42. *NATO Review*, August 1985, pp. 26–32. *Aktuell*, pp. 76–77.
43. *FAZ*, June 6, 1986.

The Federal Government's Decision to Support SDI

In a setting of sharp, often highly emotional public controversy, the Bonn government decided to take a positive stand toward SDI. This was not an easy decision, with the January 1987 parliamentary elections looming on the horizon. Also, a political impediment to any decision-making is that governments in Bonn are always coalition governments. Therefore, a decision must have the backing of two or more political parties. There are always subtile (and sometimes public) power struggles within any governing coalition. The frictions between the CDU-CSU and the FDP especially affected the debate over West German cooperation in SDI.

Christian Democrats and Christian Socials (from Bavaria) are generally in favor of SDI, although some, such as Chancellor Kohl and Research Minister Heinz Riesenhuber (and earlier Wörner) were reportedly cooler toward the project than were others, such as Todenhöfer or Strauss. In the FDP, though, it is different. Because of its perennial role as "king-maker," the FDP (popularly referred to as "the Liberals") has always been able to demand a greater number of important cabinet seats than its percentage of votes would seem to justify. The larger governing parties have resented this. Also, it always faces the unnerving prospect of falling below 5 percent of the vote in federal elections and thereby being excluded altogether from the parliament. Thus, it has carefully guarded its image as a party which moderates and controls the larger governing parties. It feels compelled to present itself as being different and a bit difficult in order to justify its continued existence. The FDP has always considered its foreign policy profile as one of its most important domestic policy assets since Hans-Dietrich Genscher became foreign minister in 1969.

It should not be surprising that the Free Democrats raised objections to their partners' views of SDI. The majority of party members were reportedly opposed to SDI, and Genscher was a half-hearted supporter at best. It raised many objections. It was worried about being out of step with its European allies, and it therefore demanded that there be "no singular participation" on the part of the FRG. It did not like the Americans' insistence that the precise terms of any accord remain confidential and that the results of research work by West German firms be subject to secrecy classifications defined by the US government. Most importantly, it was anxious about the ABM Treaty and the entire arms control process. FDP leaders demanded that "the goal of arms reduction negotiations must be to prevent an arms race in space and to end it on earth." Sometimes Christian Democrats had the feeling that the FDP was raising one problem and objection after the other, merely to sabotage West German participation in SDI. The FDP finally gave its stamp of approval to the project, but the ultimate government position was a compromise which took into account the FDP's reservations.[44]

A further source of institutional conflict in Bonn is competition among the various government ministries. Cabinet ministers are appointed by the chancellor, but they are always powerful elected figures within the ruling parties who are very sensitive to being overlooked. In comparing the relatively weak cabinet members in the US with those in Bonn, the *Frankfurter Allgemeine Zeitung* wrote on September 10, 1985, that "the responsible ministries in the German system behave like duchies, which throw their entire weight on the scales, sometimes according to the party clientel of the duke or the minister, as the case may be." The ministries of economics, research, foreign affairs, and defense have all had prominent involvement in the SDI debate and have carefully tried to protect their interests.

A new problem which emerged in the controversy over SDI was the rivalry which developed between the Foreign Office (directed by Genscher—FDP) and Horst Teltschik, CDU, the director of the department within the Chancellor's Office responsible for foreign and security affairs. Unlike his predecessors, Teltschik has not kept a low, bureaucratic profile; he has assumed a role as the key foreign and defense policy advisor to the chancellor. He is ambitious and conscious of his power, and he has been sent out as the chancellor's spokesman in matters dealing with SDI. Thus, the problem of rivalry between the national security advisor and the foreign minister, so familiar in Washington, has become an important factor in Bonn as well.

On top of these institutional and party political impediments, there are important foreign political considerations. Since its foundation in 1949, the FRG's internal politics have been powerfully influenced by outside powers. The relationship with the US has always been crucial, but the reconciliation of Germany and France has been a cornerstone of West Germany's rehabilitation in Europe. The facts that SDI was an important American initiative and that France publicly opposed it were bound to create difficulties for the FRG.

No country in the world can disregard the Soviet Union's military weight and resultant diplomatic power. The FRG, as a divided nation on the line between East and West, can least of all disregard Soviet views. With the backing of its allies, the FRG can stand firm against Soviet demands, as it did with regard to INF deployment in 1983, but it can do so only after much soul-searching. During 1985 Bonn was subjected to considerable diplomatic

---

44. For the FDP's official position on SDI, see the resolutions on SDI adopted by its Directorate at Neuss in July 1985 and by its Party Congress on May 23–25, 1986. See also *Zeit*, October 11, 1985; *FAZ*, June 4, 1985; *SZ*, November 15, 1985; *Spiegel*, Nr. 13, 1986.

pressure from Moscow to foreswear any participation in SDI. The Bonn government is aware that Moscow has never stopped working on strategic defenses and that Soviet propaganda tends to be aimed against the American program of SDI, rather than against the fundamental principle of strategic defense. Nevertheless, Bonn was made to feel the heat from Moscow when it was deliberating over SDI. It was told that SDI is an "antimissile shield for the aggressor and a limited nuclear war for the Europeans." The Federal Republic had also allegedly made itself "an accomplice" to undermining the ABM Treaty and damaging the chances for successful arms control results; support for SDI would violate "the spirit of Geneva." Bonn would bear "a heavy responsibility for participating" in hostile American actions against the Soviet Union and must be aware that "the Soviet Union has no alternative to drawing appropriate conclusions."[45] Bonn is uncomfortable when it receives such messages. However, it knows when to ignore such propaganda. It has less control, however, over West German public reaction to Soviet threats. Many citizens take such warnings from Moscow at face value and put pressure on their government to do the same.

In important ways, 1985 was the "year of SDI" in Europe. The West German government had heard of SDI only after the president's March 1983 speech and was understandably irritated that it had not been informed in advance about this important project. Having just survived an energy- and will-sapping debate over INF deployment, it did not relish another public fight. At first it tried to maintain a low profile. In 1985, though, it felt compelled to take a stand, and it took a whole year to make up its mind.

The prod was a speech given by President Reagan in early January 1985. In this speech he declared that SDI was not designed to undermine deterrence but to reinforce it. For that, a 100 percent protection was not necessary. "It must only create sufficient uncertainty in the mind of a potential aggressor regarding the chances for the success of his attack." More important for German ears, he repeated that with SDI the US did not seek military supremacy nor political advantages over the Soviet Union. As a step toward finding means for reducing the dangers of atomic war, Reagan assured that negotiations with the Soviet Union would be pursued seriously. Finally, he called on America's allies to take part in the research, which would be fully funded by the US. That was a call to which the Bonn government had to respond.[46]

On February 9, 1985, Chancellor Kohl outlined the German interests in SDI: maintenance of the strategic unity within the alliance (i.e., no decoupling); avoidance of any strategic instability during a transition phase; close consultations within the alliance and with the United States; preservation of

45. *FAZ*, February 25, 1985, and April 7, 1986. *SZ*, December 30, 1985. *WP*, November 14, 1985. Davis and Pfaltzgraff, *Strategic Defense*, p. 22.

46. *Fakten*, pp. 24–25.

a technological link with the US; and a stimulus for the Soviet Union to negotiate. He stressed that neither superpower should seek military superiority, that both sides should continue to observe the ABM Treaty until an agreement could be reached which would clarify the relationship between offensive and defensive weapons, that any such solution be a cooperative one between the US and the Soviet Union, and that the US enter into negotiations with the Soviet Union before developing or deploying space weapons. Kohl never backed away from these defined interests or conditions.[47]

On March 19, 1985, Secretary of Defense Weinberger formally invited America's allies to join in the research effort, which would be fully funded by the US. He gave the allies sixty days to reply to his invitation. As Uwe Nerlich pointed out, this invitation was quite unusual: "The normal case would be an autonomous American development with subsequent pressure to adjust to it." Few people in West Germany were flattered, however; the sixty-day deadline appeared to be an ultimatum. The Kohl government quickly announced that it would not be able to meet that deadline because too little was known about SDI. SPD parliamentary leader Vogel reminded Washington that the allies are not "vassals" to Washington. Also, British Foreign Minister Sir Geoffrey Howe gave a thoughtful and important speech crystallizing European concerns about SDI. Within days, Weinberger assured Europeans that his invitation was not an ultimatum and should not be misunderstood. He sent General Abrahamson to Europe at the end of March to report on the goals of SDI and where the research stood at the time.[48]

On the twenty-seventh of that very feverish month of March, Chancellor Kohl noted in a speech to parliament that the ABM Treaty does not forbid research. He announced that his government basically approves of the research effort under the conditions he announced in February, to which he added: that the strategy of flexible response remain intact as long as there were no effective alternative to it, that the US not be decoupled from the defense of Europe, and that the conventional force imbalance in Europe be eliminated. He also indicated that his government would evaluate the economic and technological aspects of SDI. Finally, he would attempt to achieve a common European position toward SDI. These conditions continued to reflect the attention which he had to pay to the FDP's more skeptical view of SDI.[49]

Kohl's most important speech supporting SDI was delivered before parliament on April 18, 1985, in the midst of a firey debate. It was remarkable

---

47. *Aktuell*, pp. 81–82.
48. *Protokoll*, p. 149. *SZ*, March 29, 1985. See Howe's speech in the appendix of this book.
49. *Aktuell*, pp. 82–83. *NZZ*, March 21, 1985.

for the way it united the pro-SDI position of the CDU-CSU with the more skeptical position of the FDP. He asserted that the American research program was justified, politically necessary, and in the security interest of the West. "Every possibility of moving away from the depressing and threatening image of a nuclear holocaust as the ultimate means of preventing war deserves a conscientious examination. . . . Should this path prove to be possible, one would have to attribute to Ronald Reagan a historical service. . . . For me the resolution and the moral claim of the American president in this question are beyond any doubt." He asserted that the West is faced with a serious military threat. "Whoever says no today will not do away with the risk for the alliance and will not be able to utilize this real opportunity."

The chairman of the CDU-CSU parliamentary party and possible future defense minister, Alfred Dregger, noted in the debates that "those persons who are protesting against SDI today are the same ones who earlier protested against the concept of atomic deterrence. . . . To research a defense system against offensive missiles is not only morally justified, but also politically necessary. . . . Let us imagine for a moment that the West would refuse to look into a missile defense system, while the Soviet Union, which has never yet refused any move which could increase its military strength, would research and develop such a missile defense system alone."

After strongly supporting the general concept of SDI, Kohl elaborated the conditions and reservations which reflect the positions of his coalition partners. He said that he would discuss SDI with other European allies. He would also discuss the economic and technological possibilities of SDI with the business community and with scientific experts. He himself was optimistic and indicated that it is no exaggeration that SDI would stimulate wide-ranging technological innovations. However, he emphasized that economic-technological motives alone would not determine Bonn's decisions; political and strategic ones would. It would be important that Bonn be granted a fair partnership and a free two-way exchange of knowledge and technology. The FRG, as far as possible, should be given responsibility for its own field of research within SDI and thus have an influence on the project as a whole. He also stressed that the research phase must not automatically lead to the actual deployment of space-based defenses. Instead, it would be imperative that cooperative solutions be sought in order to improve strategic stability and drastically reduce the offensive potential of nuclear weapons. The decisive criterion "is the question: can this initiative make peace in freedom more secure for us?" "For us, the central point of SDI is arms control. . . . Our top priority is a drastic reduction in the nuclear offensive systems on both sides."[50]

50. *Dokumente*, CDU/CSU, April 18, 1985.

Kohl's April speech was an important compromise between different views within the government. Of course, the opposition parties lambasted the government's position in the debates. However, as the *Süddeutsche Zeitung* wrote the day after the speech: "Helmut Kohl's problem is not the fundamental opposition from the SPD and the Greens, but the faint, but continuous murmurs of dissent from the Liberals."

After the chancellor's key speech on SDI, the FRG became a steady supporter of SDI in international forums. In the spring 1985 meeting of the Nuclear Planning Group, it joined other members in agreeing that research resulting from SDI could stabilize the strategy of deterrence; it maintained the same position in the 1986 meeting in Würzburg. At the Bonn summit in May 1985, Kohl joined British Prime Minister Margaret Thatcher in supporting SDI, although French President Mitterrand blocked endorsement of the program. The government's support was also spelled out in the *White Paper 1985*. This official statement was followed by a publication for laymen explaining the rationale for Bonn's support for SDI.[51]

The US government made good its promise for continued consultations. In June 1985 Dr. Edward Teller reported at the German Strategy Forum that there was no clarity exactly what a strategic defense would look like. He noted that the US could introduce strategic defenses incrementally, even before a global concept had been developed. Even though he could not foresee what possible solutions might be found, he was sure that at least one of the possible approaches being studied would lead to the goal. Finally, he assured his audience that Europe would by no means be decoupled from the US as a result of SDI.[52]

In September 1985 the Bonn government sent a delegation of experts selected from various ministries and from industry and research institutes, led by Horst Teltschik, to the United States in order to explore the possible ways of participating in SDI. The delegation was met by all leading figures from the Department of Defense, SDIO, and Department of State. The Germans were struck by the facts that people in Washington well understood Bonn's domestic and foreign political problems and that no one pressured them in any way. American openness was, in one sense, a bit surprising: Bonn had just been shaken by serious revelations of East German spies in very high places. It can be assumed that more than one American was uncertain just how many secrets should be shared with this ally. The delegation returned to Bonn realizing that questions relating to the feasibility, financing, and possible vulnerability of a defensive system would not be answered until the end of the 1980s at the earliest. For Teltschik, it

51. *White Paper 1985*, pp. 31–33, 52, 62–68. *Fakten*. The North Atlantic Assembly approved a resolution favoring SDI in October 1985. See *NATO Review*, Nr. 6, December 1985, pp. 16–21.

52. *FAZ*, July 1, 1985.

was very important that the FRG have something to say about any agreement which the two superpowers might reach concerning possible stationing of defensive systems. It was a question, in his view, of whether Bonn would become an object or a subject in any such development: "to be a subject means that we must bring our influence to bear wherever and however this is possible. We must try to be co-architects."[53]

The Reagan-Gorbachev summet meeting in Geneva in November 1985 strengthened German confidence that arms control negotiations would proceed. As a result, the American president's usually low popularity in the FRG rose somewhat. There was still much public hue and cry during the fall of 1985 over SDI. On December 9 and 10 a parliamentary hearing was held in which the broad spectrum of West German views on SDI were presented and debated. Although there was clearly no consensus in the FRG behind SDI, the government officially decided on December 18, 1985, to enter formal negotiations with the US in order to reach some form of agreement governing West Germany's participation.

Its communique revealed the same kind of necessary compromises which earlier statements on SDI had contained. It stressed that the government seeks no state participation in the research program and was not planning to allot funds for it; of course, the American government had not asked any allies to help pay for the program. The goal of the negotiations would be limited: to improve the legal position of German firms accepting SDI contracts. The FDP wanted a lower form of agreement than a treaty; therefore it was decided that only "letters of understanding" would be hammered out with the US government; there would be no diplomatically significant documents signed on the subject of SDI.

Since the exchange of letters was not a part of a formal agreement, there would be no binding commitments. The very signing of any document was arguably unnecessary because the Bonn government could not prevent German firms from competing for SDI contracts anyway. However, the government's decision to sign letters of agreement was basically a decision to give some tangible political support to the research program of its most important ally.

The government chose the leader of the FDP and Economics Minister Martin Bangemann to go to Washington and negotiate an agreement. His selection was influenced by the fact that he is a Free Democrat within the cabinet, and FDP support remains crucial. Also, by sending him rather than the defense minister, the government could appear to be stressing economic objectives rather than military ones. Because of the political connotations, Bangemann insisted that the FRG approach participation in the SDI program with "civilian intentions."[54]

53. *FAZ*, September 5, 1985. *Aktuell*, pp. 10–11.
54. *WP*, April 25, 1986.

The talks began in January 1986, but they bogged down over details. West German officials complained that the US team, led by Richard Perle, was too inflexible on the general subject of whether German firms could use the results of SDI research for commercial purposes. As the Easter deadline approached, Chancellor Kohl wrote to President Reagan requesting greater flexibility, and he met with Secretary Weinberger at the Grafenwöhr maneuver grounds to work out the remaining snags. Kohl's successful efforts irritated the FDP, which suspected him of stepping into the difficult negotiations at the last minute and stealing the glory from its party leader.[55]

On March 27, 1986, Weinberger signed a Memorandum of Understanding, which regulates the research participation of German firms in SDI, and a Joint Understanding of Principle, which deals with the general exchange of technology. He had wanted to sign the letters with his counterpart in Bonn, as he had done with the British defense minister, in order to underscore Bonn's political endorsement for the objectives of SDI's military objectives. The German government refused. But at America's insistence liaison of bids and contracts is handled by the West German defense ministry.[56]

Bangemann emphasized that these agreements do not mean that the West German government is officially participating in the research. The extent of cooperation would be determined solely on the basis of private sector interests in both countries. The agreements do not guarantee a minimum level of SDI orders for West German firms, although they do include a "Berlin clause," which assures that West Berlin firms will not face discrimination in the awarding of contracts. The accords indicate that the US government will retain ultimate control over the use of all technologies developed under SDI contracts. Also, the US will exercise exclusive authority in deciding what kinds of research would be kept secret and what kinds could be used in the civilian sector. The US had even insisted that the texts of the two letters be kept secret. However, within days they were leaked to the Cologne newspaper *Express*. Some persons within the Kohl government were outraged by this leak. Others were reportedly pleased since the entire question of secrecy had long been controversial in Bonn.[57]

The signing of these letters will not end the debate over SDI. Critics have not been silenced. For example, Theo Sommer lamented in *Die Zeit* on April 4, 1986, that the Bonn government "has mounted a horse which is running in the wrong direction." The *Frankfurter Allgemeine Zeitung* on March 29, 1986, admitted that "no one will be fully satisfied with the two technology agreements . . . [which are] marked by a compromise achieved with great difficulty." German television commentary on March 27 stated

55. *The Economist*, March 29, 1986, pp. 39–40.
56. Ibid. *Week*, April 4, 1986.
57. *WP*, April 25, 1986.

the obvious: "The signing of the agreements in Washington is far removed from an unconditional, or even an unhesitating, German yes to SDI." Despite the compromise nature of the final agreement, many Germans are convinced that Bonn made a wise decision to support SDI. In the words of the *Braunschweiger Zeitung,* writing on March 29, 1986, the FRG "did not sign a blank check for its participation. On the other hand, it did not, as suggested imprudently by the opposition, issue a blanket rejection of the future security policy of the West's leading power. That would have led to the isolation of the Federal Republic and consequently, to the surrender of our security."

Bonn's commitment to SDI was put to a test in the aftermath of the October 11–12, 1986, superpower summit meeting in Reykjavik, Iceland. Those Germans who were not in the streets that weekend demonstrating against SDI and the "arms race" were watching the negotiations as closely as one could. Press reaction in the days after was neither universally critical nor laudatory. There were, of course, voices of dispair. Christoph Bertram wrote in *Die Zeit* on October 24: "The chance of the century to land the greatest disarmament package in the atomic age has been lost in one great throw of the dice. . . . It was too good to be true. A once-in-a-lifetime opportunity which never comes again" was fumbled by Reagan, the "provincial politician," because of his fixation on "the hokus pokus of missile defense in space." On October 16, in an acrimonious parliamentary debate, the SPD and the Greens held Reagan and SDI responsible for the "fiasco."

The *Frankfurter Allgemeine Zeitung* had correctly predicted on October 14 that "the dispute will now be transferred to the propaganda level, concentrating on Western Europe." The *Süddeutsche Zeitung* warned on October 18 that "Europeans must be even more careful now not to let themselves be used by Moscow against Washington," and the *Stuttgarter Zeitung* wrote two days later that "the U.S. must seriously reckon with the danger that projects like the SDI could turn out to be divisive instruments in the alliance." This danger was by no means ignored by Chancellor Kohl. In response to the concerted Soviet propaganda offensive which began within hours after the summit, he commented on the Soviet leader in a now-famous interview in *Newsweek:* "I'm no fool: I don't consider him to be a liberal. He is a modern communist leader who understands public relations. Goebbels . . . was an expert in public relations, too." Of course, Kohl was stating the obvious about Gorbachev's public relations talents. But his comparing him to Goebbels as a propagandist brought down a thunder of criticism from Germany and the Soviet Union that the beleaguered chancellor felt compelled to state publicly that "it was not my intention to insult the Soviet General Secretary." His remark elicited renewed

Social Democratic and Green demands that he resign.[58]

The chancellor had no such intentions, especially since the superpower discussions could affect Western European security in such a profound way. Kohl was the first NATO leader to confer with the American president after the summit. As he revealed in advance to the *Newsweek* interviewer, his message was: "Ron, be patient. Don't allow yourself to be pushed. You're in a good position." During his Washington visit in late October, he supported Reagan's refusal to scrap SDI for such a sweeping arms control agreement as was discussed on the spur of the moment in the Icelandic capital. He admitted that the prospect of destroying strategic nuclear weapons within ten years is "an attractive vision." However, he pointed out that without American nuclear power, long-standing Soviet superiority in conventional military power would make war in Europe "more feasible and probable again." Flexible response would lose all its credibility. He also pointed to the growing danger of Soviet short-range missiles targetted against the FRG which seemed to have been forgotten in Reykjavik. It is no wonder that his security advisor, Teltschik, admitted that "the talking was brought to a halt just at the right time in Reykjavik. Time is now needed to reappraise it all and strike a balance."[59]

The German ally did utter mild words of criticism in the Oval Office. Not only does the Kohl government favor including conventional armament in overall disarmament discussions, but he stressed that the number of offensive missiles remaining after deep cuts must "necessarily determine the extent of future defensive systems." Kohl did not agree with Reagan's insistance that there must be an "insurance policy" in the form of SDI to insure Soviet compliance to a future drastic arms reduction agreement. As the *General Anzeiger* put it on October 18: "In contrast to the president, Kohl prefers to look at the gigantic space venture as an eventual instrument of pressure which could be sacrificed on the alter of world peace once there is global disarmament."[60]

---

58. "Kohl to Reagan: 'Ron, be Patient,'" in *Newsweek*, October 27, 1986, p. 29. For the reaction to Kohl's interview, see *Der Spiegel*, No. 43, 1986, pp. 19–24, and No. 44, pp. 17–21, and *WP*, October 25, 1986. "Rückfall in the Dampfplauderei," *Die Zeit*, November 14, 1986, p. 1.

59. *Newsweek*, October 27, 1986, p. 25. *Week*, October 17 and 24, 1986. *Stuttgarter Zeitung*, October 18, 1986, and *Der Spiegel*, No. 44, 1986, p. 20.

60. *Week*, October 24, 1986. On the general topic of West German views toward SDI, see the following: Robert A. Monson, "Star Wars and AirLand Battle: Technology, Strategy, and Politics in German-American Relations," *German Studies Review* 9(October 1986); and the entire *Beilage zur Wochenzeitung Das Parlament*, B 43/86, October 25, 1986. Capturing 53 percent of the votes, Kohl's government easily won the 1987 elections. However, the CDU/CSU dropped more than 4 percent, while the FDP gained 2 percent, thereby increasing its influence within the coalition. The Greens made the most dramatic gains, winning more than 8 percent, while the SPD fell to 37 percent. SDI was not a major issue in the campaigns, but the electoral results could influence the government's position toward SDI.

17

Canada and the Future of Strategic Defense

Joel J. Sokolsky and Joseph T. Jockel

THE possibility that the United States will, in a matter of years, deploy new ballistic missile defenses (BMD), either ground- or space-based, poses a series of acute dilemmas for Canada unlike those faced by any other American ally. Quite simply put, a BMD system could involve, either directly or indirectly, Canadian airspace, Canadian territory, and the Canadian armed forces. Even the deployment of a space-based antimissile system with no installations on Canadian soil could place a premium on the expansion of antibomber and anticruise missile forces located in Canada, whether or not they were manned by Canadians. This leads to worries which are sharply different from the European fear that a Strategic Defense Initiative system will "decouple" the US strategic nuclear deterrent from European defense. Canadians worry that geography already has coupled them too tightly. Yet, it is highly ironic that space developments could have precisely the opposite effect on Canada. The development of space-based systems to detect bombers and cruise missiles could lead to a diminution in the importance of Canadian territory and, if those systems are operated by the United States alone, make Canada irrelevant to its own defense.

Over three decades ago, when Canada and the United States were in the midst of erecting massive air defenses, Ottawa coolly came to the conclusion that "it may be very difficult indeed for the Canadian Government to reject any major defence proposals which the United States Government presents with conviction as essential for the security of North America."[1] That conclusion still holds. But Canadians cannot sit back, hypnotized, awaiting technological developments occurring elsewhere and conclusions reached outside their borders, which may, or may not, lead to a new North American strategic posture. Decisions have to be made now about the defense of Canada and the protection of its sovereignty.

1. Department of External Affairs, confidential study, "Continental Radar Defence," October 3, 1953. Brooke Claxton papers, vol. 102. Public Archives of Canada.

Further complicating and differentiating the Canadian situation from that of other American allies is the fact that since 1957 Canada and the United States have maintained the binational NORAD—North American Aerospace Defense Command. NORAD, whose commander is a US Air Force general and deputy commander a Canadian general, has operational control over all US and Canadian air defense forces. In NORAD, Canada directly serves a fundamental national interest by contributing to the protection of the US strategic nuclear forces upon which deterrence rests. Through NORAD participation, as well, Canada guarantees its access to the US strategic information and plans which affect Canadian interests and secures at least the strong say of a junior partner in the development of those plans. It is also the cheapest way for Canada to surveil its airspace: the United States pays for 90 percent of the costs for the command's headquarters at Colorado Springs as well as 60 percent of the costs of the new detection systems being erected for continental air defense (including those located on Canadian soil which will be discussed in detail below). The United States also provides Airborne Warning and Control Systems (AWACS) aircraft for use in Canada and the fighter-interceptors which in an emergency would back up the two squadrons of CF-18 aircraft which the Fighter Group of the Canadian Armed Forces will be deploying on Canadian soil for air defense purposes.

Canadian governments have never seriously considered, since the dawn of the atomic age, alternative arrangements to cooperation with the United States in continental defense. But the costs of such cooperation have always been recognized as significant. It has meant the placement of Canadian forces under the control of a USAF general (albeit one wearing a binational hat and seated next to a Canadian deputy), the location, in the past at least, of some US military facilities on Canadian soil, and the establishment of provisions to admit US forces to Canada in an emergency. All these features of Canadian-American cooperation have led to the attendant worry that Canadian sovereignty—that ill-defined but still ardently cherished attribute of Canadian nationhood—has suffered in the bargain. Much of the uneasiness has resulted from simple unhappiness at being the very junior, non-nuclear partner linked to the nuclear behemoth to the south.

Still, from 1963 (the last year of the Canada-United States defense "crisis" arising from a refusal by the Canadian government of the day to honor its commitments to equip its defense forces, including air defense forces, with nuclear weapons) to the early 1980s, the defense partnership has been generally without rough moments. During those years the NORAD arrangements were of decreasing importance to the security of both countries. The shift in the threat to the continent from bombers to intercontinental and submarine-launched ballistic missiles precipitated the wholesale dismantling of many of the Canadian and US antibomber defenses. These included those weapons in Canada eventually equipped with nuclear warheads, which had been the source of friction in earlier years. To the vast relief of the Canadian government and public alike, the 1972 US-USSR Anti-Ballistic Missile (ABM) Treaty removed the possibility that Canada, through its NORAD membership, would be linked to an ABM system.

Since the early 1980s, the Canada-US defense relationship has once again been subject to uncertainty and strain. NORAD participation raises for Canadians the vague, but still menacing specter of commitment to SDI. The Canadian government's position on SDI can be fairly characterized today as one of skeptical agnosticism. There are few enthusiasts in Canada for the eventual deployment of SDI-type defenses. The old concern over loss of Canadian sovereignty through US facilities being located in Canada, and the old uneasiness at being linked to the nuclear superpower have only intensified since President Ronald Reagan made his celebrated and condemned March 1983 address on SDI. There is in Canada an all but universal attachment to the doctrine of mutual assured destruction (MAD). This leads to fears that deployment of new BMD systems would invariably trigger an intensification of the arms race, into which Canadian territory and armed forces would inevitably be dragged. As Joe Clark, secretary of state for external affairs in the Progressive Conservative (PC or "Tory") government of Brian Mulroney, told the House of Commons in January 1985, "actual development and deployment of space-based ballistic missile defence systems by either side would transgress the limits of the ABM Treaty [with] serious implications for arms control."[2] He reiterated a year later that "Canada firmly supports the regime created by the ABM Treaty and the existing SALT agreement on limiting strategic forces. Our stance toward SDI research is rooted in the need to conform strictly with the provisions of the ABM Treaty."[3] Ottawa rejected the US invitation, tendered in the spring of 1985 to allied governments, to participate directly in the SDI research program, although it left the door open for private Canadian firms and institutions to compete for SDI contracts which the Pentagon would be offering.

Still, the Mulroney government has not been prepared to condemn the US research program out of hand—as long as it remains a research program. Both the prime minister and the secretary of state for external affairs have proclaimed such research as being "prudent in light of significant advances in Soviet research and deployment of the world's only existing ballistic missile defence system."[4] Nonetheless, it seems fair to conclude that the ambivalent Canadian response was shaped as much by the overall approach of the Mulroney government to relations with the United States, as by strategic considerations. Since coming to power in September 1984,

2. Canada, House of Commons, *Debates*, January 21, 1985, p. 1502.
3. Ibid., January 23, 1985, p. 10101.
4. Prime Minister's statement, September 7, 1985. Prime Minister's Office.

the Mulroney government has been pursuing a policy of "refurbishing" Canada's relationship with the United States. "Good relations, super relations with the United States will be the cornerstone of our foreign policy," Mulroney had said on the campaign trail. The faltering Canadian economy, the Tories hoped, was to be partially revitalized by a removal of the barriers to US investment which had been put in place during the years of Pierre Trudeau's Liberal party rule, and by a pursuit of an enhanced trading relationship with Canada's most important trading partner, the United States. A central element of the new approach to Washington was the cementing of a strong personal relationship between the prime minister and Ronald Reagan. Dealing with Ronald Reagan meant addressing the president's cherished SDI program.

If there were any doubts in the Canadian public's mind about the importance the US president attached to SDI and in particular to the invitation to participate in SDI research, these disappeared during the March 1985 "Shamrock Summit" in Quebec City where Reagan, true to his pledge to hold a Canada-US summit annual meeting, met the prime minister. There, in a speech televised to the Canadian people, the president delivered a lengthy and heartfelt plea for Canadian participation in the SDI program. "The possibility," he said, "of developing and sharing with you technology that could provide a security shield, and someday eliminate the threat of nuclear attack, is for us the most hopeful possibility of the nuclear age."[5]

It is rare for Canada and the United States in the management of their relationship directly to link bilateral issues. Concessions in the environmental area, for example, are rarely linked to trade matters. Defense tends to be handled the same way. Nonetheless, because Mulroney had staked so much on his personal rapport with the president in the "refurbishing" of the relationship, the Canadian responses to the SDI invitation had to be handled with great care, to say the least, lest the good will of the Reagan administration, and of the president in particular, be diminished.

The government's professed agnosticism on SDI, and the president's obvious interest in involving Canada in the program has made the Canadian participation in NORAD quite sensitive. Unfortunately, from the point of view of the Mulroney government, two major decisions had to be taken concerning continental defense during 1985 and 1986. These highlighted the dilemma posed, on the one hand, by the desire for close defense cooperation with the United States and, on the other, by the desire to avoid any commitments—*or even the appearance of commitments*—to SDI-type defenses. These decisions were, first, the upgrading of air defenses, and secondly the renewal of the Canada-US NORAD agreement itself.

---

5. Presidential address, Quebec City, March 17, 1985. White House text.

By the late 1970s, it was clear that North America's air defense capabilities had seriously deteriorated. To be sure, the manned bomber had long become the weakest leg in the Soviet strategic triad, constituting some 11 percent of Soviet delivery systems and about an equal percentage of deliverable nuclear weapons. Yet this decline had been matched by an even more precipitous decline in NORAD's capabilities to defend against the bomber. Moreover, while the number of Soviet bombers was not increasing, their sophistication was. It also appeared that the USSR would follow the US in the development of air-launched cruise missiles (ALCM), having already taken a lead in sea-launched cruise missiles (SLCM). As a report to Congress observed, Pentagon planners had paid little attention to, and Congress had provided little money for, "defenses that could ward off ALCM, SLCM and bomber attacks against the United States."[6]

In 1979 Congress directed the Department of Defense to develop a plan for improving continental air defenses. The resulting Air Defense Master Plan (ADMP) was used as a basis for negotiation with Canada on the joint modernization of NORAD's capabilities. Those negotiations led to the announcement at the 1985 "Shamrock Summit" that the two countries would collaborate on a far-reaching modernization program that would entail the establishment of a surveillance and warning system around the perimeter of the continental mainland, capable of detecting aircraft and cruise missiles penetrating the warning net at any altitude.

A detection system, to be called the North Warning System (NWS) is to provide surveillance of the key transpolar routes of attack, and will replace the obsolete Distant Early Warning (DEW) Line, built in the mid-1950s. It will consist of thirteen minimally attended long-range radars (eleven in the Canadian Arctic) and thirty-nine unattended short-range radars (thirty-six in the Canadian Arctic). Unlike the DEW Line, which is operated by a private firm under contract to the Pentagon, the NWS will be operated by Canada. Covering the rest of the perimeter will be an Over-the-Horizon-Backscatter (OTH-B) network consisting of one site in Alaska and three in the continental United States. These very-long-range radars will provide coverage of the eastern, western, and southern approaches to North America. However, they cannot provide northern coverage because of the effect of the aurora borealis over Canada's north; hence the need for the conventional NWS sites.

It will also remain necessary for NORAD to operate fighter interceptors to identify intruders and to determine their intent. Indeed, fighter coverage is to be "pushed northward," a move which will be facilitated by construction of certain airstrips in the Canadian North and their designation as

6. John M. Collins, *U.S.-Soviet Military Balance 1980–1985*. (New York: Pergamon-Brassey's, 1985), p. 154.

Forward Operating Locations (FOLs) for use by Canadian, and in an emergency American, interceptors. Other northern Canadian airstrips are to be designated as Dispersed Operating Bases (DOBs) for use by American AWACS aircraft. The United States will be footing most of the bill. The total cost of the various elements of modernization has been estimated at approximately seven billion Canadian dollars, with Canada paying about 12 percent. The costs of the NWS will be apportioned on a 60-40 US-Canada basis.

The announcement of the North American air defense modernization agreement was met immediately with charges that the new airstrips would lead to a diminution of Canadian sovereignty, because they could be used, in time of crisis, by US forces. The Mulroney government was by and large successful in beating back that charge. The Tories explained that the enhanced radar coverage in the North, to be obtained by a system operated entirely by Canadians, would in fact lead to an improvement of Canada's ability to surveil its northern airspace and hence protect its sovereignty. Far more difficult to answer were the persistent charges made by the opposition parties in the House of Commons that either consciously or inadvertently the Mulroney government, by entering into the modernization agreement, had dragged Canada into SDI. In particular, opposition spokesmen repeatedly—and quite incorrectly—claimed that the NWS, the FOLs, and the DOBs were the first step toward inevitable BMD deployments in Canada. The president's plea for Canadian participation in SDI, made only hours before the signing of the modernization agreement, fueled suspicions that luring Canada into SDI constituted the hidden agenda behind Canada-US relations and that the Progressive Conservatives were not telling the public the whole story. Those suspicions were further intensified by Secretary of Defense Caspar Weinberger. During a television interview broadcast from the Quebec summit, he said that in the long run deploying new antimissile systems in Canada was indeed a hypothetical possibility. United States and Canadian officials scrambled to issue clarifications, lest the happy mood of the summit be spoiled.

Government spokesmen, including the prime minister, the minister of national defence, and the secretary of state for external affairs, were obliged to explain again and again in response to the opposition in the House of Commons that the North Warning System could detect only cruise missiles and aircraft—not ballistic missiles; that SDI was only a research program that might or might not yield results, which in turn might or might not lead to the actual deployment of new defenses against missiles; and that in any event the Canadian government still supported the ABM Treaty. But the Mulroney government, worried that the public might not grasp the difference, also over-reacted, claiming that there was no linkage between SDI and the NWS. In reality, however, if new antimissile systems are ever deployed, they will take their place, along with the NWS, as part of the North American continent's integrated defenses.

The need to integrate air defense and BMD is implicit in one of the most frequent criticisms made of SDI. As former US Secretary of Defense Harold Brown has pointed out, BMD deployments by the US could compel the Soviets to increase their bomber and cruise missile forces. To have BMD without air defense would be akin to having a house with a roof but no walls.[7] The SDI program already recognizes this possibility. While noting that the current improvements in air defense have only a "minimal" relationship to SDI and that a "robust" air defense system would only be deployed in conjunction with an "effective defense against ballistic missiles," the 1985 Department of Defense report to Congress on SDI went on to state:

> Study efforts will not ignore the relationship between the research of the SDI and strategic air defense. Strategic air defense requirements are currently under review, and continuing progress in the area of the SDI will permit the addressing even more comprehensively of the interrelationship between SDI and strategic air defense.[8]

In fact, this relationship is already being studied. A few months before President Reagan launched SDI, the US undersecretary of defense for research and development initiated the Strategic Defense Architecture 2000 (SDA 2000) study. Its purpose is to "develop a concept for integrated defense against bombers, cruise missiles and ballistic missiles" which would serve as a planning annex to the Air Defense Master Plan. Phase I of SDA 2000 looked only at bombers and cruise missiles and was concluded in April 1985. Canadian officials participated in this phase. They have not yet accepted the US invitation to participate in Phase II, which will be looking into missile defense.[9]

The link between SDI and a more robust air defense is also indicated by the US Air Force's Air Defense Initiative (ADI). Still only a modest research program, with a budget of only $53 million, ADI is exploring whether in fact air defense "walls" can be built to complement the BMD "roof" promised by SDI.[10] But Phase II of SDI 2000 and ADI, like SDI

7. Harold Brown, "The Strategic Defense Initiative: Defensive Systems and the Strategic Debate," *Survival,* March-April 1985, p. 56.

8. United States, Department of Defense, *Report to Congress on the Strategic Defense Initiative* (1985), p. C-21.

9. The history of the SDA 2000 study was described in a letter from the Canadian Department of National Defence to the Standing Committee of the House of Commons on External Affairs and National Defence, November 20, 1985.

10. On the ADI see David Lynch, "U.S. Considers Air Defense Shield," *Defense Week,* June 23, 1986.

itself, are long-term undertakings whose end results are by no means certain. Deployment of the NWS and associated improvements is necessary even if SDI never results in actual deployments, because the aging DEW Line and other warning facilities can no longer detect all forms of the air-breathing threat. As noted earlier, discussions between Canada and the US on air defense improvements had begun even before the president elevated BMD research to such high visibility and importance by announcing SDI.

Unfortunately, the complex nature of the relationship between SDI and air defense has been lost in Canadian discussions. Although the issue has been before the public for months, considerable confusion remains—much to the chagrin of the Mulroney government. The press generated still more confusion. No major news organization in Canada has a full-time defense correspondent, and for years the press, along with Parliament and the public, has paid scant attention to Canadian defense policy. Suddenly, though, defense was a hot topic. The press scurried to cover and, in many cases, simply to understand defense issues—not always with success.

The opposition, in short, scored a major success by highlighting the SDI issue. This made the renewal process of the NORAD agreement, which was to expire in May 1986, all the more sensitive. Two problems faced the Tories as they prepared for renewal. The first was that at the time of the last renewal in 1981, the then-ruling Liberal government accepted two changes in the text which seemed at the time to be unremarkable. The first was a change in NORAD's name from the North American Air Defense Command to North American Aerospace Defense Command. Because the 1981 document had not included a BMD mission for NORAD, the renaming of the command was both incorrect and confusing. Strictly speaking, NORAD, if it was to be renamed at all, should have been called the "North American Air Defense and Aerospace Surveillance Command."[11] The command was tasked and equipped to provide warning of air-breathing, ballistic missile, and space threats. The only active defense it could (and can) mount is against the air-breathing threat. Even here, given the lack of capability to intercept cruise missiles until the NWS and related improvements are in place, it still lacks the capability to intercept cruise missiles. As the 1986 renewal process got underway, critical voices in Canada asked whether the renaming foretold changes in NORAD's mission to include BMD and whether Canada had begun to reconsider its opposition to antimissile defenses. Those in Canada who harbored such suspicions could point to a second change which had been made in the 1981 agreement: the dropping of the so-called "ABM Clause."

11. Canada, House of Commons, Standing Committee on External Affairs and National Defence, *Report on Canada-U.S. Defence Cooperation and the 1986 Renewal of the NORAD Agreement*, February 1986, chairman's forward.

The ABM clause had been inserted in 1968, at a time when the United States was moving toward deployment of a limited ABM system. It provided that NORAD participation did "not involve in any way a Canadian commitment to participate in active ballistic missile defense."[12] Renewals in 1973, 1975, and 1980 retained the clause until it was dropped in 1981. Various official explanations have been given as to why Canada agreed to drop the clause. Ironically, these had to be given by the Progressive Conservative government, which assumed office in 1984, in response to questions raised in the House of Commons by the Liberals, now in opposition, but whose party had been in power when the decision was made in 1981. Joe Clark explained that the clause was deleted because the 1972 ABM Treaty was still considered to be in force. Therefore, the ABM clause in the NORAD Agreement was redundant. The government of Canada wanted to avoid any suggestion that either Canada or the United States would even consider taking any actions contrary to the treaty. Moreover, the United States had not even maintained the one ABM system allowed it under the treaty.[13]

Critics in the opposition New Democratic party (NDP), the now-chastened Liberal party and disarmament groups saw the workings of another dark and sinister plot to drag Canada, via NORAD, into SDI. They called upon the government to reinsert the clause in the 1986 renewal. A Tory-dominated parliamentary committee suggested another course of action: "that the government consider inviting the United States to issue at the time of the renewal of the NORAD agreement a joint declaration reaffirming both countries' commitment to deterrence and strategic stability, as well as their support for the integrity of the ABM Treaty and a negotiation process leading to variable reduction of armaments."[14] The Mulroney government, adhering to its claim that the clause was unnecessary, ignored the recommendation. President Reagan and Prime Minister Mulroney signed the renewal at the White House in March 1986. It was accompanied by another round of firm statements by the Progressive Conservatives that Canada was not obliged in any way to participate in any eventual SDI deployment.

Canada is far from being off the hook, through, in facing decisions about its participation in space-based systems related to North American defense. Events are proceeding too rapidly in the United States to allow that luxury. On October 1, 1985, the US established a unified Space Command (USSPACECOM), whose commander in chief (CINCSPACE) wears two other hats, that of commander of the US Air Force Space Command (USAFSPACECOM) and that of NORAD's commander in chief (CINCNORAD).

---

12. *Agreement between Canada and the United States of America,* Canada Treaty Series 1968, no. 5.

13. Canada, House of Commons, *Debates,* February 4, 1985, p. 1961.

14. *Report,* p. 78 (see footnote 11).

The USSPACECOM, the USAFSPACECOM, and NORAD are all to be jointly headquartered in a new building near the current NORAD complex outside Colorado Springs. For some Canadians, this puts Canada, via NORAD, all too close to American space activities, even though USSPACECOM, unlike NORAD, is an entirely American command, with no Canadians on the staff. They worry that CINCSPACE will be given operational responsibilities for antisatellite (ASAT) operations and BMD systems to the extent that those systems are fully deployed. In their view, NORAD, as USSPACECOM's twin, will still prove to be SDI's Trojan horse. But in testimony before the Canadian parliamentary committee looking into NORAD renewal, the current CINCNORAD, General Robert T. Herres, stressed that even if CINCSPACE were given ASAT and BMD responsibilities, those missions would not become NORAD missions unless there was a change in the NORAD terms of reference. He also emphasized that Space Command, as an operational command, is to be distinguished from the Pentagon's Strategic Defense Initiative Organization (SDIO) which, for the time being, is purely a research organization. Space Command would have been created regardless of the president's March 1983 initiative.[15]

For the Canadian military and others who are worried about Canadian access to space developments, the problem is not that binational NORAD and all-American USSPACECOM are twins, but rather that they are not linked closely enough. As a report by the Working Group on US-Canadian relations of the Georgetown University Center for Strategic and International Studies put it, the consolidation of US space assets under USSPACECOM is "bureaucratically . . . a moving train already picking up both speed and momentum. . . . The effective time window for a meaningful Canadian conceptual or organization input, consequently, is fast closing."[16] This concern was echoed by Major General L. Ashley, at the time chief of air doctrine and operations at National Defence Headquarters, and currently commander of the Canadian Armed Forces Air Command. As he put it, "my main concern now is that the division of responsibility may eventually make it difficult to participate in ventures that should be of interest to Canada. These would include space-based radar surveillance of the atmosphere."[17]

Canada has a vital interest in such activities. If feasible, space-based air surveillance could greatly enhance Canada's ability to monitor its sovereign airspace, and may prove cheaper than land-based radars, such as the North

15. General Herres testimony, Canada, House of Commons, Standing Committee on External Affaris and National Defence, *Proceedings*, 54: 29–30.

16. George A. Carver, Jr., ed., *The View from the South: A U.S. Perspective on Key Bilateral Issues Affecting U.S.-Canadian Relations*, Significant Issues Series, vol. VII, No. 4 (Washington, D.C.: Georgetown University Center for Strategic and International Studies, 1985), p. 46.

17. General Ashley testimony, SCEAND *Proceedings*, 34: 31.

Warning System. A Special Committee on National Defence of the Canadian Senate noted:

> Canada could control the use of its own satellites and make sure that they remain dedicated to passive detection and surveillance needs. Canadian military satellites over the North could also provide Canadian civil authorities with much useful information about activities in the Arctic and frontier regions. They could, for example, help monitor many forms of air, land, and sea movements across the North, keep track of oil spills and other dangers to the environment, or document the impact of development. They could improve communications with remote settlements and facilitate search and rescue operations, while at the same time enable Canadian industry to aim at the forefront of worked technological development in the space field.[18]

Canada is a world leader in some aspects of space technology, especially those related to telecommunications satellites. But as a relatively small country of twenty-five million people, it does not possess the resources to develop the full range of technology required for surveillance. It would not like a situation in which the US alone would develop, own, and operate the satellites which would be so useful for Canadian purposes. The Canadian Department of National Defence has been thus most eager to cooperate with the United States. Most notable is Canadian involvement in the American TEAL RUBY program which is aimed at developing satellite detection of bombers and cruise missiles. Before the Space Shuttle tragedy in January 1986, it had been expected that the US would put into orbit an experimental TEAL RUBY satellite in 1986. Canadian aircraft were to have flown within its surveillance area, and Canada was to have access to the results of the test. While Canada is paying nothing for the immediate development of the satellite, it provides a defense scientist who is working at the TEAL RUBY project office in California, and it is conducting studies on high Arctic cloud cover. The latter is important because the infrared rays used by the satellite cannot see through clouds. Canadian funds are also being provided for the construction of a data processing station which will analyze the results of the experiment.[19]

Yet here, too, the specter of SDI arises. TEAL RUBY currently falls under the US Defense Advanced Research Projects Agency (DARPA), not the SDIO. But it may be difficult to maintain a strict noninvolvement in SDI-related research if the US combines programs for space-based air

---

18. Canada, Senate, Special Committee on National Defence, *Canada's Territorial Air Defence* (Ottawa, 1985), p. 40.

19. See the testimony of Dr. D. Schofeld, Chief, Research and Development, Department of National Defence, in Canada, House of Commons, Standing Committee on External Affairs and National Defence, *Proceedings*, 34: 31, and before the Senate Special Committee on National Defence, *Proceedings*, 4: 24.

defense surveillance research with similar research into missile surveillance under the SDI program. Already there is "coordination" between DARPA and the SDIO on possible SDI applications of TEAL RUBY technology. The press has also reported that the TEAL RUBY "project has been described as a crucial element of the SDI program."[20]

Another possibility is that the impact of the Gramm-Rudman deficit-cutting legislation may compel the Pentagon to transfer research programs into the SDIO, which is being protected from automatic cuts. Already, opposition spokesmen in the Canadian House of Commons are calling for the termination of Canadian participation in the project, because of its links to SDI. The Mulroney government, having been compelled to acknowledge Canadian government participation in an SDI-related project, has hedged on its earlier statement proclaiming no government-to-government involvement. Quite justifiably, it has pointed out that Canada should not ignore valuable technology just because its development in the US may be linked to SDI research.[21] Nevertheless, coming on the heels of previous attempts to put distance between Canada's air defense needs and SDI, such statements have only fueled opposition criticism and press cynicism.

In the short run, Canada faces the difficult task of trying to guarantee its access to certain aspects of US space technology while continuing to emphasize the distance between the Canadian military and SDI. The task becomes all the more difficult as the US defense establishment contemplates the prospect of Canadian territory and airspace becoming less important as a result of space developments. As John Hamre, a senior staffer for the US Senate's Armed Services Committee has observed, "technology and program choices for upgrading North American air defenses appear to be eroding the necessity of Canadian participation. . . . I believe Americans continue to view continental air defense as a joint venture. However, in the future that joint venture will likely be more a matter of choice than necessity."[22] Even ADI, which envisions an elaborate air defense system to complement SDI, appears to be concentrating on surveillance and interception capabilities that, being located in the US and in space, would reduce the importance of Canadian contributions.[23] While they were once warmly welcome, Canadian defense officials are already beginning to feel a distinct chill in the precincts of US aerospace defense policy and operations.

Although ignored in the sizable mountain of commentary and analysis which has grown up around SDI, Canada could well be the first of Ameri-

20. *Boston Globe*, March 5, 1986, p. 4.

21. *Ottawa Citizen*, June 17, 1986, p. C3.

22. John Hamre, "Continental Air Defence, United States Security Policy, and Canada- United States Defence Relations," in R. B. Byers, et al., *Aerospace Defence: Canada's Future Role?* Wellesley Paper 9/1985 (Toronto: Canadian Institute of International Affairs, 1985), p. 27.

23. See footnote 10.

ca's allies to have to make major decisions regarding SDI if and when it moves out of the laboratory to actual deployments. Already the impact of SDI has been felt on the Canadian government as it attempts to manage its security relations with the United States. Yet for Canada, perhaps more than for any other ally, the implications of SDI remain uncertain, and troublingly so. At the heart of this uncertainty is the inability to predict with any surety whether BMD deployment would mean more or less American reliance on Canadian territory and more or less Canadian participation in the air defense efforts that would accompany defense against missiles. Indeed, there can be no roof without walls.

Canada therefore faces several different strategic futures, with quite different implications. Most trying for Canada and Canadian-American relations would be BMD deployments coupled with the need, from the US point of view, for new air defenses located in Canada. Less trying would be BMD deployments linked with an American ability to conduct the necessary air defenses from space. Optimal from the Canadian perspective would be no BMD deployments, but the deployment of space-based air surveillance systems over Canada which Canada operated either jointly with the US or on its own. The trick for Canada is how to opt out of some aspects of space defense without closing the door entirely on others.

18

The Pros and Cons of Strategic Defense: Discussions at the Virginia Military Institute

Technical Feasibility

QUESTION: *Can an entire SDI system ever be made to work?*

BARRY M. BLECHMAN: I have no doubt that American scientists, over time and given enough money and effort, will be able to develop each of the components of the Strategic Defense Initiative, each of the single types of weapons or sensors or capabilities which are required. But in considering the very important question of whether we should deploy SDI, and whether we should change our approach to strategy and so forth in order to do it, one really has to consider the overall systems-effectiveness. The question is: how will all these components work when there is an enemy, not only attempting to evade it, but attempting to attack it—when it has to deal, in a very short period of time, with a very large number of objects. The question of overall systems-effectiveness is really the decisive one in determining whether, even in the best case, even in the event that we successfully develop the technology, that all the components work individually, the system can ever be effective enough for us to rely upon.

QUESTION: *Can we test the capabilities of SDI, even if we can physically construct a system?*

GEORGE W. RATHJENS: We can test a lot of the components. We can do a lot of simulations. And we could probably do a pretty good job of simulating a terminal defense component. It is going to be harder to simu-

Unless otherwise indicated, all questions were posed by cadets at the Virginia Military Institute. These discussions were held April 7–8 and October 16, 1986.

late the whole system in an effective way. And so you have to make judgments about how well large systems will work based on tests of parts of it. That is the best you can hope for.

LOUIS MARQUET: We really can. But I would add that the same statement can be made about offensive retaliatory systems which are currently in place. We are depending on the credibility of our being able to absorb an overwhelming first strike against our command centers, our communications links, our retaliatory forces, our submarines and airports. We are assuming that we can rise from the dust of that strike and that all our systems can be reconstituted and mobilized to deliver a retaliatory strike. This is the grounds for deterrence in the first place. So we are making the same kind of assumption in terms of testing the components of that system in a very similar strategic context.

QUESTION: *What are some of the countermeasures to SDI, and if the Soviet Union can produce them cheaply, why should we even begin developing our own SDI?*

MARQUET: We have been inventing many of these countermeasures, if not most of them, as we try to address the question of what it is going to take to leapfrog these countermeasures. Clearly, this problem is different from going to the moon because we have to concern ourselves with the response, and we have to be convinced that we can come up with an effective and cost-effective technical and operational approach to the problem. There is far more than just technology involved in a military system.

Let me mention a few of our detractors' favorite countermeasures and perhaps a few words about responses to them as well. One of them that I hear a great deal about is fast-burn boosters. The idea is that if the opponent can reduce the time required for the boost phase, which is currently on the order of three hundred seconds, down to perhaps fifty to a hundred seconds and get the boost phase completed before the boosters are out of the atmosphere, it would obviously stress the time-lines and the performances of that boost-phase intercept system and, in particular, of directed energy. It turns out that if you can make the boosters fast enough, you can probably eliminate kinetic energy weapons as a potential candidate for that intercept. So, what are you going to do about this? Well, first of all, most of the forms of directed energy we are looking at penetrate the atmosphere, so the fact that the boost phase is completed before it gets out of the atmosphere is not going to be a detriment. What is a detriment, of course, is that we have to be able to achieve our kill in a very short period of time.

In order to offset this absentee ratio (meaning how many of your weapons platforms are really in the battle at any one time) and to make the situation as favorable as possible for you, you want to have a long-range weapon, a very high brightness system, which pushes you toward shorter and shorter wavelengths. The nature of light being what it is, the shorter the wavelength, the less spread of the optical field for the given dimensions of your transmitter. You would like to get very short wavelength systems—high brightness with long ranges. Quite frankly, our technical goals are quite substantial. We are trying to get absentee ratios on the order of 3:1 to 5:1, rather than the 20:1 to 50:1 that sometimes gets quoted. The kill rate—that is the time that we actually dwell on the targets—is less than a second. Therefore, if we can meet these technical objectives, which is what our goals are, we will be able to carry out intercepts on fast-burn boosters without a great deal of additional effort.

Fast-burn boosters also have a penalty associated with them: they cannot carry as many RVs, and it is more expensive per RV delivery. That is probably the most useful measure of currency in comparing costs: how much does it cost to deliver an RV, not how much does it cost to build one or to build a missile. If you compare the cost of RVs delivered from our MX, which launches ten of them, to the Midgetman, it costs almost ten times as much for each delivered RV for the Midgetman. It is just not an efficient way of delivering an offensive attack. Midgetman has other virtues; I am not going to denigrate it: it is mobile; it is perhaps more survivable. Nevertheless, it is much more costly.

My second most favorite countermeasure is the ground-based laser assault against our space-based assets. The argument goes as follows: if we are going to build these great bright laser beams and be able to deploy them in space or to relay them in space through mirrors in order to intercept boosters on the other side of the earth, then surely it is going to be straightforward and take much less stressing technology to put it on the ground and zap our satellites, whether they are weapons, surveillance, or relay-mirror satellites. Space systems, it is argued, are more vulnerable than boosters. I am not willing to say categorically that this is not true, but I believe that there are many reasons to doubt that this may necessarily be true. A booster has a very challenging job that it tries to perform, namely putting a lot of weight into space as it burns through the atmosphere at high accelerations. On the other hand, it is true that today's satellites are flimsy, vulnerable structures, for the simple reason that no one has ever had to protect them against laser radiation or kinetic-energy interceptors. This does not mean that they cannot be hardened, that we cannot put up decoys, that we cannot conceal the deployment of some of these objects and maneuver them in order to avoid attacks. We can do all of these things, if we must, in order to insure survivability.

Most of the things I just mentioned cannot be done for boosters. Boosters are very easy to detect because they have large bright plumes. They are under considerable stress. It is hard to imagine hanging shields around the outside of them, whereas you could perhaps do that sort of thing with space assets. Thus, it is an asymmetric situation. It is not at all obvious that you cannot harden your space systems to survive a ground attack.

In order for a laser beam on the ground to intercept and destroy a space target, you must compensate for the atmosphere's aberration. We carried out some low-power experiments at Maui that demonstrated that we do have the fundamental understanding to do this. But that was done in a very special way. In order to carry out that compensation, you must measure what the atmosphere's aberration is dynamically, in real-time, on a time scale of a millisecond, and correct for that aberration. That is done by placing a beacon satellite, a reference mirror, if you like, on our relay systems. It is our satellite and we can hang any tracking aids we like on it and can provide that reference to compensate for aberrations. If you are going against a hostile satellite, he certainly is not going to make the job easy. He is going to do everything he can to destroy your ability to find out where he is and to measure that atmospheric aberration. It is an extremely challenging job. We do not know how to do it.

RATHJENS: You need not destroy satellites from the ground; you can do it from space-based systems. Assuming a real competition, where both of us have space-based systems that are designed to destroy adversary boosters in the boost phase or in mid-course, it is very likely that those same systems could be used to destroy the other fellow's satellites. Take one example, though not one that is my favorite by any means: if we had a space-based system that involved an X-ray laser system—nuclear-driven X-ray lasers—to intercept the adversary RVs, it would be an easier task to use those same space-based assets to destroy adversary space-based assets. The great incentive, incidentally, if we ever got to the point of deploying these things, is for each side to strike first at the other's space-based assets. They become the target of choice in any situation of real crisis instability, assuming, that is, that you have systems that are deployed in the first place, which I very much doubt.

I wish to make a more general comment about countermeasures which involves why we should worry about SDI. The countermeasures game is one that we will play. We always have in the past. Every time the Soviet Union has deployed defenses, we have developed counters to them. The same applies for our side. Every time we improve our anti-submarine warfare capabilities, they build deeper-diving, quieter, faster submarines, longer-range missiles, and so on. It is competition. However it comes out, one of the more worrisome things about going down the SDI path is that it will be an arms race. As we look at new options, as we do interesting R&D, long before we get to the point of deployment, they—if they act the way they have in the past—will begin to think of ways of countering what we are doing, and we will do the same. Since the time from early R&D to production and deployment is probably a decade or more, there is plenty of time for both sides to react and to counterreact. Therefore, we are going to have a lot of action-reaction that is likely to consume at least billions of dollars per year, possibly tens of billions. I am not worried about crisis instability, because I do not think we will ever get to the point where we can deploy systems that are good enough for either side to do it really on a significant scale. But I am troubled by the fact that, because of this countermeasure problem and constant competition, we are headed down the path of at least moderate expenditures and accelerating competition which will lead nowhere except to both sides' having spent a great deal of money.

The primary problem is still whether or not the countermeasures will stay ahead. Let us go back to the important business of the cost-exchange ratio. The argument is that, yes, we can develop pretty good technology, perhaps systems which could be technically interesting. But if the adversary could offset them through substantially lesser expenditures, it is probably not a very practical system to deploy. That is, ours could be countered at relatively lower costs.

What it will cost will depend very much on what the objective will be. If we are talking about a very effective defense—90, 95, to 100 percent effective—offense is going to continue to have the lead. On the other hand, the cost-exchange ratio does change dramatically if the defense is interested in assuring the survival of only a relatively small number of targets and is willing to let a large number go. In that case, defense may look more attractive. It might then be interesting to deploy defenses, at least at modest levels, to defend a few targets. Or it might be interesting to deploy defense against lesser adversaries. We once talked about deploying a defense against the Chinese. Indeed, we began going down that path.

QUESTION: *What are some countermeasures to SDI?*

JACK MENDELSOHN: There are a number of ways of defeating an adversary. Let me just mention a few. There are responses to all of them, and there are penalties involved in instituting these. But there are also penalties in trying to respond.

On the boost phase, you could reduce the burn time of the booster. You could multiply the number of hot spots which sensors might look at, to try to confuse, overwhelm, or staturate the sensing system. You could harden the booster. You could rotate the boosters. You could release the front ends of the boosters sooner rather than later in order to get the multiple targets out as soon as you can. In the mid-course, you can either make things look like something else or try to keep them from looking like what they really are. You could either add decoys, or you could mask objects. You want to complicate the sensing and acquisition process. In the terminal phase, you can try to maneuver. You can try to concentrate your attack in certain ways. You can try to confuse or overwhelm the defenses. You can try to blind the defenses. You can simply try to attack the defenses, possibly by means of space mines. Mirrors are very fragile.

One very simple political maneuver, which is in the back of everybody's mind, is the simple expedient of the Soviet Union declaring any of its air space off limits to weapons. We are in a very fragile environment where both nations have agreed to allow relatively benign intelligence satellites to overfly. This is the right of innocent passage. The Soviet Union could easily say that the right of innocent passage exists, but the right of harmful or lethal passage does not, and the first space mirror or whatever which turns up over the Soviet Union would be considered a target. It is reasonable. It may be an act of war, but it is not an act of war in the same sense as shooting down an aircraft might be. It will not precipitate an exchange. And if the Soviet Union wishes to declare the presence of attack systems over its nation as an act of war, we would be likely to do the same.

QUESTION: *How could the electrical systems required for defensive satellite systems be protected from electro-magnetic pulses or microwave systems, which could disrupt their operations?*

MARQUET: We have been looking at a number of technologies. First are hardened electronics, which are intrinsically several orders of magnitude less sensitive to electronic countermeasures. In addition to this, we are developing methodologies using fault-tolerant types of architectures, computing processing which, in the event that there are malfunctions, may be localized and possibly moved around. It is one of the most critical areas because it is virtually impossible to prove either that your systems will work or that it will not. The uncertainty resides on both sides. Our goal is to prevent these systems from having to be used in the first place. Therefore, if the Soviets, for example, were counting on taking out our electronics by a high-altitude burst or a microwave attack weapon aimed at one of our satellites, hoping to disengage its electronics, then they are clearly taking enormous risks because the vulnerability of our electronics is a very subjective issue. It depends on how you harden, how you design, and how you have packaged the systems. Then it may not be at all obvious, once you have done that, that you have killed it, so what do you do next?

RATHJENS: One can probably argue that some kinds of satellites or instruments in space can be protected against nuclear weapons attacks. A single nuclear burst might not necessarily destroy more than one satellite. But we certainly cannot protect them against a nuclear weapon directed against a given satellite. Probably by protecting against electro-magnetic pulse and other effects, we could have a system which would require the adversary to dedicate a weapon to each satellite he wants to knock out. A possible exception is the use of X-ray lasers as a means of attacking multiple satellites.

MARQUET: That sort of thing can be countered by a relatively thin sheet of material that will absorb the X-rays. They do not penetrate material very far at all. By deploying thin sheets to protect these satellites from those X-ray lasers, you can provide them with a protective shield against them. Thus, even directing an X-ray laser against the satellite does not guarantee its demise.

QUESTION: *Submarine-launched cruise missiles could be launched close to our shores and with low trajectories could underfly SDI. Is that true?*

MARQUET: There has been considerable ambiguity on what the objectives of the program have been. SDI is currently addressing the issue of intercontinental ballistic missiles and submarine-launched ballistic missiles. That does not mean that every kind of threat—bombers, cruise missiles, and the so-called "rusty freighters" or "suitcase threat"—is going to be eliminated in one fell swoop. It is clear that there is still something to be concerned about—terrorist actions and smaller-scale attacks. All would have to be addressed in order to fill in the gaps.

The reason that the administration is first addressing the problem of intercontinental ballistic missiles, which take only twenty to thirty minutes to get to their targets, is that they are clearly the most destabilizing kinds of weapons which exist on earth today. They take such a short period of time that there really is not enough time to talk on the "hot line" between Moscow and Washington. They cannot be recalled, unlike aircraft, cruise missiles, and "rusty freighters." Therefore, in terms of crisis management, they are extremely destabilizing.

It is also true that the program is fundamentally different. The president asked us to explore the possibility of moving away from a strategic posture based on MAD, to a posture based on denial of effectiveness by defense. That is to say, it would still be deterrence, but it would be deterrence by denial of military objectives through defense, versus the denial of attack by the threat of massive retaliation. Frankly I think the question which the president raised was really the cogent one: Is it not better to defend our people than to avenge them? Can we, in fact and over the long term, continue to exist in a world where very suspicious and hostile nations stare at each other over cocked guns? That is the objective.

The program does not currently address the threat of cruise missiles. There are many programs in the Defense Department that do. Certainly many of the technologies that are developed under the SDI program, such as space directed energy, could conceivably intercept and destroy cruise missiles. They would presumably be much more vulnerable than boosters might be. The difficulty might be finding them in the first place. But we consider them a secondary order of factor in the overall strategic posture. We need to look for the quarter down the street in the dark, if you like, rather than under the lamp, where it might be easier to look. The most distressing problem is ballistic missiles, and there is no point in handling the strategic air defense problem until we have solved that one. That is why we have essentially dismantled our air defense system.

QUESTION: *Why has more attention not been paid to ground-based defensive systems, rather than to more complex space-based systems?*

MARQUET: If the objective is to protect our missile silos, versus moving away from dependence on retaliation, then we would have a very different kind of program. We want to deny the offense the opportunity to pick his targets and to design structured attacks. We addressed this extensively in the 1960s and 1970s and came to the conclusion that conventional defense of terminal systems really were extremely difficult and, at that time, infeasible because the advantages really were with the offense, which calls the tune. I certainly agree that dealing with the booster during its boost phase is the most challenging part of the problem. However, I also believe that the emerging technologies, particularly directed energy, offer us the hope that we can solve it.

QUESTION: *The control systems for SDI will rely on computer systems which have yet to be developed. But given the software requirements and short flight time, what would the role of human decision-making be in SDI?*

MARQUET: I would like to make a couple of points about the reliability of control systems. First of all, I want to contrast the potential consequences of a malfunction of an offensive-designed system versus a defensive-designed system. Let us imagine, for whatever reason, that our current retaliatory system is triggered. This could perhaps result from a computer malfunction. We have gone through some simulated engagements in the past where the people in Cheyenne Mountain were convinced that they were under attack. Computers do fail, and in the best of all possible worlds, they will always fail. I do not think we can get perfection in anything, human or machine. But we are not talking about that perfection here. The consequences of an offensive mistake are rather obvious and incredible because we would unleash a retaliatory strike.

Now let us consider a malfunction in a defensive system. Imagine that somehow one of our space-based systems is activated, and it shoots down a missile, which is launched from the ground. Certainly the consequences of shooting down, say, a shuttle full of astronauts would not be terribly pleasant. It is no worse than shooting down an unarmed civilian Korean airliner, but it is still not pleasant. It does not, however, amount to starting World War Three. So, number one, the consequences are very different; they are not symmetrical.

Number two, there will be an enormous amount of computer processing. We will always have to ask the question, will man be in the loop, will the president be in the loop. Clearly he will increasingly be in the loop as the system is raised to a higher and higher level of activation. If you like, it can be set on automatic only in the most extreme crisis, and setting it on automatic might be a very stabilizing aspect of crisis management. Automatic action really only needs to be done in a layered defense when you have a massive attack.

There will be a threshold. In other words, there will be that first sixty to a hundred seconds in the boost phase where you need to activate the space component of your system. It is only necessary under a massive attack. Then there are the subsequent layers where you have minutes or tens of minutes in which to get a more thorough assessment of the system to be invoked. So there is a threshold before various levels of automation would be invoked. Likewise, there is an asymmetry of consequences of error.

RATHJENS: I think that the problem is very different, again depending on what your objective is. When you talk about a multi-layered system which includes boost-phase, mid-course, and terminal defense, the management problems are enormous. You have to pass data from one part of the defense system to the other in a very complex way. Whether you can ever debug a system and get the right program, at least one you can be sure works, is very debatable. You never get a chance to practice in a real sense with one of those things.

On the other hand, consider terminal defense of point targets. The separate components of it do not have to be orchestrated or all tied together. From the battle-management standpoint, the software problem would be immeasurably easier. It is another reason why, if you are really interested in enhancing deterrence, you probably want to forget about all this space management and focus on the terminal defense problem where you have a more manageable management problem.

QUESTION: *Do you agree with the objections voiced by the Union of Concerned Scientists?*

MARQUET: My problem with their conclusions is that they are very simplistic. They really do not tell about one-half of the story. They point out the problems which we ourselves identify, and then they take a dogmatic approach that these problems cannot be solved. I believe that it is characteristic, and I think it is unprofessional for scientists to take such a dogmatic position. I have problems with their very title: the Union of Concerned Scientists.

QUESTION: *What effect has the space shuttle tragedy on January 28, 1986, had on SDI research?*

MARQUET: Of course, the space shuttle disaster has an impact on our research program. We have been counting on using the space shuttle to carry out an important number of space experiments which we believe are going to be essential during this research portion of our program. For example, in my office we have been planning a very ambitious space-based pointing and tracking experiment to determine whether or not we can actually track the plume of a booster: to acquire that booster itself, hand that track over to a precision pointing system, and hold a laser beam stably along the side of the booster. Of course, we are not talking about using lasers which have a lethal capability. We are talking about very low power lasers. But from the standpoint of precision pointing and tracking requirements, it is going to be a very challenging experiment. That was one of the first experiments we have been wanting to carry out. It was scheduled for late fiscal year 1987.

At this point in time, we do not know exactly when the shuttle will be back in operation. We suspect that it will cost us some money and time. But it is important to point out that we view the space shuttle as a mechanism for carrying out a research program. If and when it is decided that we will move into the deployment stage, the space shuttle is totally inadequate from the standpoint of supporting the kinds of systems that we have been visualizing. Some types of simplified, much less costly, perhaps unmanned, more robust systems of space transport will be required to reduce the costs of getting material into space. We can realistically reduce costs significantly by using much of the space shuttle technology base, by taking man out of the system, and by changing the infrastructure that provides the support for that one system. The space shuttle system is not intended to play a direct role in the deployment or maintenance of any potential SDI system.

QUESTION: *All past technologies have become obsolete. Will not this pattern apply to SDI as well?*

MARQUET: I would like to see it first apply to ballistic missiles.

QUESTION: *Aside from their desirability or feasibility to defense applications, what are the implications of new technologies on a new era of weapons design and development?*

MARQUET: We are in a new era of weapons designs and developments, but let me just talk about tactical technologies. If the technologies which we have undertaken and collected together for investigating within an SDI—not just directed energy, but advanced sensors, advanced processors for the computer systems, communications links, and so forth—come anywhere close to achieving the goals that we have set for them, then the appropriate question is: "What is the implication on the technology for the next generation of tactical weapons?" If you consider the competition between the

United States and the Soviet Union and look at the way that competition has gone, it is very clear we cannot out-man, out-gun, out-tank, out-plane the Soviets. We certainly have the industrial capacity, if we chose to use it, if we decided to use 20 percent of our GNP for military support, as the Soviets are obviously able to do.

Where do we have the advantage? It is intrinsic in our free-enterprise system, in our innovation and advanced technologies. The Soviets have, in fact, been forced to develop a discipline of their own: copying Western technology. They are getting pretty good at that. This is not to suggest that they cannot do innovative work of their own; they certainly can when they choose to do so. But the rewards and punishments, in terms of economic progress and advanced technologies, are very different in a controlled society than in ours. We do have that advantage; that is our secret weapon. We must, in fact, continue to exploit this advantage in order to hold them at bay until some day in the future when their system will evolve to the point where we can put together more tranquil kinds of exchanges.

Specifically, what might we expect from our new tactical technologies? I believe that the days of the manned aircraft, the tank, and the surface ship are limited, as we go to more sophisticated electronic systems. It is very likely that instead of our using a tank to engage a Soviet tank, we will have much more reliable, cheaper, and more precise weapons that can destroy tanks. Conceivably, space-based systems could be used to detect and designate conventional munitions and their delivery systems, the interceptors on the ground, the artillery shell, for example, which homes in on the laser beam illuminating from a space satellite to the top of a tank. Possibly an aircraft could be intercepted doing the same kind of thing. With the advances in our technologies, computers, satellites, sensors, interceptors, and so forth, within a time frame of twenty or thirty years, we are really going to see some very rapid changes in our perceptions of what conventional forces are all about.

GREGORY M. SUCHAN: I note that the Soviet Union seems to recognize this changing evolution as well. Remember the article that General Nikolai Ogarkov wrote a few years back, which many people believe was the reason he lost his job as chief of the general staff. He indicated that the revolution in weapons technology is such that in the future, people are going to have to rely on certain things other than nuclear weapons and larger and heavier tanks as the basis of military power. He argued that greater emphasis will have to be placed on precision weapons and things for which accuracy is more important.

Strategic Doctrine

QUESTION: *How would the development of defensive options fit into the United States's overall strategic doctrine?*

STEVEN A. MAARANEN: We cannot stand still. We have come to a certain point with an offensive force strategy and posture which places very heavy responsibilities on our forces. Because of a variety of changes in technology and in the way the Soviet Union has proceeded in conducting its policy, we are coming to a dead end, in terms of the continuing prosecution of that particular policy.

There are a lot of ways forward. There may be ways forward with offensive nuclear forces alone. There may be a way in terms of proceeding to a kind of mutual assured destruction policy, which abandons limited nuclear options. There may be technical options for proceeding with limited options and the flexible response strategy we have now. But there are technical requirements associated with these programs, and economic costs associated with programs like that, that are going to be very difficult to satisfy and very difficult to complete with success.

Then there is a variety of options for the employment of defenses, either defenses by themselves or, more likely, defenses in combination with a variety of offenses. Thus, if we do not yet have the answer to the question, "which of these strategies is the right one to choose?", there are persuasive reasons why we should be looking at defenses, but at a spectrum of defenses in combination with offense. Ultimately, there is an argument I find persuasive to move progressively in the direction of ever more effective defenses over the long run. I do not think that we can be sure this is the right answer, but we should not cut off the debate about strategic defensive forces. I am quite confident that we do not yet see our way clear to a satisfactory strategy in the absence of defenses.

QUESTION: *Why would a person not support a weapons system, including one in space, which could possibly protect our citizens?*

MENDELSOHN: There is no reason not to put something up in space if you are certain that it would protect us. But that is not what we have got with SDI. Nobody knows if it is going to work. Nobody knows how to implement it if it does work. If we had a defensive system which could protect the nation, I would say "Fine, let's do it." I think the argument really is: "Do we have that possibility in the mid- or long-term?" There is a great deal of question about that. If, indeed, we do not, then we will only precipitate a long chain of events where we destabilize the relationship. In other words, if one side believes its forces are being threatened in some way, then it may precipitously attack in a crisis situation. Or you may stimulate additional offensive deployments. I would be the first to concur that it is a great idea to defend yourself. However, unless you know that that is really the case, and I argue that we do not and can not know, then you will unleash a chain of events which could be less apt to defend you in the long term.

RATHJENS: There is at least a possibility of an overreaction. If one side appears to be deploying defenses which it thinks might work, and the other side thinks might work, then it is prudent to overreact, if you believe in deterrence. To make sure those defenses are not going to work, you improve your offensive capabilities, possibly including more than enough offenses to offset the defenses. We have seen that happen.

The Soviet air defenses are an example. As they improve their air defenses, we really do overreact. What are we doing now? We have built the B-1 bomber, and we are developing the stealth bomber. We are building cruise missiles in very large numbers. The net result will likely be that if the Strategic Air Command ever does attack the Soviet Union, that country would be more damaged by US aircraft than it would have been if it had not built defenses in the first place. By building defenses which can be relatively easily offset, you are almost issuing an invitation to the other side to overreact. Therefore, you put yourself in a worse position than you started with.

QUESTION: *Is there an alternative without MAD and without SDI?*

RATHJENS: I do not think so. We are stuck with nuclear weapons for as long as the knowledge of how to build them exists. We could disarm totally, and if we got into a crisis situation with the Soviet Union or any other power, presumably the first thing we would do is to start turning out nuclear weapons again. We have a situation akin to the mobilization in World War One. So we are stuck, I am afraid, with having those capabilities. The best we can hope for is to reduce the likelihood of conflict so that that becomes less relevant. I do not despair. My biggest trouble with SDI is frankly not the arms race or crisis instability, but the diversionary aspect of holding out the hope that through a technical fix we are going to get around the dual problem that is inherent in a world in which we have a difficult adversary and nuclear weapons. As long as we hold out that hope that the solution is to be found through technology, I think that we live in peril, and unnecessarily so, to some degree.

If we would focus more on political solutions, I suspect we could make some progress. People say that that is hopeless. How can you ever hope to achieve anything in terms of reducing the emphasis on conflict with the Soviet Union or the likelihood of preventing a crisis from getting out of hand? It does seem difficult, but there is a bit on the record which provides a little more hope in finding political solutions, rather than military solutions. I will just give three examples, all of which may not necessarily be apropos to the situation at hand, but which may have something to them.

Through the Camp David agreements, bitter antagonism between Israel and Egypt has been defused to some degree so that conflict between those two nations is probably less likely. This had nothing to do with any particular tinkering with arms or trying to solve the problem technically. It was a political solution.

Under Nixon and Kissinger, we changed our relations with China dramatically. I recall the day when Melvin Laird, the secretary of defense, was arguing that we needed more weapons to deal with the Chinese than we did to deal with the Soviet Union. Before then another secretary of defense, Robert McNamara, suggested that we had to deploy a defense against China. All that changed very dramatically almost overnight. When we decided that it was in our mutual interest to change the relationship, we did so. It is argued that we did that only because the Soviet Union was a mutual enemy. I grant that that may be a decisive and mitigating circumstance. But I do not despair of changing things with the Soviet Union.

I will give you one other example of the kind of agreement that had a small effect, but nevertheless a constructive one. In 1955 we negotiated a political agreement with the Soviet Union on the status of Austria. In effect, we declared it out of bounds and an area about which we were not going to fight. We decided to agree that Austria would not be a source of contention. I think we are better off, and I think the Soviets are better off. I know that the Austrians are better off.

Thus, it is not impossible to make political progress, but it will not get us away from deterrence. The fact is that we cannot get away from the existence of nuclear weapons. We can just hope that through political behavior we can get into a situation in which deterrence is less relevant, where, incidentally, defense would be much less relevant as well. Above all, I am troubled that in SDI we are holding out the hope of the ultimate technical fix when I do not believe that there is a technical fix to the problem of deterrence.

MENDELSOHN: MAD is not simply based on the existence of nuclear weapons. The capability to punish malevolence is going to remain the primary means of deterring or stabilizing, as it were, a relationship. It has nothing to do with irrationality, but it seems to work when dealing in adversarial relationships. Nuclear weapons and human nature seem to me to be immutable, at least for the foreseeable future.

One of the reasons that SDI is enormously popular is that most people give an intuitive answer to the question, "Isn't it a good idea to protect yourself?" The best answer to this question is really non-intuitive. The non-intuitive response is that defenses may not be good for you because they entail a series of political reactions that would put you in a worse situation than if you did not have defenses. That is a non-intuitive argument that 95 percent of the population either will not consider or will reject because defenses seem like the right thing to do. I would ask that you reexamine what seems right—that we defend ourselves—and consider it may really entail more problems than one thinks.

SUCHAN: It is basically counter-intuitive to say that to defend yourself is more threatening than to threaten somebody. It is true that some kinds of defenses could be precisely that threatening. You can imagine circumstances under which defenses could be extremely threatening. The considerations of the people who worked out the 1972 ABM Treaty were important ones. The kind of systems we were looking at in 1972 were exactly that kind. They made very little sense except as an adjunct to a first strike and for protecting yourself against a ragged retaliation. But what we are looking at now with SDI is qualitatively different. It has the possibility of changing the rules of the game to where you are no longer offensively dominated.

There have been periods in history where defenses have been stronger than offenses. The entire Middle Ages were an era of defense dominance. Basically, the whole time from about the American Civil War to World War One was a period of defense dominance. It is not at all impossible that defenses could again be dominant over offenses. It would probably be a much better world if that were true.

Basically, it is not necessary to equate deterrence and mutual vulnerability and to say that in order to have deterrence, you have to be vulnerable to the other guy's nuclear weapons. Neither side thought that, say, in the 1950s and early 1960s, when both sides maintained significant air defenses against a predominantly bomber threat from the other side. It is only for a period now of about twenty to thirty years that we have faced mutual vulnerability, at least temporarily, as an inevitable fact of life. Some people want to make it into a virtue and say that it is a good thing. This is not necessarily true at all. If SDI can find a way of making it no longer necessary to be vulnerable to this threat of offensive nuclear weapons, then we will all be better off in the end.

QUESTION: *Can we get away from our reliance on deterrence?*

BLECHMAN: It is indeed the case that the Soviet Union has a very different nuclear doctrine than the United States does. It was assumed by Americans, in the 1960s particularly, that the Soviets somehow were accepting our declared doctrine as their own. Part of Soviet doctrine that we, until recently, did not share, was that one should plan to have land-based missile forces capable of taking out the opponent's land-based forces in a first strike. They have developed those capabilities, or at least we believe thay have to a reasonable extent. They do not have the most accurate missiles, but they have enough megatonnage on their missiles. Also, we have few enough targets that they are likely to be able to take out our missiles.

However, the two sides' strategic forces are asymmetrical. They have two-thirds of their capabilities in land-based forces; we have about 20 percent of ours. We have emphasized bomber forces; they have virtually no bomber forces to speak of. Perhaps this will change in the next decade; they are developing a bomber capable of carrying air-launched cruise missiles. But to this point they have had no bomber force to speak of. They have a smaller and less capable force of strategic submarines in whose survivability they have limited confidence. It was only recently that we began to speak publicly of our intention to go after their strategic submarines, but we were developing the capabilities and exercising in ways of which they were well aware long before that. Thus, we have two asymmetrical forces, and each side can develop its doctrines as a derivative of its opportunities, of its military traditions, and of the resulting distribution of its nuclear capabilities.

Doctrines aside, the key question which has to be examined is whether it is feasible or imaginable that one can get away from deterrence; whether it is possible through the development and deployment of defenses, be they space-based directed-energy defenses, or extensive land-based defenses, using traditional technologies, which the Soviets have certainly been emphasizing; whether it is possible through those technical developments to get away from deterrence. The president himself first spoke of a great shift in doctrine toward a comprehensive defensive strategy. Now he says that we will continue to depend upon deterrence, that we will have effective deterrence with defenses.

In my view, that is not the smart way to go. It would lead inevitably to an unsafe, unstable, and accelerated arms competition. The smart way to go is to try to strengthen deterrence on the two sides and to avoid unnecessary expenditures, or a great diversion of resources, and the instabilities that would be associated with defenses. We should be trying to strengthen the ABM Treaty. We should be insisting that they dismantle or at least cease working on the radar they are building which violates the treaty. The administration has complained about the radar, but for years it has not done anything about it. We should be seeking additional amendments to the treaty, such that it could be abrogated only after, say, five years advance notice instead of only six months, which is now required. We should be seeking more precise definitions of the type of development work which would be permitted under the treaty. This way we could strengthen deterrence. We would have a more stable relationship and a greater chance of avoiding nuclear war.

MAARANEN: One thing we should remember is that we have placed a very heavy burden on our nuclear strategy by guaranteeing extended deterrence, that is, by extending the nuclear capabilities of the United States to support US alliances overseas. This is an onerous responsibility which places heavy demands on our forces. It is responsible for much of the development of the characteristics both of overall US strategic policy and limited nuclear options. It is also responsible for the specific military capabilities that we have believed necessary to build into our nuclear offensive forces in order to execute that policy.

Can you get away from deterrence? Of course you can! You can stop developing and deploying the military capabilities that allow you effectively to assure deterrence. That is all we have to do to get away from deterrence. Is that a good idea? Obviously not.

Are there things we can do to maintain deterrence? The Soviet Union has done a lot of things to deny us the capability of executing our nuclear strategy. The Soviets do things in a positive sense in order to execute their offensive strategy. There are a number of things they have done in a defensive sense to deny us the capability of successfully executing our offensive nuclear strategy. There are lots of developments which would be necessary in our nuclear force capabilities and in the development of the planning process for applying our forces in order to maintain the kind of deterrence that we have known in the past and which many people, by implication, would like to rely on in the future. I would like to hear them supporting those kinds of activities. I do not think that Congress is likely to support many of the requirements. I do not think many of us would like the kinds of requirements which would be necessary to maintain deterrence as we have known it in the past.

Therefore, let us look at strategic defenses. Maybe it might be a less threatening and much more satisfactory way to begin to move away from an offensive-only deterrent posture. It might be one which would give us the confidence that we can maintain our commitment to our allies, but also get away from the difficulties that are now arising in our posture which are very difficult to resolve any other way.

QUESTION: *How will SDI affect the use of tactical nuclear weapons on the battlefield? It is now believed that they are not likely to be used because of the danger of escalation to the level of an international exchange. But with SDI that could be avoided. How would our strategic doctrine change to deal with that?*

MAARANEN: If one looks at the kinds of weapons systems which are a threat to the European theater, which is the largest threat, there is a combination of theater ballistic missiles, many of which would be vulnerable to the kinds of technologies being developed within the SDI. There may be a significant capability for the SDI kinds of components, which are being developed for global deployment, to handle some of the more threatening European ballistic missile threats.

There is significant interest on the part of our allies as well. They increasingly recognize that there is a growing threat to the escalatory linkage, to the deterrent forces of the NATO alliance. This is coming from very accurate, shorter-range or medium-range ballistic missiles in the European theater. There is even the prospect down the road that the nonnuclear warheads on board those missiles would significantly threaten our nuclear capability to respond and escalate in the European theater. Thus, our allies are interested in supplementing anything that the United States is doing which would provide more effective defenses in the European theater against longer-range ballistic missiles.

BLECHMAN: If there were strategic defenses erected, there would be a somewhat greater likelihood of using tactical nuclear weapons. I think we have the priorities reversed. If anything, we should be emphasizing defense against shorter-range systems, particularly in the European theater so as to maintain a higher threshold for strategic exchanges, rather than making it more likely that nuclear weapons would be used in the European theater.

QUESTION: *How would SDI fit into an entire defense strategy? It would seem that SDI would have little impact on low-intensity conflicts. Given the high potential costs of SDI, would not a better strategy for defending the United States be a kind of Marshall Plan for assisting Third World countries, rather than SDI?*

BLECHMAN: There is a very serious problem with SDI: its effect on other budget priorities. I really object to the administration's unwillingness to make serious estimates of the cost of this program and to discuss its impact on other priorities. I am of the opinion that it would be much more in our interest to spend more on aid programs. However, in this context, SDI's primary effect will be on other military requirements. We are talking about a program which is going to be pushing harder and harder against other priorities within a fairly constant budget ceiling. The administration has given SDI top priority in its own budget. It is now at a relatively low level, but if SDI continues, it will be pushing against our ability to maintain credible forces in Europe or to develop capabilities to deal with contingencies in Asia, the Middle East, or wherever. This is not being debated.

The administration is saying that it is too early, and that we do not know exactly what SDI will be. That is really nonsense, and I think it is irresponsible not to face up to the facts and to debate SDI on its merits. Maybe we should be going into SDI; maybe the technology is promising. But at least let us debate and allow the public to look at it and say it would cost so-and-so much to put in a system over a twenty-year period and to state where the money will come from. Are we going to have an incremental tax for it? Is it going to come out of aid programs, social programs, or other military programs? These are very serious questions, and they deserve very serious answers.

MAARANEN: Many of the areas where we have seen developments unfavorable to the United States are not directly related to our association with the Soviet Union. But the real question is: Do you believe that the strategic balance is now satisfactory, on the basis of offensive nuclear weapons alone and is going to remain satisfactory indefinitely in the future on the basis of that strategy? Do you believe that it is an affordable and an acceptable strategic approach? I do not believe that it is. I think we are going to have to face some very hard choices, and it is going to be expensive. We do not know the exact costs, but they may be the kinds of costs which are inevitably down the road.

QUESTION: *How would the implementation of SDI affect modern-day conventional forces?*

RATHJENS: It depends on what your expectation is of SDI. If it is a really good defense against ballistic missiles that can deliver nuclear warheads, that is one thing. In that case, it would probably make the delivery of conventional ordnance by the same kinds of missiles even less attractive. You do not have to achieve the same levels of attrition. This is relative to the European context when we talk about defense against missiles which can carry nuclear, conventional, or even chemical warheads. In that case, it would make some difference whether SDI were developed.

If you have a perfect SDI, the extreme, what I would characterize as the "president's wish," which would protect against the delivery of all nuclear weapons, you would make the delivery of any kind of conventional ordnance impossible, except for very short ranges. A system which will intercept 100 percent of the nuclear weapons can surely intercept more easily 99 percent of conventional weapons. You can probably tolerate that 1 percent, so it changes the name of the game. Perfect SDI defense would put us back to the trench warfare of World War One, without any ability of either side to deliver conventional ordnance any way or anywhere.

BLECHMAN: There will be at least a short- and mid-term effect in that implementation of SDI is squeezing out of the budget those measures which are supporting conventional modernization. The administration gave SDI exemption from cuts last year. It instructed that the budget this year give it the highest priority. It requested virtually the same amount this year as was projected last year, despite last year's cuts. As a result, once the budget is brought down to a realistic figure, and if SDI continues to be protected, then the funds will have to be taken from other programs, probably from conventional modernization.

In the next decade, when SDI begins imposing serious budgetary demands, and if we begin engineering development, then the effect would be much more serious than we are seeing now. This will be true unless the country makes a decision that it really wants this thing and is willing to pay for it, and that it is willing to increase substantially the overall defense budget to do that. That might be a rational choice, but it is a choice which this country really should confront. It should decide yes or no, rather than to try to accommodate this new initiative within the existing budgetary confines and pushing out a lot of other things.

QUESTION: *Do you have any particular concern about SDI?*

MARQUET: The main point is that the objective is to deter the attack in the first place. Put yourself in the position of the potential attacker. Look at the system, however effective it may be, and try to predict the consequences of your attack. Any defense system, even one that is considerably less than perfect, is going to have an enormous leverage on your ability to make that calculation with any precision. Clearly it is going to have an enormous leverage on the assurance with which you might carry it out. Our objective is to prevent it in the first place, wherein everybody survives.

European Defense

QUESTION: *Does SDI make conventional war in Europe more possible? If so, after spending large sums of money on SDI, would we have to spend large sums of money in order to upgrade our conventional forces in Europe?*

PIERRE M. GALLOIS: This would be to accept defeat. The Soviet Union has almost three hundred million people under a single political power, whose political and social status are almost totally different from ours. To face conventional war under these conditions is to court defeat. We cannot accept such a proposition. Many people often say that we Europeans are three hundred million people who can resist. One cannot compare a geographical entity which is Europe with a theater of operations. As far as I know, there is not a defense of Africa, or a defense of the Americas, or a defense of Asia. This would make no sense, and it is the same for Europe.

Let us consider a specific example. Should the Russians attempt to take over a part of Norway in order to protect the Kola Peninsula, do you think the Turks, Greeks, Spanish, and Portugese are going to unite their forces in order to recover this territory? You know very well that the answer is no. And should the Russians try to take over a portion of Turkey, do you think that the Danes, the Norwegians, the Belgians, the Dutch, and the Greeks—the Greeks!—are going to fight for the Turks? You know very well that the answer is no.

What is European defense? The Russians are not stupid enough to attack all these fourteen nations. They would more probably select one victim and concentrate their force against him. The others would risk an attack and would do nothing. Hence the idea of a defense of Europe, which has been talked about so much in the past, makes no sense in reality. There remains the fact of France and Germany, which may represent a certain force in central Europe. But these two nations are very different. In case of war, the

Germans risk a conventional occupation of their territory, while the risk to France (and to Germany) is atomic annihilation. Hence also, such an association of these two nations facing the power of Russia is very difficult to consider. And I am sorry to say that accepting conventional warfare in Europe would be the loss of Europe.

MAARANEN: Soviet military strategy in Europe, if the Soviet Union should ever go to war, calls for a rapid and effective offensive across Western Europe from their positions in the Warsaw Pact. In Soviet eyes, the success of that offensive would depend on the prompt destruction of some two hundred to six hundred critical military targets in Western Europe. In the past we have assumed that those attacks, following the initiation of local hostilities, would be made by Soviet nuclear forces, both air-carried weapons and ballistic missile forces. If you look at the nuclear forces of the Soviet Union deployed in the Western European theater, it looks like their size and design is precisely to do that job. If they could suppress the nuclear retaliatory capabilities of NATO, then potentially the Soviet Union would be able to effect the rest of its offensive strategy and possibly preclude a nuclear response on the part of the NATO alliance.

The critical element is the ability confidently to destroy two hundred to six hundred targets. The Soviet ballistic missile threat to those targets is becoming much more critical than it has been in the past. The deployment of the intermediate-range SS-20 missile, the short-range SS-21, SS-22, and SS-23 missiles and the potential deployment of nonnuclear warheads on some of those systems may provide sufficient accuracy to destroy those forces even without employing nuclear forces. That is a matter of very great concern to NATO planners and NATO allies. In the early stages of the deployment of a theater ballistic missile defense, if you are defending precisely those critical targets, that would reduce the Soviets' high confidence in a ballistic missile attack on them, therefore making the whole Soviet offensive highly risky and improbable of success. That looks like the way we would want to begin to deploy theater ballistic missile defense in the European theater. Way down the road, if you get to a point where you have an extremely effective defense against the full spectrum of nuclear threats, then it may be the time to address the question of whether the conventional force capabilities of the alliance would have to be increased. In the near term, antiballistic missiles may pose some real opportunities for limiting some of the most critical weaknesses in NATO strategy.

QUESTION: *If the president's program of strategic defense succeeds, would France lose anything in terms of its own independently controlled nuclear forces?*

GALLOIS: This is a difficult question. I think that the statement of President Reagan may be valid for America, possibly also for Russia, insofar as making these weapons obsolete is concerned. Also it is possible that American strategy could change. But not for us. If there were a Russian SDI working one day, we would be obliged—we are already obliged—to increase our inventory, try saturation, develop short-range artillery weapons, or change our strategy. We would not take as hostages targets far inside of Russia any more, but try to plan a strategy to take as hostages the forces that the enemy needs and is moving toward the West. Then we would have to face a difficult problem: to intercept and to destroy weapons which have a very short, depressed trajectory, five to ten minutes' flight time, and multiple warheads. We would have to penetrate their defense. This is the only strategy which we can rely on. The difference of force and the difference of determination between a democracy and an autocracy are such that we have to rely on deterrence and not to rely on any form of combat, which would be lost in advance.

I do not think that anybody really believes that the Russians may decide one day to launch the very risky operation of starting war in Europe against Western Europe without having prepared all the conditions of their victory. If they met any resistence, they would take all available means to overcome such resistence, including atomic weapons if necessary. I do not see their advance being checked somewhere prompting them to tell their troops to get back to their barracks, sending a telegram of excuse to the Elysee and to the White House. This is not the picture we have of Russian determination, should they start a war.

The Russians have two methods of expansion: When they push their land frontier to the west, the south, or the east, they use their Red Army, and in that case, what they gain is not reversible. On the other hand, when they try to project their influence far from their borders, the Red Army is not employed as organic elements. They use only "volunteers" from Moscow, from Czechoslovakia, or Bulgaria. In that case, they can tolerate withdrawal and accept failure. They did it in Somalia; they did it in the Sudan; they did it in Egypt. But in Europe, because we are where we are, without being able to sever our geographic connection with them, if they start something, it will be to achieve victory at any cost. Therefore, we must be prepared to deter them at any cost.

QUESTION: *What is the impact of SDI on conventional warfare and the defense of Western Europe?*

MARQUET: Let me talk specifically about military applications. We are already beginning to see some very, very interesting opportunities for applying some of the technologies that are coming out of SDI in the tactical ballistic missile arena. Particularly in Western Europe, where the tactical ballistic missile is a very serious problem, we have to ask ourselves the question: if there are no nuclear weapons, if there are no intercontinental ballistic missiles, have you left the world safe for conventional warfare? And that is a very important problem. I think that the answer to that is that if we have the technical capability to indeed make nuclear ballistic missiles impotent and obsolete, think of how that surgical instrument to which General Gallois has alluded will perform in terms of tactical ballistic missiles, tanks, aircraft, and other conventional munitions. From the standpoint of the defenders again, it will have a very major impact in protecting Western Europe.

QUESTION: *In the short term, any strategic defense will be an imperfect defense. How would a mutual American and Soviet hard-target defense affect US flexible response strategies, particularly the use of American strategic nuclear capacity in the defense of NATO?*

JACQUELYN K. DAVIS: By virtue of the enhanced survivability of its nuclear forces, a nation theoretically would be able to threaten in a more credible fashion the use of such weapons in defense of vital interests. That assumption is inherent in NATO strategy of flexible response. Of course, any discussion of strategic choices of specific weapons employment, especially with regard to the escalatory chain and the US willingness to escalate, is very scenario-dependent. This is true in the sense that it is assumed that if a crisis were to escalate into a wartime engagement between the United States and the Soviet Union, it would be global in nature. It is more than likely that a war-initiating crisis would not start in Europe, which is probably the most stable theater in terms of East-West balance of forces.

The perception in Europe is that a protected United States or an enhanced deterrence credibility of US strategic forces would underpin the willingness of the United States to come to the support of NATO Europe. Of course, the Europeans have always assumed that our troops which are deployed forward would be engaged immediately in any kind of crisis response. The issue is the extent to which the United States would go all out in the defense of Europe if it were not initially involved, or if its homeland were not threatened directly. The president's conceptualization of the Strategic Defense Initiative is that a protected deterrent makes more credible the United States's ability to intervene. It therefore shores up the deterrent potential of the alliance because of the perceived strengthened position of the United States.

The Soviet Union and Strategic Defense

QUESTION (posed by Tom Diaz of the *Washington Times*): *The Soviet Union has built up an extremely accurate land-based, large missile force of which the SS-18 is the epitome, which could be seen as a first-strike capabil-*

ity. Two, the Soviets have super-hardened those silos and their command and control functions in which the privileged part of the society would be protected from all but the most accurate and capable warheads from us. Three, they most assuredly have the world's only deployed missile defense system around Moscow, and there is much debate right now as to whether they are preparing either to break out of the ABM Treaty or have some nation-wide system of defense. They are also engaging in an extensive strategic defense program of their own which includes directed-energy research which is ahead of ours. Taking all that together, the question is: is it not reasonable for people on our side to conclude that the Soviet Union is preparing to have a first-strike capability against the United States? Would it not also have both a defense capability which, assuming our response would be to be ready, could protect that part of the society which they value most, and some civil defense capability which could take care of the rest of the people, within acceptable limits, given their history of losses in the Second World War? In this case, the question is: if one believes that that is what is happening, then does it make any sense for us, intuitively or counter-intuitively, to sit on our hands?

MENDELSOHN: What sense can be made out of the complex of activities that the Soviets have been involved in? None of us knows what Soviet intentions are, and the complex of their activities together may or may not be analysed in any reasonable way. The questions are: What are the real capabilities with which we are dealing? What military threats do these really represent? The fact that the Soviets have a large, accurate ICBM system targetted on our smaller and also accurate ICBM system is an interesting asymmetry of forces. We can make somewhat less out of it in terms of what impact it would have on the retaliatory capability of the United States. Were we to lose virtually every one of our ICBMs, fully two-thirds or more of our strategic force would still be viable. The Soviets know that. I think those kinds of scenarios are not credible.

The hundred ABM interceptors around Moscow are a little bit like nine thousand antiaircraft or air-defense weapons that are available to the Soviets. We do not take them very seriously. We have a very elaborate program of our own to ensure penetration of any kind of ABM system. I would be very interested to see what the discount factor in our own targetting is or what we think the impact on our attack would be of those hundred ABMs. I suspect we count them as inflicting very low, if any, attrition on our nuclear warheads.

Another indication of how little we deal with the ABMs around Moscow is the British program. The British have spent a lot of money and time designing penetration aids and actually off-loading RVs in order to put on penetration aids. They feel, as we do, quite confident of being able to penetrate or neutralize any Soviet defenses.

The Soviets are, as we should be, involved in a very vigorous and intensive research program. There is no question about that. I do not think that is going to change unless we agree between the sides that we want to stop that. That is not likely, and I do not think it is advisable. I do not know what the Soviets' intentions are, but I know that some of their activities actually amount to less than ours.

QUESTION: *Is it true that the Soviet Union has developed strategic defenses? If so, how effective are they?*

SUCHAN: The fact of the matter is that the Soviet Union now does have strategic defenses deployed to a much greater extent than the United States does. It has a much more advanced air defense than the United States does, and it has an existing ABM system, which the United States does not. It is also true that there are technological advantages which the United States has. Over the long term, these could be significant, but until you actually get down to bending metal and getting a weapons system out there, how well you can build a computer does not matter.

MENDELSOHN: Although the Soviet Union has an ABM system deployed, it is generally considered to consist of technologies which the United States developed or began to develop in the 1960s and abandoned in the 1970s. It is true that the Soviets have done some hard work and have achieved our level of a decade ago.

In my opinion, our Air Force believes that it can penetrate Soviet defenses, to a very high degree. It is not a terribly sophisticated plan. You simply blast a hole through them. That is what all the defense-suppression weapons are for. So let us try to put all this Soviet defense issue into some kind of context.

The president and the secretary of defense speak at a very general political level. They are, in effect, trying to mobilize support. One has to look at their views critically and understand that they are in a context; they are not misspeaking, but they are speaking for a purpose.

SERGEI KISLYAK: I would like to take up the notion of a strategic defense and an inner defense. Yes, we have a missile defense system around Moscow, recognized in an agreement with the United States which permits the US to build its own analogous system. I remind you that this American system was not dismantled. It is not now operational, but it has been conserved. If it were needed again, it could be reactivated.

The basic difference between our activities in the ABM field and the SDI "research" is that ours is limited by the ABM Treaty regime, for example for the protection of the capital of the Soviet Union. What is meant by SDI is the creation of a shield to cover the territory of the United States, the very idea of which is opposed to the ABM Treaty.

As far as the air defense of the Soviet Union is concerned, yes, we have an air defense. It was a reaction to what we faced in the way of a threat from the outside. Take a look at the political map of the world and you will see that the Soviet Union is surrounded by American bases. They are in Europe, in the Indian Ocean, in Japan, and in South Korea. Everywhere there you will find forward-based weapons of the United States. There is no secret that all of them are targetted against the Soviet Union. The major portion of them are aircraft, including American strategic aviation, which is by far bigger than ours. The United States has consistently placed more emphasis on strategic aviation than we do. Of course we have to take countermeasures, which we did. We have created an air defense system. You might not even need such an air defense system as we do, because we do not have foreign-based aircraft around your territory. We do not pose as much of a threat to your territory as you do to ours. But this air defense system has nothing to do with the SDI concept. This would be mixing up very different things.

QUESTION: *Must we have some kind of strategic defense just because the Russians have one?*

SUCHAN: I do not argue that because the Soviet Union has a strategic defense program, we must also have it. It is certainly true that some kind of ABM effort is required on the part of the United States as a safeguard against a possible Soviet abrogation of the ABM Treaty. Given the Krasnoyarsk radar and other activities, that is certainly legitimate. The Soviet Union itself does understand well the importance of strategic defense, and it does whatever is possible, within certain restraints, including the ABM Treaty.

The Soviets point out that they need air defenses because the Soviet Union confronts a large number of American aircraft. But if one believes that all one has to do to maintain deterrence is to rely on one's own offensive systems, who should care? The Soviet Union has nine thousand nuclear warheads on its strategic ballistic missiles right now. What do a couple of hundred bombers matter in that sort of environment? The Soviet Union has plenty to deter the United States from using any of those aircraft.

With reference to the Cuban Missile Crisis, one of the few nice things that one can say about the postwar period is that both sides have behaved much more cautiously in their relationship toward each other than have two dominant powers ever before in history. Nuclear weapons are probably the cause for that, rather than unilateral restraint. We do know that crises can happen. In my view, the 1973 October War was a lot scarier than the Cuban Missile Crisis. The fact that nuclear weapons themselves were not directly involved is not particularly relevant. We came very close to an armed conflict with the Soviet Union, in a strategic environment where the Soviet

Union had achieved parity on a strategic nuclear level. I would like to think that such a crisis is unlikely to happen. I have no empirical basis for hope that it never will happen.

QUESTION: *Is the Soviet Union interested in developing nuclear systems for actual use in warfare, rather than for deterrence? Is the Soviet Union committed to stabilizing the military balance?*

MENDELSOHN: I think that the Soviet Union appreciates as the United States does that it is not in its interest to destabilize the balance in a way which could precipitate any kind of an exchange; it is not in their interest to move in that direction. It is in their interest, as it is in ours, to develop as capable a set of systems as they believe they can, or that they can deploy. If you look at both the posture of the Soviet systems, their state of readiness, and their capabilities, the likelihood that the Soviet Union intends to attack the United States out of the blue just is not there. There are scenarios which are very popular about a first strike against American nuclear systems and the possibility of carrying out major destruction scenarios against the United States. Again, if we look very carefully at the capabilities of Soviet systems and to the real posture of its forces, these scenarios do not look terribly credible.

Recent studies by the Congressional Research Service and others indicate that even in the case of an all-out attack against American fixed silos—meaning that the Soviets would use every one of their land-based ICBMs—the US would still have about 60 percent of its available nuclear forces. That is a fairly reasonable assumption of what the Soviets might be able to do. I suspect that the Soviets think that they can do even less than that estimate. I question that the Soviets really do believe that they can pull off a disarming first strike.

QUESTION: *Does the Soviet Union believe in MAD?*

RATHJENS: First of all, it is a mistake to characterize MAD as a matter of doctrine for anybody. It is a sad statement of our condition. It is a consequence of the fact that we and the Soviet Union deploy a large number of nuclear weapons and have the means of delivery at our disposal. The question is: "Why is the Soviet Union interested in civil defense? Doesn't it believe in MAD?"

The Soviet Union is a country that has a greater defense tradition than we do, and I use defense here in the particular sense. It is a country which has been invaded many times; we have not. Therefore, it is committed to defense, to civil defense, to air defense. My judgment is that the Soviets fear nuclear war, like most of us do. But it is their view that it might happen, and if it does, they ought to do what they can to reduce the consequences. This is a perfectly reasonable position. It is the intuitive reaction. It is counter-intuitive to suggest that trying to do something about defense is counter-productive. In this respect, the Soviets have been less willing to accept these counter-intuitive arguments than we have. Thus, they have historically made greater efforts in defense almost every year than we have.

QUESTION: *In the high-technology world we live in today, is it not a moot point how far we might be ahead of the Soviet Union in terms of technology?*

MENDELSOHN: In terms of technology, we are ahead of the Soviet Union in practically every key technology. But if the system does not work, it does not matter whether you have a Mercedes or a Pontiac. Second, if it does work to some degree, because we do have superior technology, it can be defeated with a lower degree of technology.

ROBERT L. WALQUIST: The Russians have continually surprised the United States in terms of their technological capability. As a result, we need to be very careful that we do not underestimate that capability. On the other side of the coin, it is very interesting that the Russians have always been awed by what the US technology can achieve. In general, it appears that our technology is ahead of the Russian technology. That, on the one hand, can give you some comfort; we can field systems which are more sophisticated than the Russians can field.

But you should not sleep well with that thought in your mind for a couple of reasons: One, if you do not have the technology available to you, you try to solve the problem in some other fashion. You take more of a brute force approach. One such approach is that you try to overwhelm a problem with quantity rather than quality. Two, the United States takes an awful lot longer from the time that we come up with an idea in the laboratory and start with the initial research and technology to the time that we put a system out in the field. The differential between what it takes us to do that and the best information we have about what it takes the Russians to do that is that we take about an extra three to five years.

As best we can determine, our technology lead on the Russians is approximately three to five years. They continually pick up our technology. There are various ways to get it; one is to capture equipment when we do get it out in the field. Thus, the problem with our maintaining a technological lead is not to get ahead of the Russians but to stay even with them because of their shorter time cycle in fielding equipment. In several areas they have achieved technological advances that we have not. They are also working on many of the types of systems that we are working on. They are working on directed-energy weapons, just like we are in support of SDI. They are building and have deployed large phased-array radars. They have deployed complex satellites in space.

The key aspect is two-fold. We have really disrupted the Soviet planning cycle when we introduced SDI. They like to lay out five-year plans, ten-year plans. They tend to have a structured approach to how they run their economy, how they run their military. They were on a particular course of action based upon what they saw us doing and upon what they needed to do in the military area. They were trying to keep their costs under control and to start bolstering their civilian economy. Then the US came in and said to the Soviets that we are going to do something different in the strategic area; we are going to look at developing a strategic defense system which may make all your current planning obsolete. This will cause you to do differently things with your current ballistic missiles because they will be very vulnerable to SDI. It will also cause you to spend money on developing a defensive system like the US is looking at developing. This put quite a bit of turmoil into the Soviet thought process. It also meant that if we continue to go ahead, then they feel—and they apparently do feel—that they must respond to our going ahead. It is going to cause a big shift in their planning, in what they planned to put into their military rather than their home civilian market area.

The second thing is that the Russians have always been concerned with the United States's ability to do what it says it is going to do. One of the things that SDI has done is to focus on a lot of little areas of technology which were going on in the United States somewhat undirected. Each was very useful and was moving forward, but each sort of doing its own thing. When these were combined under SDI, it gave a focus to all that research. There has been a terrific synergistic interaction in that research. The country has made about ten years progress in technology in about three years because of a lot of things which have come out of the woodwork; little areas of technology which were related to other areas and could bolster the development and growth of those other areas. As a result, the Russians are very concerned that when we put our minds to an SDI system, we are probably going to be able to do it.

MARQUET: One has to be very careful to differentiate what the Russians say and what they do. The Russians have said that defenses are destabilizing, and they have made a big case of this. They clearly do not believe that; the Russians have never believed that. They have never bought the idea of mutually assured destruction. They have proceeded to deploy the only operational ballistic missile defense system. This is something which is within the constraints of the treaty, but it does exist, and they are continually upgrading it. They have continued to pursue single-mindedly their research in advanced technologies to continually upgrade the performance of that system. They have also made substantial advances in some of the most exotic technologies, and we have learned some things from them. Thus, they have never had to be convinced of the value of defense. What they find disturbing is that for the first time since 1972, we in the United States are beginning to see that there is perhaps some value in defense as well. Of course, that is clearly upsetting their apple cart. Defenses are not destabilizing, especially if you are the only one who has it.

Arms Control

QUESTION: *What would be your idea of compromise with the Soviet Union on strategic defenses?*

SUCHAN: I do not think we are talking about something where we must consider where the compromise can be drawn. SDI is a program to discover whether something is possible. If we end up determining that effective defenses are not going to be feasible, then there is not going to be any disagreement among the people involved; we will trash the idea. On the other hand, if we determine that effective defenses are feasible and that something can be deployed in a survivable, cost-effective manner and at a price that the United States decides it can afford, then we are not going to be talking about where it is going to be compromised. There will be different policy choices as to how to proceed, but it is not going to be a matter of splitting the difference.

There seems to be a temptation on the part of many, particularly in the political field, to say: "Why don't we split the difference between no defense and effective defense and come up with something like a terminal defense of our ICBM silos?" My own preference would be to do nothing rather than to get involved in the terminal defense field. It is something that has proven historically not to be cost-effective at the margin. Offense unquestionably has the long-term advantage against this kind of defense. I cannot imagine that America would break an arms control treaty and pay a huge financial cost in order to protect a bunch of ICBM silos, rather than Chicago. This is a question which is incredibly premature.

RATHJENS: To my mind, there is no room for compromise here in terms of deployment. I agree that we have to do some R&D. If we are interested in the defense of selected targets, it is not a question of compromise. Then we will want to take a hard look at whether defense is the way we want to go. We ought not to decide something like this without looking at alternative ways of accomplishing similar objectives. I cannot imagine addressing the question of compromise without addressing those other questions of what we want to do and what are the other ways of getting this done? If it turns out that we really think we need those limited defenses of military assets or other targets, and if we can defend them effectively and meet the Nitze criteria—is it cost-effective (can the other side offset what we are doing at lesser cost) and is the defense vulnerable?—then I do not see any problem with going ahead. But I think those are very tough criteria to meet. We should not spend any money beyond R&D.

MENDELSOHN: There is no prospect of an arms control agreement that will lead to a deployment of SDI or which will restrain offensive weapons systems while SDI is being deployed. There may be some kind of interim INF agreement, but the likelihood that there will be a strategic arms control agreement while we are beginning to deploy defensive systems is just not in the cards.

QUESTION: *Given the fact that the American administration has dismally failed to reach an arms control agreement and the fact that the Soviet Union is so aggressively opposed to SDI, how would the US go about getting a cooperative agreement during the interim with the Soviet Union while the US tries to get an SDI up in space?*

SUCHAN: Ambassador Nitze has said that a cooperative transition would, at best, be a tricky proposition. I think that it is unlikely that we can achieve Soviet agreement to a cooperative transition unless the Soviet Union is convinced that, in the absence of cooperation, the US would proceed with the deployment of effective defenses and that under a non-cooperative scenario, the Soviet Union would end up comparatively worse off. If it came to that conclusion, the Soviet Union would have every incentive to cooperate with the United States, both on the offensive and the defensive sides of a cooperative transition.

When we talk about defenses, it is not just a question about our putting in defenses and doing it according to some sort of schedule. We are also talking about offense elements in this regime as well. It would mean that, for example, if we could reach a START agreement to go down to the level of forty-five hundred strategic ballistic missile warheads and that there would be additional reductions following that at equal levels, we would hopefully have more stability.

The Soviet Union does not have a veto over whether the United States goes to defenses. We want to cooperate with the Soviet Union, but a decision on its part not to cooperate does not mean that effective defenses will not happen. What puzzles us right now is that the Soviet Union realizes that defenses may happen whether it wants to cooperate or not. It seems to have come to the conclusion that the transition period should begin with high and unstable levels of offensive nuclear weapons, rather than at relatively low and more stable levels, which would apply if the arms control regime were already in place. The United States obviously prefers the latter scenario. That is the way we would hope to begin this transition.

QUESTION: *Is the Soviet Union interested in working together with the United States to establish a transition to a cooperative defense-dominant regime?*

KISLYAK: The argument that no matter what the Russians do at Geneva, the Americans will proceed with SDI reminds me of one person telling another: "If you want to talk to me, shut up!" This is not a way of dealing with each other, especially when you are talking about nuclear weapons. I think that if we want to negotiate seriously, we have to take into consideration the security needs and the security perceptions of each of the countries. The approach that whatever is done in Geneva, we will continue, is not the way to deal with the problems.

QUESTION: *Should the US share SDI technical information with the Soviet Union? Are there dangers in sharing?*

MENDELSOHN: One of the things you do not want to do is to give blueprints or details of your operating systems to an adversary. The reason is not so much because they might replicate those systems and turn them against you, but more likely because they would obviously seek to find whatever weaknesses or chinks there might be in the designs or ways to defeat or subvert the design in some manner. The statement that we would share the technology with the Soviets is an outrageous idea.

The political genesis of the offer to share technology is worth recollecting. The problem was posed: what do the Soviets do the day before the system goes into space? They feel threatened, and they are prone to launch because they feel threatened. Then the idea came up that a way to keep them from feeling threatened is to share this technology with them. People were beginning to try to address some of the issues: how does the other side react in the face of some kind of a defense with any kind of effectiveness?

MARQUET: We are an open society. We are sharing with the Soviets our debates and our opinions about defensive systems and our general state of technology. But in a more realistic sense, the way things really do work in an open society is that there is a massive effort to infiltrate and to get information from our society, both technical and otherwise. Truckloads of legally and illegally obtained documents are shipped back to the Soviet Union constantly. So the fact of the matter is that, whether we intend or not, we will share this technology with the Soviets. In designing the systems, any particular flaws that may be determined by trying to understand details of technologies (such Achilles heels do exist), we have to assume that the information will be available to the Soviets within a maximum of five to ten years after it is deployed anyway.

QUESTION: *How should the United States respond to alleged violations of the ABM Treaty?*

BLECHMAN: The credibility of our complaints against the Krasnoyarsk radar is cheapened by comparing it with a long series of other allegations, many of which may or may not be the case. The radar is a very clear-cut case. Most of the other things are much more ambiguous. By grouping them together, one cheapens the seriousness. But most importantly, although it has raised this question and complained about it, the administration has been willing to continue negotiations as if it were no serious problem. The president should have sent a very private emissary to the Soviets and made it clear that the radar is different; that it is overt and undeniable; that it is an indisputable violation of a very serious agreement. Of course, we all know why the administration has behaved in this way; it has to do with politics. Still, by publicizing these charges, and then by not taking any effective action, by not demonstrating that we are very serious about it, we only encourage this kind of violation in the future and lessen the possibility of a real arms-control regime.

SUCHAN: Regarding the ABM Treaty and the illegal radar in Siberia, we have to make a very big thing of this, and they are willing to discuss it with us. One of my Soviet colleagues in Geneva said that he was sick and tired of hearing about Krasnoyarsk, and the world was too. He complained that every child in Africa had heard about Krasnoyarsk. Short of putting diplomatic pressure on the Soviet Union, there is not an awful lot we can do directly about Krasnoyarsk.

PART TWO

19

Address on the Strategic Defense Initiative

Ronald Reagan

MY predecessors in the Oval Office have appeared before you on other occasions to describe the threat posed by Soviet power and have proposed steps to address that threat. But since the advent of nuclear weapons, those steps have been increasingly directed toward deterrence of aggression through the promise of retaliation.

This approach to stability through offensive threat has worked. We and our allies have succeeded in preventing nuclear war for more than three decades. In recent months, however, my advisers, including in particular the Joint Chiefs of Staff, have underscored the necessity to break out of a future that relies solely on offensive retaliation for our security.

Over the course of these discussions, I've become more and more deeply convinced that the human spirit must be capable of rising above dealing with other nations and human beings by threatening their existence. Feeling this way, I believe we must thoroughly examine every opportunity for reducing tensions and for introducing greater stability into the strategic calculus on both sides.

One of the most important contributions we can make is, of course, to lower the level of all arms, and particularly nuclear arms. We're engaged right now in several negotiations with the Soviet Union to bring about a mutual reduction of weapons. . . . Let me just say, I'm totally committed to this course.

If the Soviet Union will join with us in our effort to achieve major arms reductions we will have succeeded in stabilizing the nuclear balance. Nevertheless, it will still be necessary to rely on the specter of retaliation, on mutual threat. And that's a sad commentary on the human condition. Wouldn't it be better to save lives than to avenge them? Are we not capable

---

This is an extract of President Reagan's address over nationwide radio and television on March 23, 1983.

of demonstrating our peaceful intentions by applying all our abilities and our ingenuity to achieving a truly lasting stability? I think we are. Indeed we must.

After careful consultation with my advisers, including the Joint Chiefs of Staff, I believe there is a way. Let me share with you a vision of the future which offers hope. It is that we embark on a program to counter the awesome Soviet missile threat with measures that are defensive. Let us turn to the very strengths in technology that spawned our great industrial base and that have given us the quality of life we enjoy today.

What if free people could live secure in the knowledge that their security did not rest upon the threat of instant US retaliation to deter a Soviet attack, that we could intercept and destroy strategic ballistic missiles before they reached our own soil or that of our allies?

I know this is a formidable, technical task, one that may not be accomplished before the end of the century. Yet, current technology has attained a level of sophistication where it's reasonable for us to begin this effort. It will take years, probably decades of efforts on many fronts. There will be failures and setbacks, just as there will be successes and breakthroughs. And as we proceed, we must remain constant in preserving the nuclear deterrent and maintaining a solid capability for flexible response. But isn't it worth every investment necessary to free the world from the threat of nuclear war? We know it is.

In the meantime, we will continue to pursue real reductions in nuclear arms, negotiating from a position of strength that can be ensured only by modernizing our strategic forces. At the same time, we must take steps to reduce the risk of a conventional military conflict escalating to nuclear war by improving our non-nuclear capabilities.

America does possess—now—the technologies to attain very significant improvements in the effectiveness of our conventional, non-nuclear forces. Proceeding boldly with these new technologies, we can significantly reduce any incentive that the Soviet Union may have to threaten attack against the United States or its allies.

As we pursue our goal of defensive technologies, we recognize that our allies rely upon our strategic offensive power to deter attacks against them. Their vital interests and ours are inextricably linked. Their safety and ours are one. And no change in technology can or will alter that reality. We must and shall continue to honor our commitments.

I clearly recognize that defensive systems have limitations and raise certain problems and ambiguities. If paired with offensive systems, they can be viewed as fostering an aggressive policy, and no one wants that. But with these considerations firmly in mind, I call upon the scientific community in our country, those who gave us nuclear weapons, to turn their great talents now to the cause of mankind and world peace, to give us the means of rendering these nuclear weapons impotent and obsolete.

Tonight, consistent with our obligations of the ABM Treaty and recognizing the need for closer consultation with our allies, I'm taking an important first step. I am directing a comprehensive and intensive effort to define a long-term research and development program to begin to achieve our ultimate goal of eliminating the threat posed by strategic nuclear missiles. This could pave the way for arms control measures to eliminate the weapons themselves. We seek neither military superiority nor political advantage. Our only purpose—one all people share—is to search for ways to reduce the danger of nuclear war.

My fellow Americans, tonight we're launching an effort which holds the promise of changing the course of human history. There will be risks, and results take time. But I believe we can do it. As we cross this threshold, I ask for your prayers and your support.

20

SDI, Arms Control, and Stability: Toward a New Synthesis

Paul H. Nitze

THE primary security objective of the United States is to reduce the risk of war while preserving our liberty and democratic political system. Over the past twenty-five years, the United States has pursued this objective through two related means. We have sought to deter war by maintaining a force structure adequate to convince potential adversaries that the risks and costs of aggression would far outweigh any possible gains. Simultaneously, we have sought to limit the nature and extent of the threat to the United States and to stabilize the strategic relationship with our principal adversary, the Soviet Union, through arms control agreements.

The United States is now engaged in research to find out if new technologies could provide a more stable basis to deter war in the future by a shift to a greater reliance on strategic defenses. Arms control could also play an important role in designing a more stable strategic regime in the future. I propose to examine the relationship among SDI, arms control, and stability. I hope to show that our SDI research and arms control policies, as currently defined, provide a cohesive and firm basis for enhancing strategic stability in the future and ultimately for reducing the risk of war.

Arms Control and Stability

Two important corollaries to the objective of reducing the risk of war are the objectives of assuring overall functional equality between the capabilities of the two sides and of assuring crisis stability. Crisis stability implies a

---

Mr. Nitze delivered this address to the *Time* Magazine Conference on SDI, Washington, D.C., June 3, 1986. It was published by the US Department of State, Bureau of Public Affairs, Current Policy No. 845 (Washington, D.C.: GPO, June 1986).

situation in which no nation has an incentive to execute a first strike in a serious crisis or, in peacetime, to provoke a crisis that might lead to a military confrontation. This situation obtains if no significant advantage can be achieved by initiating conflict. Equivalently, crisis stability also implies that a potential aggressor perceives that he could end up in no better a military position after expending a major portion of his forces in executing the attack and then absorbing a retaliation than would the defender after absorbing the attack and retaliating. These two goals—assuring overall functional equality and crisis stability—are closely interrelated. The United States cannot tolerate either significant inequality or substantial crisis instability.

Trends in the strategic balance over the past fifteen years lend new meaning and importance to these classical goals. The growth of Soviet capability to destroy hardened targets—such as ICBM silos in an initial strike, with their large, land-based, MIRVed ballistic missiles—has created a serious force structure asymmetry and a growing danger of instability in a crisis. Soviet strategic defense activities, coupled with a military doctrine that stresses the importance of offensive and defensive force interactions to achieve Soviet aims in any conflict, have likewise been threatening.

Both the United States and the Soviet Union recognize that it is the balance between the offense-defense mixes of both sides that determines the strategic nuclear relationship. The Soviet Union must realize that a successful "creepout" or "breakout" in its strategic defense capabilities, or conversely, unilateral restraint by the United States in this area, would further shift the strategic nuclear balance in its favor and potentially undermine the value to the United States and its allies of US deterrent forces. Through its ongoing overt and covert defense activities and its arms control policies, the Soviet Union has been attempting to foster such a shift. Currently, in the arms control arena, the Soviet Union seeks to protect the gains that it has achieved in the strategic nuclear balance by limiting and delaying US defense programs, especially SDI. This focus on SDI reflects Soviet concern over the fact that they are no longer alone in their exploration of the defensive potential of advanced technologies and over the prospect of having to divert resources from proven ballistic missile programs to high-technology programs in fields where we are likely to have a competitive advantage.

US arms control efforts are oriented toward achieving strategically significant and stabilizing reductions. For example, we seek to lower the ratio of accurate warheads to strategic aim-points and reduce a potential attacker's confidence in his ability to eliminate effective retaliation. I should note that while the role of arms control in enhancing US security and in bringing about a more stable strategic relationship is important, it is secondary to what we are able and willing to do for ourselves. US strategic modernization programs provide the necessary foundation on which our deterrence and arms control policies must rest. SDI should be understood in the context of the goals of our modernization and arms control policies and the dangers inherent in the future possibility of having deterrent forces inadequate to respond to, and thus deter, the threat.

We should make no mistake about the fact that Soviet offensive and defensive capabilities pose real threats to the security of the West. Our work in SDI is, in part, a reaction to the unabated growth of this threat, especially during the last fifteen years. Through SDI, we seek both new capabilities and a new approach to rectify the deteriorating strategic balance.

The ABM Treaty and the Origins of SDI

The president's March 1983 speech expressed his strongly held belief that we should reexamine the basis of our deterrent posture to see if we could deter aggression through a greater reliance on defense rather than relying so heavily on the threat of devastating nuclear retaliation. This belief reflects both our disappointment in the deterioration of the strategic balance since the signing of the SALT I agreements and our hope that new defensive technologies can mitigate adverse developments in the area of strategic offensive weaponry.

The United States in the early 1970s had proceeded from the assumption that the strict limitation of defenses in the ABM Treaty would provide the basis for significant reductions in offensive weaponry. The theory was simple: if both sides had survivable retaliatory nuclear forces at about the same level of capability and both sides were otherwise effectively defenseless against the nuclear capability of the other, then neither side would have an incentive to strike first, regardless of the circumstances. Therefore, significant reductions to equal levels of capability, tailored so as to enhance security, would improve the security of both sides.

However, the Soviets showed little readiness during the SALT negotiations to agree to measures which would result in meaningful limits or cuts in offensive nuclear forces. Within the framework of the SALT I interim agreement and SALT II, the Soviets deployed large numbers of MIRVed ballistic missiles of sufficient throw-weight and accuracy to pose an evident threat to the survivability of the entire land-based portion of US retaliatory forces. This violated a basic premise of the SALT process. The growth in Soviet nuclear capabilities, in general, and in the asymmetry in counterforce capabilities, in particular, is fundamentally inimical to the security of the United States and its allies.

Despite erosion of the value of the ABM Treaty through Soviet noncompliance and through the absence of comparable Soviet restraints on offensive systems, the United States is and will continue to remain in full compliance with its ABM Treaty obligations. A principal factor leading to that accord was the conclusion reached in the United States during the

ABM debate of the late 1960s that defenses, at the then-existing level of technology, could be overwhelmed at less cost by additional offensive systems than would be required to add balancing defenses. Therefore, we were concerned that the deployment of a relatively ineffective territorial ABM system on either side could prompt a proliferation of offensive nuclear forces and cheap but effective countermeasures. An ABM system based on then-current technology would not have been militarily effective, survivable, or cost-effective at the margin.

By contrast, our interest in SDI research is premised on the judgment that new technologies may now be available that could reverse our judgments of the late 1960s about the military ineffectiveness, vulnerability, and cost-ineffectiveness of strategic defenses. It is important to keep in mind that these three requirements are as relevant today as they were sixteen years ago; it is the capabilities of the technologies that may have changed.

The SDI Decision Criteria: A Path to Stability

The President's Strategic Defense Initiative, published in January 1985 as the most authoritative description of the president's vision, discussed these requirements for an effective defense. These criteria are posited as necessary for maintaining stability.

To achieve the benefits which advanced technologies may be able to offer, defenses must be militarily effective. Defenses must be able, at a minimum, to destroy a sufficient portion of an aggressor's attacking forces to deny him confidence in the attack's outcome, in general, and in particular, to deny him the ability to destroy a significant portion of the military target sets he would need to destroy.

The exact level of defense system capability required to achieve these ends cannot be determined at this time, since it depends on the size, composition, effectiveness, and inherent survivability of US forces relative to those of the Soviet Union at the time that defenses are introduced. However, in addition to the requirement of military effectiveness, two other necessary characteristics of an effective defense have been identified and constitute current presidential policy as put forth in a recent National Security Decision Directive. They are survivability and cost-effectiveness at the margin.

Survivability is defined not in terms of system invulnerability but the ability of a system "to maintain a sufficient degree of effectiveness to fulfill its mission, even in the face of determined attacks against it." The president's analysis characterizes survivability as "essential not only to maintain the effectiveness of a defense system, but to maintain stability." Vulnerable defenses could, in a crisis, provide the offense with incentives to initiate defense suppression attacks to gain a favorable shift in the offense-defense balance as a prelude to a first strike.

Similarly, in the interest of discouraging the proliferation of ballistic missile forces, the defensive system must be able to maintain its effectiveness against the offense at less cost than it would take to develop offensive countermeasures and proliferate the ballistic missiles necessary to overcome it. This is the concept of cost-effectiveness at the margin. It describes the stability of the competitive relationship between one side's defensive forces and the other side's offensive forces—that is, whether one side has major incentives to add additional offensive forces in an effort to overcome the other side's defenses.

The term cost-effectiveness is expressed in economic terms. While this concept has valid application not only for strategic defenses but for other military systems as well, the United States understands the criterion of cost-effectiveness at the margin to be more than an economic concept.

In particular, we need to be concerned, in our evaluation of options generated by SDI research, with the degree to which certain types of defensive systems encourage or discourage an adversary to attempt to overwhelm them with additional offensive systems and countermeasures. We seek defensive options which provide clear disincentives to attempts to counter them with additional offensive forces.

Our continued adherence to these criteria indicates the deep interest that the United States has in maintaining and enhancing stability. The United States is demonstrating this interest in other ways as well. In particular, our goals related to a possible transition to greater reliance on defenses, together with our view of SDI as a means of enhancing deterrence and stabilizing the US-Soviet balance and not as a means of achieving superiority, underscore our concern for stability.

Assuring Confidence in Our SDI Research

President Reagan personally assured General Secretary Gorbachev at the November 1985 summit that the United States seeks to enhance peace and that we are pursuing SDI as part of our effort to enhance deterrence and global stability. In this regard, as we have repeatedly made clear, the United States is conducting research only on defensive systems, with primary emphasis on non-nuclear technologies. While it is difficult to be certain of capabilities of potential systems based on technologies not yet developed, defenses based on the new technologies we are investigating would not have the role of striking targets on the ground.

Despite Soviet unwillingness during the first four rounds of the nuclear and space talks to engage in meaningful dialogue in the defense and space negotiating group, the United States has consistently demonstrated in our statements and actions that we do not seek to gain a unilateral advantage from strategic defense. This openness stands in marked contrast to the closed nature of Soviet strategic defensive activities, the intentions of which we must extrapolate from an operationally offensive Soviet military doctrine with heavy emphasis on the unabated growth in Soviet nuclear weapons capabilities.

Consistent with our traditional emphasis on verification, the United States does not expect the Soviet Union to accept our assurances on faith alone. On the contrary, in Geneva we have made concrete proposals which would enable the United States and the Soviet Union to assess the defensive nature of the research being conducted by each side.

If and when our research criteria are met, and following close consultation with our allies, we intend to consult and negotiate, as appropriate, with the Soviets pursuant to the terms of the ABM Treaty, which provide for such consultations on how deterrence could be enhanced through a greater reliance by both sides on new defensive systems. It is our intention and our hope that, if new defensive technologies prove feasible, we—in close and continuing consultation with our allies—and the Soviets will jointly manage a transition to a more defense-reliant balance. A jointly managed transition would be designed to maintain, at all times, control over the mix of offensive and defensive systems, thereby assuring both sides of the stability of the evolving strategic balance. An implicit goal of a jointly managed transition would be to identify in advance potential problems in, for example, the stability of the mix of offense and defense and to act to resolve such problems.

Of course, arms control would play an important role in such a transition. Properly structured cuts in offensive arms are not only worthwhile in their own right but they could also facilitate the shift to a more defense-reliant posture. Unilateral modernization measures can enhance transition stability. Improving the survivability of our offensive forces, for example, would especially contribute to stability in an early transition phase.

Our interest in pursuing a cooperative transition with the Soviets should not be seen, however, as granting them veto power over US decision making. Any US decision to develop and deploy defenses would still reflect the same goals of peace and enhanced deterrence through a stable transition, even if our good faith efforts to engage the Soviets in a cooperative transition were to fail. I am convinced, however, that a successful SDI research phase proving the feasibility of survivable and cost-effective defenses would provide compelling incentives for the Soviets to consider seriously the advantages of a jointly managed transition. In Geneva, we seek to provide a forum for such consideration.

Balancing Offense and Defense in Geneva

The Soviet approach in Geneva has been to advance the self-serving and unacceptable concepts of "a ban on space-strike arms" and "a ban on purposeful research," both impossible to define in meaningful and verifiable terms. They would like to limit US capabilities and stop US research while avoiding constraints on their own weapon systems and research through definitional ploys.

The United States is committed to the SDI research program, which is being carried out in full compliance with all of our treaty obligations, including the ABM Treaty. Indeed, the United States seeks to reverse the erosion of existing agreements, including the ABM Treaty, caused by Soviet violations. In seeking to stop or delay SDI, the Soviet Union also talks about strengthening the ABM Treaty. However, their approach for doing so has so far been based on artificial distinctions such as that between "purposeful" and "fundamental" research.

The Soviets maintain that deep cuts are only possible, and that stability can only be preserved, if the United States agrees to halt substantive work on SDI. The United States cannot accept this thesis. We propose, instead, a serious discussion on the offense-defense relationship and the outlines of the future offense-defense balance. Were the Soviets to work with us in a meaningful exploration of significant reductions in START and INF, we could examine how the level of defense would logically be affected by the level and nature of offensive arms.

The ABM Treaty marked the beginning of an arms control process which, in retrospect, has been profoundly disappointing. The offensive reductions which were supposed to accompany it have not materialized, and the Soviets are in fundamental violation of one or more of the treaty's key provisions. Consequently, we are working to halt the treaty's erosion by the Soviet Union and persuade them that full compliance with its terms by both sides is in our mutual interest.

The United States does not believe that there is reason now to change the ABM Treaty. Through our SDI research, we wish to determine whether or not there is a better way to ensure long-term stability than to rely on the ever more dangerous threat of devastating nuclear retaliation to deter war and assure peace. If we find there is, and if at some future time the United States, in close consultation with its allies, decides to proceed with deployment of defensive systems, we intend to utilize mechanisms for US-Soviet consultations provided for in the ABM Treaty. Through such mechanisms and taking full account of the Soviet Union's own expansive defensive systems research program, we will seek to proceed in a stable fashion with the Soviet Union. In this context, we must remember that the ABM Treaty is a living document.

Articles XIII and XIV provide for consultation with the aim of appropriate amendment of the treaty to take account of future considerations, such as the possibility of a new—and more stable—strategic balance.

Toward a New Synthesis

Current United States SDI research activities and arms control policies are designed to provide a basis for securing stability in a future strategic regime. The goal of stability can be guaranteed only if we maintain our commitment to the standards and criteria consistent with it.

The United States is committed to achieving strategic stability and, therefore, to a predictable and stable arms control process to complement our strategic programs to assure our primary security objective of reducing the risk of war.

21

SDI: Progress and Promise

Ronald Reagan

OUR country's security today relies as much on the genius and creativity of scientists as it does on the courage and dedication of those in the military services. It also relies on those with the wisdom to recognize innovation when they see it and to shepherd change over the obstacles and through the maze. . . .

There are three stages of reaction to any new idea, as Arthur C. Clarke, a brilliant writer with a fine scientific mind, once noted. First, "It's crazy; don't waste my time." Second, "It's possible, but it's not worth doing." And, finally, "I always said it was a good idea." . . .

Clearly, intelligent and well-meaning individuals can be trapped by a mindset, a way of thinking that prevents them from seeing beyond what has already been done and makes them uncomfortable with what is unfamiliar. And this mindset is perhaps our greatest obstacle in regard to SDI.

We're at a critical point now on national security issues . . . . Many of our citizens are still unaware that today we are absolutely defenseless against the fastest, most destructive weapons man has ever created—ballistic missiles. Yet, there are still those who want to cut off, or severely cut back, our ability to investigate the feasibility of such defenses. Congressional action on the defense authorization bill is coinciding with increasing diplomatic activity with the Soviet Union. Yet, at the same time, we're in the midst of a budget fight which could take away the very leverage we need to deal with the Soviets successfully.

Back in 1983, I challenged America's scientific community to develop an alternative to our total reliance on the threat of nuclear retaliation, an

---

Extracts from President Reagan's remarks at a briefing on the Strategic Defense Initiative, Washington, D.C., August 6, 1986. Department of State, *SDI: Progress and Promise,* Current Policy No. 858 (Washington, D.C.: GPO, 1986).

alternative based on protecting innocent people rather than avenging them; an alternative that would be judged effective by how many lives it could save, rather than how many lives it could destroy.

All of you know that during the past three decades deterrence has been based on our ability to use offensive weapons to retaliate against any attack. Once an American president even had to make the excruciating decision to use such weapons in our defense. Isn't it time that we took steps that will permit us to do something about nuclear weapons, rather than simply continue to live with them in fear? And this is what our SDI research is all about, and there can be no better time than today, the 41st anniversary of Hiroshima, to rededicate ourselves to finding a safer way to keep the peace.

Many people believe the answer lies not in SDI but only in reaching arms control agreements. Trust and understanding alone, it is said, will lead to arms control. But let's not kid ourselves; it's realism, not just trust, that is going to make it possible for adversaries, like the Soviet Union and the United States, to reach effective arms reduction agreements. Our SDI program has provided a historic opportunity; one that enhances the prospects for reducing the number of nuclear weapons. Technology can make it possible for both sides, realistically, without compromising their own security, to reduce their arsenals. And the fear that one side might cheat—might have a number of missiles above the agreed upon limit—could be offset by effective defenses. Clearly, by making offensive nuclear missiles less reliable, we make agreements to reduce their number more attainable. Particularly is that true where one side now is an economic basket case because of the massive arms buildup that it's been conducting over the last few decades—the Soviet Union.

There has been progress. There's a serious prospect today for arms reductions, not just arms control; and that by itself is a great change, and it can be traced to our Strategic Defense Initiative. SDI can take the profit out of the Soviet buildup of offensive weapons and, in time, open new opportunities by building on today's and tomorrow's technologies.

I say this fully aware of the Soviet campaign to convince the world that terminating our SDI program is a prerequisite to any arms agreement. This clamoring is nothing new. It also has preceded steps we've taken to modernize our strategic forces. It was especially loud, for example, as we moved to offset the unprovoked and unacceptable Soviet buildup of intermediate-range missiles aimed at our allies by deploying our Pershing IIs and cruise missiles.

Today, we continue to negotiate with the Soviets, and they are negotiating with us. In fact, their recent proposals—in stark contrast to those gloomy predictions—are somewhat more forthcoming than those of the past. . . . Forecasting is not useful, but, let me just say again, I am optimistic. It is demonstrably in the interest of both our countries to reduce the resources that we commit to weapons. . . .

As for SDI, let me again affirm, we are willing to explore how to share its benefits with the Soviet Union, which itself has long been involved in strategic defense programs. This will help to demonstrate what I have been emphasizing all along—that we seek no unilateral advantage through the SDI.

There's been some speculation that in my recent letter to General Secretary Gorbachev, I decided to seek some sort of "grand compromise" to trade away SDI in exchange for getting the Soviets to join with us in the offensive reductions. Now, to those who have been publicizing what is supposed to be in that letter, I hope they aren't offended to find out that they don't know what's in that letter because no one's really told them. I know. Let me reassure you right here and now that our response to demands that we cut off or delay research and testing and close shop is: no way. SDI is no bargaining chip; it is the path to a safer and more secure future. And the research is not, and never has been, negotiable. As I've said before, it's the number of offensive missiles that needs to be reduced, not efforts to find a way to defend mankind against these deadly weapons.

Many of the vocal opponents of SDI, some of them with impressive scientific credentials, claim our goal is impossible; it can't be done, they say. Well, I think it's becoming increasingly apparent to everyone that those claiming it can't be done have clouded vision. Sometimes smoke gets in your eyes. And sometimes politics gets in your eyes. If this project is as big a waste of time and money as some have claimed, why have the Soviets been involved in strategic defense themselves for so long, and why are they so anxious that we stop? . . .

I'm more than happy with the strides made in our ability to track and intercept missiles before they reach their targets. The goal we seek is a system that can intercept deadly ballistic missiles in all phases of their flight, including and, in particular, the boost phase—right where they're coming out of the silos. Our research is aimed at finding a way of protecting people, not missiles. And that's my highest priority and will remain so.

And to accomplish this, we're proceeding as fast as we can toward developing a full range of promising technologies. I know there are those who are getting a bit antsy, but to deploy systems of limited effectiveness now would divert limited funds and delay our main research. It could well erode support for the program before it's permitted to reach its potential. . . .

We and the other free people of the world are on the edge of a giant leap into the next century. That turning point in 13½ years will not only mark the end of a century but the beginning of a new millennium. And the free people of the world are ready for it. Our research on effective defenses helps to point the way to a safer future. The best minds from some allied countries are already working with us in this noble endeavor, and we believe others will join this effort before too long. In SDI, as elsewhere, we've put technology that almost boggles the mind to work—increasing our productivity and expanding the limits of human potential. The relationship between freedom and human progress has never been more apparent. . . .

If we cut back on our own forces unilaterally, we will leave our adversaries no incentive to reduce their own weapons. And we will leave the next generations not a safer, more stable world but a far more dangerous one. The future is literally in our hands. And it is SDI that is helping us to regain control over our own destiny.

Just one last little incident, if you aren't aware of it already, that might be helpful to you and some people that you might be discussing this subject with. Back when Fulton was inventing the steamboat and it came into reality, there was an effort made to sell it to Napoleon in France. And the great general, with all his wisdom, said: "Are you trying to tell me that you can have a boat that will sail against the tide and the currents and the winds without any sails?" He said: "Don't bother me with such foolishness." Well, we know where the foolishness lay, and let's not make the same mistakes. . . .

I'll just leave you with this thought once again. When the time has come and the research is complete, yes, we're going to deploy.

22

Reykjavik Summit: The American View

Ronald Reagan

GOOD Evening. As most of you know, I have just returned from meetings in Iceland with the leader of the Soviet Union, General Secretary Gorbachev. . . . The implications of these talks are enormous and only just beginning to be understood. We proposed the most sweeping and generous arms control proposal in history. We offered the complete elimination of all ballistic missiles—Soviet and American—from the face of the Earth by 1996. While we parted company with this American offer still on the table, we are closer than ever before to agreements that could lead to a safer world without nuclear weapons. . . .

Allow me to set the stage by explaining two things that were very much a part of our talks, one a treaty and the other a defense against nuclear missiles which we are trying to develop. Now you've heard their titles a thousand times—the ABM Treaty and SDI. Those letters stand for: ABM, antiballistic missile; SDI, Strategic Defense Initiative.

Some years ago, the United States and the Soviet Union agreed to limit any defense against nuclear missile attacks to the emplacement in one location in each country of a small number of missiles capable of intercepting and shooting down incoming nuclear missiles, thus leaving our real defense—a policy called Mutual Assured Destruction, meaning if one side launched a nuclear attack, the other side could retaliate. And this mutual threat of destruction was believed to be a deterrent against either side striking first.

Printed here are extracts from President Reagan's October 13, 1986, address to the nation and from his October 14 remarks to officers of the State Department and the Arms Control and Disarmament Agency. These transcripts were released by the Office of the Press Secretary, The White House, Washington, D.C.

So here we sit with thousands of nuclear warheads targeted on each other and capable of wiping out both our countries. The Soviets deployed the few antiballistic missiles around Moscow as the treaty permitted. Our country didn't bother deploying because the threat of nationwide annihilation made such a limited defense seem useless.

For some years now we have been aware that the Soviets may be developing a nationwide defense. They have installed a large modern radar at Krasnoyarsk which we believe is a critical part of a radar system designed to provide radar guidance for antiballistic missiles protecting the entire nation. Now this is a violation of the ABM Treaty.

Believing that a policy of mutual destruction and slaughter of their citizens and ours was uncivilized, I asked our military a few years ago to study and see if there was a practical way to destroy nuclear missiles after their launch but before they can reach their targets rather than to just destroy people. Well, this is the goal for what we call SDI, and our scientists researching such a system are convinced it is practical and that several years down the road we can have such a system ready to deploy. Now, incidentally, we are not violating the ABM Treaty which permits such research. If and when we deploy, the treaty also allows withdrawal from the treaty upon six months' notice. SDI, let me make it clear, is a non-nuclear defense. . . .

On Saturday and Sunday [October 11 and 12], General Secretary Gorbachev and his Foreign Minister Shevardnadze and Secretary of State George Shultz and I met for nearly ten hours. . . . We discussed the emplacement of intermediate-range missiles in Europe and Asia and seemed to be in agreement they could be drastically reduced. Both sides seemed willing to find a way to reduce even to zero the strategic ballistic missiles we have aimed at each other. This then brought up the subject of SDI.

I offered a proposal that we continue our present research, and if and when we reached the stage of testing we would sign now a treaty that would permit Soviet observation of such tests. And if the program was practical, we would both eliminate our offensive missiles, and then we would share the benefits of advanced defenses. I explained that even though we would have done away with our offensive ballistic missiles, having the defense would protect against cheating or the possibility of a madman sometime deciding to create nuclear missiles. After all, the world now knows how to make them. I likened it to our keeping our gas masks even though the nations of the world had outlawed poison gas after World War I.

We seemed to be making progress on reducing weaponry although the general secretary was registering opposition to SDI and proposing a pledge to observe ABM for a number of years as the day was ending.

Secretary Shultz suggested we turn over the notes our note-takers had been making of everything we'd said to our respective teams and let them work through the night to put them together and find just where we were in agreement and what differences separated us. With respect and gratitude, I can inform you those teams worked through the night till 6:30 A.M.

Yesterday, Sunday morning, Mr. Gorbachev and I, with our foreign ministers, came together again and took up the report of our two teams. It was most promising. The Soviets had asked for a ten-year delay in the deployment of SDI programs.

In an effort to see how we could satisfy their concerns while protecting our principles and security, we proposed a ten-year period in which we began with the reduction of all strategic nuclear arms, bombers, air-launched cruise missiles, intercontinental ballistic missiles, submarine-launched ballistic missiles and the weapons they carry. They would be reduced 50 percent in the first five years. During the next five years, we would continue by eliminating all remaining offensive ballistic missiles, of all ranges. And during that time we would proceed with research, development, and testing of SDI—all done in conformity with ABM provisions. At the ten-year point, with all ballistic missiles eliminated, we could proceed to deploy advanced defenses, at the same time permitting the Soviets to do likewise.

And here the debate began. The general secretary wanted wording that, in effect, would have kept us from developing the SDI for the entire ten years. In effect, he was killing SDI. And unless I agreed, all that work toward eliminating nuclear weapons would go down the drain—cancelled.

I told him I had pledged to the American people that I would not trade away SDI—there was no way I could tell our people their government would not protect them against nuclear destruction. I went to Reykjavik determined that everything was negotiable except two things: our freedom and our future.

I'm still optimistic that a way will be found. The door is open and the opportunity to begin eliminating the nuclear threat is within reach.

So you can see, we made progress in Iceland. And we will continue to make progress if we pursue a prudent, deliberate, and, above all, realistic approach with the Soviets. From the earliest days of our administration, this had been our policy. We made it clear we had no illusions about the Soviets or their ultimate intentions. We were publicly candid about the critical moral distinctions between totalitarianism and democracy. We declared the principal objective of American foreign policy to be not just the prevention of war but the extension of freedom. And we stressed our commitment to the growth of democratic government and democratic institutions around the world. And that's why we assisted freedom fighters who are resisting the imposition of totalitarian rule in Afghanistan, Nicaragua, Angola, Cambodia, and elsewhere. And, finally, we began work on what I believe most spurred the Soviets to negotiate seriously—rebuilding our military strength, reconstructing our strategic deterrence, and, above all, beginning work on the Strategic Defense Initiative.

And yet, at the same time we set out these foreign policy goals and began working toward them, we pursued another of our major objectives: that of seeking means to lessen tensions with the Soviets, and ways to prevent war and keep the peace.

Now, this policy is now paying dividends—one sign of this in Iceland was the progress on the issue of arms control. For the first time in a long while, Soviet-American negotiations in the area of arms reductions were moving, and moving in the right direction—not just toward arms control, but toward arms reduction. . . .

Let me return again to the SDI issue. I realize some Americans may be asking tonight: Why not accept Mr. Gorbachev's demand? Why not give up SDI for this agreement?

Well, the answer, my friends, is simple. SDI is America's insurance policy that the Soviet Union would keep the commitments made at Reykjavik. SDI is America's security guarantee—if the Soviets should—as they have done too often in the past—fail to comply with their solemn commitments. SDI is what brought the Soviets back to arms control talks at Geneva and Iceland. SDI is the key to a world without nuclear weapons.

The Soviets understand this. They have devoted far more resources for a lot longer time than we, to their own SDI. The world's only operational missile defense today surrounds Moscow, the capital of the Soviet Union.

What Mr. Gorbachev was demanding at Reykjavik was that the United States agree to a new version of a fourteen-year-old ABM Treaty that the Soviet Union has already violated. I told him we don't make those kinds of deals in the United States.

And the American people should reflect on these critical questions. How does a defense of the United States threaten the Soviet Union or anyone else? Why are the Soviets so adamant that America remain forever vulnerable to Soviet rocket attack? As of today, all free nations are utterly defenseless against Soviet missiles—fired either by accident or design. Why does the Soviet Union insist that we remain so—forever?

So, my fellow Americans, I cannot promise, nor can any president promise, that the talks in Iceland or any future discussions with Mr. Gorbachev will lead inevitably to great breakthroughs or momentous treaty signings.

We will not abandon the guiding principle we took to Reykjavik. We prefer no agreement than to bring home a bad agreement to the United States. . . .

If there's one impression I carry away with me from these October talks, it is that, unlike the past, we're dealing now from a position of strength, and for that reason we have it within our grasp to move speedily with the Soviets toward even more breakthroughs.

Our ideas are out there on the table. They won't go away. We're ready to pick up where we left off. Our negotiators are heading back to Geneva, and we're prepared to go forward whenever and wherever the Soviets are ready. So, there's reason—good reason for hope. . . .

The following day, October 14, 1986, the president made these remarks to officers of the Arms Control and Disarmament Agency and the Department of State.

At Reykjavik, the Soviet Union went farther than ever before in accepting our goal of deep reductions in the level of nuclear weapons. For the first time, we got Soviet agreement to a figure of one hundred intermediate-range missiles—warheads—for each side worldwide. And that was a truly drastic cut. And for the first time we began to hammer out the details of a 50 percent cut in strategic forces over five years. And we were just a sentence or two away from agreeing to new talks on nuclear testing. And maybe most important, we were in sight of a historic agreement on completely eliminating the threat of offensive ballistic missiles by 1996.

My guess is that the Soviets understand this, but want to see how much farther they can push us in public before they once again get down to brass tacks. So here's how I see the meeting in Iceland adding up.

We addressed the important issues of human rights, regional conflicts and our bilateral relationship. And Mr. Gorbachev and I got awfully close to historic agreements in the arms reduction process. We took discussions into areas where they had never been before. The United States put good, fair ideas out on the table and they won't go away. Good ideas, after all, have a life of their own. The next step will be in Geneva where our negotiators will work to build on this progress.

The biggest disappointment in Iceland was that Mr. Gorbachev decided to make our progress hostage to his demand that we kill our strategic defense program. But, you know, I've had some experience with this kind of thing. One of my past jobs was as a negotiator of labor agreements in the motion picture industry, and I got used to one side or another walking out of contract talks. It didn't mean that relations had collapsed or that we'd reached an insurmountable impasse, it sometimes meant that a little maneuvering was going on.

Well, it's important for us right now to see that the real progress that we made at Reykjavik and to unite so that we'll be strong for the next stage in negotiations. And if we do that, I believe that we have it within our grasp to achieve some truly historic breakthroughs.

Last week I described Iceland as a base camp on our way to the summit. Well, this week I want to report to you that I believe there exists the opportunity to plant a permanent flag of peace at that summit. And I call on the Soviets not to miss this opportunity. The Soviets must not throw this away, must not slip back into a greater arms buildup. The American people don't mistake the absense of a final agreement for the absense of progress. We made progress—we must be patient. We made historic advances—we will not turn back.

Now, America and the West need SDI for long-run insurance. It protects us against the possibility that at some point when the elimination of ballistic missiles is not yet complete, the Soviets may change their mind. We know the Soviet record of playing fast and loose with past agreements. America can't afford to take a chance on waking up in ten years and finding that the Soviets have an advance defense system and are ready to put in place more missiles—or more modern missiles—and we have no defense of our own, and our deterrence is obsolete because of the Soviet defense system.

If arms reduction is to help bring lasting peace, we must be able to maintain the vital strategic balance which for so long has kept the peace. Nothing could more threaten world peace than arms reduction agreements with loopholes that would leave the West naked to a massive and sudden Soviet buildup in offensive and defensive weapons.

Believe me, the significance of that meeting at Reykjavik is not that we didn't sign agreements in the end, the significance is that we got as close as we did. The progress that we made would've been inconceivable just a few months ago.

On issue after issue, particularly in the area of arms reduction, we saw that General Secretary Gorbachev was ready for serious bargaining on real arms reductions. And for me, this was especially gratifying. Just five and a half years ago, when we came into office, I said that our objective must be—well, it must not be regulating the growth in nuclear weapons—which is what arms control as it was known had been all about, I know—I said that our goal must be reducing the number of nuclear weapons, that we had to work to make the world safer, not just control the pace at which it became more dangerous. And now, the Soviets, too, are talking about real arms reductions. . . .

And particularly important, of course, was America's support for the Strategic Defense Initiative. Now, as you know, I offered Mr. Gorbachev an important concession on SDI. I offered to put off deployment for a decade, and I coupled that with a ten-year plan for eliminating all Soviet and American ballistic missiles from the face of the Earth.

This may have been the most sweeping and important arms reduction proposal in the history of the world, but it wasn't good enough for Mr. Gorbachev—he wanted more. He wanted us to accept even tighter limits on SDI than the ABM Treaty now requires. That is, to stop all but laboratory research. He knew this meant killing strategic defense entirely, which has been a Soviet goal from the start. And, of course, the Soviet Union has long been engaged in extensive strategic defense programs of its own. And unlike ours, the Soviet program goes well beyond research, even to deployment. The Soviet proposal would've given them an immediate one-sided advantage and a dangerous one. And I could not, and would not agree to that. I won't settle for anything unless it's in the interest of America's security.

23

Department of Defense News Briefing Following the Reykjavik Summit

Richard N. Perle

PERLE: Again and again, the president asked Mr. Gorbachev what possible objection he could have to the deployment of defenses after ten years, and after having eliminated offensive ballistic missiles. Again and again the president pressed him to explain how defensive systems, wholly lacking in offensive capability, could threaten the Soviet Union. The president never received a satisfactory answer or even a plausible response—there was no satisfactory answer.

Prior to Reykjavik, the Soviets had been making the argument that a combination of offensive and defensive forces could enable the side that had an effective defense coupled with offenses to launch a first strike. The president sought to settle that issue by combining agreement to defer deployment with agreement to abandon offensive ballistic missiles. And he took the step the Soviet side was asking: that the deployment would follow the elimination of ballistic missiles. Under those circumstances the deployment of defenses could in no way enable the side that had them to launch a first strike because the other side would have no missiles that could be intercepted on the retaliatory strike. In other words, we accepted, in a sense, and put forward a proposal that, in my view, is logically consistent and recognizes the reality that we have not yet arrived at that point in the US-Soviet relationship where we would be wise to take on trust a Soviet declaration that they had eliminated the last of their ballistic missiles, as we would eliminate the last of ours.

You know and I know that if the United States agreed to dismantle all its offensive ballistic missiles we would do so. We wouldn't hold a few back, we wouldn't produce others, but we can have no comparable assurance on the

---

This is an excerpt from Mr. Perle's briefing for reporters at the Pentagon on October 14, 1986, immediately following the Reykjavik summit meeting.

Soviet side. So defenses in that role would be highly stabilizing, protecting the most significant and far-reaching disarmament.

My own view is that the Soviets were uncomfortable with the prospect of eliminating all ballistic missiles, but they were conscious of the political and diplomatic implications of having the talks fail on the basis of their unwillingness to do that. So at the last minute they introduced a demand that we confine research to the laboratory, which went beyond any of their previous demands and made it impossible for the president to accept.

QUESTION: *This disagreement would have left, after the end of ten years, if implemented in offensive missiles, would have left stealth bombers, air-launched cruise missiles, sea-launched cruise missiles, in some form, whether reduced from present numbers or not, we do not know. Particularly, bombers and air-launched cruise missiles, at least right now, are an American strength; we put much more stock in them than the Soviets. When you couple what is an advantage in "air-breathing" leg of the triad with the prospects of a workable SDI, why doesn't that still amount to what the Soviets might see as a first strike advantage for the United States, maybe not a first strike in thirty minutes but a first strike in several hours?*

PERLE: Let's be clear about a couple of things that are inherent in that question.

First, we were prepared significantly to reduce all legs of the triad so that our remaining bomber capability would have been reduced as we would expect reductions on the Soviet side. But second and more important, the possession of the strategic defense would have no bearing on the balance between us if we each had forces restricted to aircraft and the weapons carried by those aircraft and shorter range, non-ballistic nuclear systems. If that was the Soviet concern then the proper Soviet response was to say, "we can accept your proposition, but we have to work out the terms as they would relate to aircraft."

Now the difference between aircraft and missiles is clear. The world has been living under the shadow of nuclear forces, made up significantly of ballistic missiles that reach their target in twenty minutes or so, that can't be recalled and give rise to the possibility of a first strike. With aircraft only and shorter range systems for deterrence in Europe, we would return to a situation that once existed in which deterrence was based on systems that take hours to reach their targets, that can be recalled in the event of an accident and the world would be a much safer place if deterrence depended on those systems rather than on ballistic missiles.

QUESTION: *I don't follow why you say strategic defense would have no bearing on a world in which both sides only relied on aircraft for long-range weapons?*

PERLE: Under the agreement, the Soviets would have no missiles for the defense to intercept. Strategic defenses would only impede the plans of the country that retained offensive ballistic missiles and used them.

QUESTION: *You're not saying strategic defenses would have been entirely useless against aircraft, are you?*

PERLE: With respect to defense against aircraft, the Soviets are so far ahead of the United States in that area, they have an advantage approaching a monopoly.

QUESTION: *Richard, put yourself for a moment in the Soviet officials shoes, if you can make that leap; what would be your objections to SDI?*

PERLE: Well, frankly I think the Soviets are after more than SDI. I think they are after the whole range of American research and advanced technologies, which would help explain why at the last minute in the negotiations they introduced this concept of terminating all space research outside the laboratory, which means terminating all space research. We don't know and I don't think the Soviets know how to define that term in a manner that would be verifiable and that would be restricted to research on strategic defenses. How do you test communications systems? How do you test sensors? How do you test the whole panoply of military systems in space if you can only test inside the laboratory. You simply can't do it. So what I think the Soviets were driving at was causing the United States to scuttle virtually all of our activity with respect to space.

QUESTION: *Why are they so afraid of that?*

PERLE: I have a theory as it relates to strategic defense, and that theory is that the Soviets, who have been researching this much longer than we have—there is a period of about a decade in which our own activities were severely limited while theirs were rather abundant—I think they've come to some conclusions about the potential for strategic defense; and, conceivably, because they lay great emphasis on this, they have also discovered a potential for offensive uses of space that we haven't yet discovered. But they seem concerned that we might somehow, in the course of the SDI program, stumble upon offensive technologies and they're trying to stop that. My guess is that they have already stumbled upon such technologies.

QUESTION: *We may stumble upon those?*

PERLE: It's not the purpose of the SDI program, and if our concern were to develop offensive weapons that utilize space, it would be far more efficient and far more effective to proceed directly to do that, unencumbered by any ABM Treaty restrictions. So the Soviet argument isn't a very logical one.

QUESTION: *In terms of the technologies involved, their view on who would win or lose an all-out technological race: what is your view on that?*

PERLE: I think their view is that the American, the Western technological base would enable us to field systems the theoretical concept for which they may already understand, but the manufacturing technology for which they may not have.

QUESTION: *Would they not be able to buy or steal much of that technology?*

PERLE: Well, as you know, they are energetically involved in both, but that's a pretty risky proposition. I think as our efforts to tighten up on the control of militarily relevant technology have proceeded, they've become less certain about their ability to get Western technology.

QUESTION: *Is it your view that the Soviets really could not compete?*

PERLE: This is a tricky business, this question of high technology. When we got inside the negotiations on Saturday night, they began at eight o'clock and went through until 6:30 the following morning. There was a rather long delay, and the delay was the result of the fact that we had a typewriter and a typist who could produce drafts, but we didn't have a Xerox machine at the Hofdi House. The Soviets had a supply of carbon paper. When they produced a draft they could produce it in ten copies. We were unable to produce more than a single copy. So they depended on low technology and we didn't. There may be a lesson in that. We ended up using their carbon paper.

QUESTION: *In your chronology of how events unfolded, you said at the last minute the Soviets came in with this notion of restricting SDI to laboratory research and that came after President Reagan had advanced his plan to eliminate all ballistic missiles. So prior to that were they amenable to more research, that beyond the laboratory, and is the implication of what you said that if we hadn't asked for the elimination of all ballistic missiles perhaps they would tolerate more research?*

PERLE: I can't prove this. But I think it is at least a plausible explanation that when we proposed the elimination of all offensive ballistic missiles and were prepared to agree in conjunction with that to defer deployment for ten years, and to remain within the terms of the ABM Treaty in the conduct of research, development, and testing; they looked at that package, they understood the diplomatic and political implications of rejecting the proposal to eliminate all offensive ballistic missiles, decided they couldn't live with that and they needed a device for causing the negotiations to fail at that moment but not on the issue of the elimination of offensive ballistic missiles.

I can't prove it, but prior to that the Soviet side had said that they wanted a ten-year delay and they wanted during that ten-year period a commitment that we would conduct the SDI program within the terms of the ABM Treaty. That is, that for ten years we would not withdraw from the ABM Treaty and we would adhere strictly to it. When we agreed to that, they then added this other demand. I don't know whether they . . .

QUESTION: *So perhaps they're willing to allow more than laboratory research. Prior to that there was the possibility that they would do that.*

Let me just ask one followup question. When you take the position that we would keep our research within the limits of the treaty, which interpretation of the treaty is this—the traditional one that we're observing now or the one that the administration asserts it has a legal right to adopt, the broader one? Which one is it, because they're completely different views of what the treaty means.

PERLE: They are indeed different views, and that obviously would have to be a matter for discussion between the sides.

QUESTION: *So that was not settled then?*

PERLE: That was not settled, because the Soviets went beyond the treaty. I think the Soviets understand very well that a thorough examination of the negotiations that took place between the 20th of November 1969 and the 26th of May 1972, on which date the ABM Treaty was signed, would reveal that throughout the course of those negotiations the Soviets resisted an American proposal that research with respect to defensive systems based on new technologies or other physical principles, which is the term that was then used, would validate the legally correct interpretation that the president has adopted. There has been enough discussion about this so that I think it's clear that the Soviets understand that they have to go beyond the ABM Treaty if they're going to drive a stake through the heart of the SDI program.

QUESTION: *What happens now? Where does everything stand when it comes to, say, the medium-range talks, the INF talks? Is all of that out the window?*

PERLE: Let's take the intermediate-range talks. The Soviets took the position beginning with the Summit in November last year in Geneva that they were prepared to reach an agreement on intermediate-range systems, unlinked to the outcome of the negotiations on strategic forces or space and defense. They've repeated that position in every capital in Europe. Gorbachev on his first travels to Europe, in London and then again in Paris with Mitterrand, reiterated this position. That has been the Soviet position.

That's been our view as well, and it was agreed, as I say, in Geneva last year.

. . . I don't know what the Soviets will say if and when we do table that proposal, but to revert to their earlier position on the grounds that intermediate nuclear forces are now linked to the American agreement to restrict its research program beyond the terms of the ABM Treaty seems to me on the surface an untenable position and one that entails Soviet repudiation of the position they've had on the table for about a year. So if they do that, I think it will carry a high potential price for them in Europe. . . .

QUESTION: *Can you clarify for me the final Soviet SDI offer? Would that have prohibited anything that currently would be allowed under the narrow interpretation of the ABM Treaty?*

PERLE: Yes, it would have gone far beyond even the narrow interpretation of the treaty. And frankly, I think that a close analysis of that proposal would reveal that it would have forced the abandonment of an effective program of research involving space across the board. . . .

We're going to need deterrence for some time to come. I think it's safe to say that we will still need deterrence in 1996, and for that reason the American proposal was focused on . . .

QUESTION: *Nuclear deterrence?* .

PERLE: Nuclear deterrence. The American proposal was focused on offensive ballistic missiles. So one would still have had tactical nuclear forces other than ballistic missiles in order to deter Soviet attack in Europe, and one would have bomber forces in order to maintain an overall deterrence. But it would be a much more stable relationship if we could get rid of the ballistic missiles. We think it's in our interest to do it and we think it's in the Soviet interest as well. Ten years is a short time in which to accomplish something of that magnitude, but I think that the president was determined to go as far as we could go, even if it means a radical restructuring of our deterrent forces in order to eliminate the weapons that he believes are the principal source of instability. . . .

QUESTION: *Richard, how much damage do you think a full-bore Soviet competition with the US over the SDI would do to the Soviet economy? Is that one of the things we're striving to accomplish?*

PERLE: The Soviets have said that they are not going to follow the United States in developing strategic defenses. That may be because they in fact are leading us in the development of strategic defenses. What they have said is that they will counter the defenses rather than deploy defenses of their own. I think they are probably working on all fronts.

But what we are doing, what the president is determined not to abandon is a research program that we sought to associate with the elimination of the ballistic missiles so as to allay the concerns the Soviets had expressed to us. And when we had exhausted their concerns by responding to each of them, they came up with a new concern.

QUESTION: *Considering that most analysts of the Kremlin agree that Gorbachev has a lot on the line in terms of being able to redirect resources, to enhancing the Soviet economy, is there an effort by the Reagan administration to make that even more difficult for them?*

PERLE: No, not at all.

QUESTION: *You're not trying to engage them in "break the bank?"*

PERLE: On the contrary, the elimination of offensive ballistic missiles, together with the other reductions down to levels that were discussed in Geneva, would relieve the burden on the Soviets to continue to invest funds in strategic weapons. We would seek to finance SDI out of the funds that would be saved by abandoning offensive ballistic missiles, in which we invest very considerable sums. One has to ask why the Soviets would require an extensive antiballistic missile defense if we had no ballistic missiles. As a matter of fact, the system around Moscow, the principal limitation of which is that it has only one hundred interceptors, at least as far as we know, would be pretty effective against zero ballistic missiles. So there is no reason to believe that there would be an urgent requirement on the part of the Soviets to develop an elaborate defense in the absence of an offense to counter.

QUESTION: *Apparently, Gorbachev said that he did not believe President Reagan that we would share the benefits of SDI. He said you don't share the benefits of oil and gas technology, why would you do SDI? You know a little bit about export controls. How would we actually share the benefits? Would we turn over blueprints? What would we do?*

PERLE: We didn't get into a detailed discussion of that because Gorbachev categorically rejected it. He said he wasn't interested. . . .

It is fair to say that we attempt to restrict the sale of militarily relevant technology to the Soviet Union under current circumstances in which they are massively armed against the United States. In a world in which offensive ballistic missiles have been eliminated, the sharing of defensive technology would be irrelevant to the strategic balance conserved to protect us both against third countries and against cheating, and that's a very different situation. I think we could go quite far into those circumstances. But the precise modalities . . .

QUESTION: *Would that still not be hostage to other Afghanistans or other immigration cases? I mean, might the Soviets believe that this sharing might not take place because of our objections to their actions?*

PERLE: I think our record of keeping our treaty commitments is a pretty good one. We were never bound by treaty to provide oil and gas technology, and so we felt perfectly free to restrict that in response to Afghanistan. We would be bound by treaty and we honor our treaty commitments, no matter how inconvenient they may be.

QUESTION: *But you made three references to Soviet cheating. Isn't that the basis, your basis, the president's basis, for staying . . .*

PERLE: How could it be otherwise? We're talking about very significant disarmament. We're talking about agreements that could not be verified to the degree that would leave us certain that the Soviets had not hidden or subsequently produced ballistic missiles. These things are not very large these days, and they're highly mobile. No American president, in my view, is going to accept a situation in which if the Soviets cheat on an agreement that can't be fully verified the strategic balance between us changes dramatically overnight, and they alone would have the ability in a very short period of time to attack and destroy targets in the United States.

QUESTION: *The bottom line is really that, you just don't trust them.*

PERLE: Of course we don't trust them. We have good reason not to trust them. But we're prepared to enter into agreements as long as we can ensure those agreements in the absence of the kind of trust that exists, say, between ourselves and the Canadians. . . .

QUESTION: *Is it your judgment that they are playing hardball to get the best deal they can on SDI, or that their aim is to kill SDI? In other words, is there, in your judgment, the possibility of a compromise deal on SDI or not?*

PERLE: I think there's the possibility of a compromise. Indeed we put a compromise on the table. But the president will not kill the SDI program. He's prepared to limit research, development, and testing to that which is permitted under the ABM Treaty. He was prepared to defer deployment for ten years. But he is not prepared to terminate our research program.

QUESTION: *Will they settle for less than killing in your view?*

PERLE: They wouldn't settle for less in Reykjavik, but this is one meeting in a long process.

QUESTION: *Are we prepared to offer more?*

PERLE: I think they now know that they are dealing with a president who is willing to say yes to an agreement that is acceptable, but is also willing to say no to an agreement that isn't. It's conceivable that Gorbachev is under the mistaken impression that this was a president who would not say no. . . .

QUESTION: *Can I get a clarification? You mentioned on this whole thing of turning the technology of SDI over to the Soviets. There has obviously been a lot of skepticism to that. You're saying we're willing to write that into a treaty, in negotiating part of a treaty, that we would turn that over at the end of this ten years?*

PERLE: We were prepared in our earlier proposals to share technology with the Soviets. When the general secretary said, "I'm not interested in sharing, forget about it," we then proceeded to attempt to put together a compromise that at that point did not contain a commitment to sharing.

QUESTION: *But we're willing to firmly commit to that . . .*

PERLE: We are prepared to go quite far in sharing. People wonder how we can do this, and all I can say to you is that the notion of sharing was the president's own idea. It's not a conventional idea, but this is not a conventional president.

QUESTION: *You left open a question before, you said the president put forward an offer and they rejected it, and you left open the question of can we go further in any concessions that were already made. Is that the final offer? Is it dead after that? After the one we made?*

PERLE: We're going to return to the negotiations in Geneva, I think they resume tomorrow. And we're prepared to explore with the Soviets any ideas they might have. But we believe that the elimination of offensive ballistic missiles together with a deferral on SDI deployment and strict adherence to the ABM Treaty in doing the research, development, and testing that is permitted by the treaty, should be of interest to the Soviet Union.

QUESTION: *Wouldn't an insurance policy, as the president wants, if the Soviets accepted a verification procedure that you wrote, Richard Perle wrote, would that be a good enough insurance policy?*

PERLE: No. I don't know how to write a verification regime that would enable you to be confident that in the whole of Soviet territory there was not somewhere some quantity of offensive ballistic missiles. It's just too difficult. They don't have to be deployed, they can be stored.

But let me come back to the critical point. What harm would the defensive shield do to the Soviets or to anyone else as a way of reinforcing the elimination of ballistic missiles?

QUESTION: *The harm at the moment would be stopping any kind of arms reduction. That's what the critics say.*

PERLE: I understand the Soviet position, but as I said at the outset, they have failed to give us a convincing reason for their objection to our proposal. They may have fears they're not expressing. They may be based, as I suggested earlier, on things they know that we don't yet know. I don't know how to explain their view. But until they explain more persuasively why our proposal is unacceptable and what it is that concerns them, we can't even begin to think creatively about how to respond to those concerns. We responded in Reykjavik to every concern the Soviets had previously expressed, and I believe to every concern they made intelligible in Reykjavik. If there are other fears that [we] don't know about, we can't very well craft a proposal to deal with it.

QUESTION: *You're talking about the defensive shield as though the shield were . . . going to be put in place at some date in the future, though uncertain. Isn't it still a pig in the poke scientifically?*

PERLE: Well, I wouldn't call it a pig in a poke.

QUESTION: *What have you got?*

PERLE: We have a research program, and that's all we insisted upon is a research program, and the right, should the research program succeed, and after the elimination of all offensive ballistic missiles, to deploy that insurance policy. If we don't have it, we don't have it. Then you have a period in which there are some obvious risks, but the president is prepared to take some risks. He wasn't prepared to kill the program now.

QUESTION: *If you think about it, this is actually a potentially very dangerous proposal. You say you can't know if in fact they've eliminated all their missiles, verification isn't good enough. We don't know if we'd really have a shield up after eleven years or twelve years or thirteen years. Under your proposal there would be this period of uncertainty in which they could have missiles and we could have none, and we'd be at that peril.*

PERLE: Bear in mind that the scope and nature of the defense bears an obvious relationship to the offense. A defense adequate against a presumably relatively small number of clandestine ICBMs is much less demanding than a defense against massive offenses. So we are pretty confident that over the ten-year period we could move to deploy a defense sufficient to buy us some insurance.

QUESTION: *Can you tell us what that defense is?*

PERLE: No, I can't. It's too soon in the research program. Obviously, we would look for an efficient defense that met at least our minimum requirements, and if it did better than that, so much the better. . . .

QUESTION: *If I could go back to the question of why the Soviets are opposed to the SDI. You suggested that in the context of the proposals there they were concerned about not having offensive missiles at the time. The president questioned the Soviets' motives last night in saying why they would be opposed to strategic defense. I'm still not clear on why you feel the Soviets made this proposal and why they're so opposed to SDI, at least within the context of the negotiations.*

PERLE: I'm not clear in my own mind. That's why I said that repeatedly the president attempted to elicit from Gorbachev a statement that we could work with as to why he was concerned about the deployment of defenses in the aftermath of the elimination of offensive ballistic missiles and he didn't get that. I offer just a hypothesis, that the Soviets may believe that the SDI program will produce technologies that will give the United States some decisive military advantage, and it is at least conceivable to me that they think that because they themselves have gone rather far, at least in the theoretical work that they've been doing these many years. The Soviets are awfully good theorists, and they're good scientists. Where they have difficulty is in translating even quite imaginative concepts into manufactured hardware with high reliability and high technology. So they may know something we don't know, but they're not saying what it is.

QUESTION: *Is that why the president suggested, as was reported, that Gorbachev wasn't interested in agreement?*

PERLE: I think the president was troubled by the fact that when we met the Soviets, remember, this is a dynamic situation. They're putting out proposals; we're putting our proposals. They said ten years; we agreed ten years. They said stay within the ABM Treaty; we said we'll stay within the ABM Treaty. And then having responded positively, when we added the elimination of offensive ballistic missiles, they then came back and said you've got to stop, in effect, all SDI research. So this raises the question of what the Soviets were really after.

As I say, my own belief, which I can't prove, is that the sticking point was not really defensive research. The sticking point was the elimination of ballistic missiles. But since Gorbachev had previously talked about eliminating all nuclear weapons from the world and so forth, he did not want the talks to fail on that point, so he made a new and we think unreasonable and unexpected demand with respect to limitations on research and development in the hope that the Summit would be seen to have failed not on the elimination of ballistic missiles, but on SDI.

QUESTION: *For the sake of public relations you mean, for public opinion?*

PERLE: Yes. . . .To summarize it, I think the Soviets laid out a position, possibly believing that we wouldn't accept it, and when we accepted it, they discovered that they didn't like it. . . .

QUESTION: *Is it your suspicion that the Soviets went to Reykjavik to posture and propagandize and not to make a deal?*

PERLE: I don't want to speculate on why they went to Reykjavik. I think there was a lot of give and take back and forth. We want to believe that it might have been possible to conclude an agreement. I think they went beyond what we could accept in the proposal to limit research to the laboratory, unverifiable. If the question is did they go there knowing that there was no way they were going to reach an agreement, I wouldn't say that.

24

Defence and Security in the Nuclear Age: The British View

Sir Geoffrey Howe

THE Western Alliance has always emphasised its wholly defensive character. Repeatedly, we have pledged ourselves never to use any weapons except in response to an attack. The North Atlantic Treaty commits each of us to defend ourselves and our allies. The options for doing so have been debated for the past generation. It is my purpose today to review those options; to explore the relationship between defence in its broadest sense and systems intended to defend against nuclear weapons; and to highlight the issues our alliance must tackle if its success in ensuring peace is to continue.

Forty years ago, the nature of warfare between major powers was irrevocably altered. The extent of this revolutionary shift in international affairs has taken a long time to sink in. It was Einstein himself who said: "After the dawn of the nuclear age, everything changed except our way of thinking."

Gradually we have learned to think differently; to realise as Bernard Brodie once put it, that thus far the chief purpose of a military establishment has been to win wars. From now on, its chief purpose must be to avert them. Here is the core of the paradox of deterrence. But how to avert? How to manage the existence of nuclear weapons so as to ensure that they never have to be used?

The Shield of Deterrence

Some suggested that the West should cut the Gordian knot by surrendering all its own nuclear weapons and trusting in the good intentions of the

---

Sir Geoffrey delivered this address at the Royal United Services Institute in London, March 15, 1985. That portion at the end of the address regarding US-Soviet talks has been omitted.

Comrades in Moscow. Such a strategy found little favour in the West; apparently it was never even considered in the East. A second alternative, the idea of a preemptive first strike, was and still is rejected with equal firmness by the peoples and governments of the West.

The remaining option—to maintain a level of forces, conventional and nuclear, which would present a credible defence against any attack and make aggression pay an unacceptable price—was the almost universal choice. Deterrence based on the whole spectrum from conventional to nuclear forces became the basis for the defence of this country and our allies.

We have lived with that choice, the decision to deter the enemy rather than attack or surrender to him, ever since; this Government and all its predecessors have supported it wholeheartedly. And we have enjoyed forty years of peace. Rather than die by the nuclear sword, we have lived by the shield of deterrence.

The nuclear age was born out of new technology. And the development of technology has not faltered. This is a fact of life in the twentieth century. Twenty years ago, the prospect of limited defences against the nuclear threat began to emerge. Simultaneously, even the theoretical possibility of a successful preemptive attack began to disappear. Technology created survivable forces which were capable of riding out any attack, even by nuclear weapons, and still retaliating in a second strike. In addition, the concept that negotiations between adversaries could actually limit, and eventually reduce, the levels of forces on both sides—and thus enhance the stability of deterrence—assumed a new importance.

Thus is was that in 1972 the United States and the Soviet Union reached an historic agreement, the ABM Treaty; this was the intended basis for the future strategic balance. Under the ABM Treaty the United States and the Soviet Union committed themselves for an unlimited period to restrict active defences against the nuclear threat from ballistic missiles to a very low level—so low that the United States subsequently decided that any deployments of such defences were not worth pursuing. The Russians, for their part, provided only for unlimited protection of Moscow.

The ABM Treaty reflected the agreement that there could be no winner in a nuclear conflict and that it was a dangerous illusion to believe that we could get round this reality. At the same time, the first SALT agreement set limits on the future expansion of nuclear forces.

The net effect was to eliminate the option of full-scale deployment of defensive systems, perceived in 1972 to be destabilising, costly and in any case ineffective; and to enhance the strategy of nuclear deterrence through the clear recognition of mutual vulnerability.

The common acronym for the latter is MAD. In my view the 'D' should stand for Mutual Assured Deterrence, not destruction. This Government, in exactly the same way as its predecessors, is not committed to destruction. It is committed to preserving peace by ensuring that potential aggressors are deterred from threatening it. We are confident that this will continue to be effective so long as would-be aggressors are faced with the credible prospect of unacceptable damage in return.

The Soviet Build-Up

To return to the historical analysis, the march of technology continued. As limits were placed on some elements of nuclear forces, others began to multiply. Any assumption that the Russians would match in the 1970s the degree of self-restraint demonstrated by the Americans proved ill-founded. Over the past fifteen years, the Soviet Union has been improving and modernising its strategic forces to a much greater extent than the United States. As a result the Soviet Union's weapons are in general much newer. It is currently developing and testing a new generation of ICBMs and SLBMs, a new strategic bomber, and a complete new family of cruise missiles. While the United States perhaps retains a slight advantage in overall warhead numbers, the Soviet Union has more delivery systems. It also has more warheads on its missiles; and it has nearly three times as much throw-weight.

There is no doubt that the present Soviet capabilities cause justifiable concern on our side. Their build-up over the past decade goes far beyond the reasonable requirements necessary for the defence of the Soviet Union. It is in the light of this Soviet programme that the United States and its allies felt compelled to take steps to modernise their own forces.

In Intermediate-range Nuclear Forces (INF), there is still a marked imbalance in favour of the Warsaw Pact, despite the belated deployments by NATO of a limited number of ground-launched cruise missiles and Pershing IIs. The continuing build-up of SS-20s remains of particular concern. Major improvements are taking place in Soviet shorter range dual-capable systems; improvements are also under way in Soviet battlefield nuclear artillery.

In 1985 we now face once again a range of options to preserve our defence into the next century. We could disarm and throw away our weapons. To do so would be to rob ourselves of the chance to negotiate balanced reductions with our adversaries, or to exploit the new advantages which technology may have to offer. Alternatively, we could indulge in a new orgy of armament. This would mean seeking a dubious safety in levels of forces far above the minimum we now believe necessary to prevent attacks upon us. The Government rejects both these courses.

Evidently, we still have the third option: to maintain sufficient forces to deter any aggression against us and our allies; and to seek at the same time balanced reductions in these forces on both sides. Deterrence must remain credible, while the political, strategic and financial costs of maintaining such levels is reduced. Despite weaknesses in some areas, I believe that present and planned US and NATO nuclear forces provide a stable balance of deterrence and will continue to do so. I emphasise that the Western modernisation programme is an essential factor in maintaining this balance. It is no good accepting the principle of deterrence and then failing to will the means needed to maintain the effectiveness of the deterrent.

Strategic Defense Initiative

And now, with President Reagan's Strategic Defense Initiative, a fourth possible option has been reintroduced, after more than a decade in limbo. The strategic debate is focussing on new possibilities for active defence against the nuclear threat.

In our approach to these new issues the Government believes that certain well-tested premises retain their validity: the need for strong and credible military forces, the need to maintain deterrence, and the need to pursue arms control measures. The four points agreed last December at Camp David between the prime minister and President Reagan, and reaffirmed during their Washington discussions last month, are of prime importance in this context. These points are by now familiar. But as they represent the basis for the British Government's approach to the strategic future, their restatement is in order.

As the prime minister has said, she agreed with the president that:

— the US and Western aim is not to achieve superiority but maintain balance, taking account of Soviet developments;
— SDI-related deployment would, in view of treaty obligations, have to be a matter for negotiation;
— the overall aim is to enhance, not undercut, deterrence; and
— East-West negotiation should aim to achieve security with reduced levels of offensive systems on both sides.

Thus our policy as we face the new challenge continues to be consistent. But what are the real problems to be resolved and the new questions to be answered?

President Reagan, in the historic address which launched the Strategic Defense Initiative almost two years ago, spoke of his vision that new technology might make it possible to create comprehensive defences against nuclear attack. These could render ballistic weapons impotent and obsolete; they could free the peoples of the world once and for all from the threat of nuclear annihilation.

From the start, such a vision was always recognised as subject to uncertainty. As the president himself said in March 1983, it will take years, probably decades of effort on many fronts. There will be failures and setbacks just as there will be successes and breakthroughs. Subsequent statements in Washington have underlined the tentative nature of the venture.

Weapons in Space

Nonetheless, the president's vision has already made a decisive impact in several respects. It has focussed interest on military activities in space and on new weapons systems which might theoretically be deployed or aimed there. It has also drawn to public notice the very considerable research under way in the Soviet Union on a range of potential defensive measures.

Not enough attention has been paid to this Soviet research. It is extensive and far-reaching and has been going on for many years. Any discussion of future Western strategies must take full account of it. To ignore or to dismiss what is happening in the Soviet Union would be not only myopic; it would be dangerous.

Given the dimensions of space, it is hardly surprising that imaginative schemes for its military exploitation have been almost infinite; so too have been the misconceptions and distortions of what is happening now and may happen in the future. To use terms such as Star Wars about either US or Soviet intentions is to distort the very real problems and their potential solutions.

Equally, in Soviet calls for the demilitarisation of space I see more propaganda than substance. Activities in space with military relevance are not by definition evil. It is neither feasible nor desirable to try to preclude all of them. Current Soviet rhetoric makes a less than serious contribution to a most serious debate. Greater precision and deeper thought are required. For a start we must distinguish between present military activities in space and those that may at some far-off point in the future achieve reality. And at all times we must keep in mind the key question: will new developments enhance or undercut deterrence? At present, space is used by a limited number of military systems, on both sides.

First, communications and surveillance satellites which add significantly to the effectiveness and credibility of Western defences, and thus to their deterrent effect. Efficient and cost-effective, they provide a unique contribution to the defence of the West.

Second, reusable launchers. These pack-horses of the space age are equally valuable. Nor by their nature do they pose a real threat of aggression; the shuttle is too limited, too costly and too vulnerable a platform for that purpose.

Third, there is the potential use of space for the delivery of nuclear warheads by ballistic missiles. We must seek to ensure that this will always remain an unrealised potential.

Lastly, we face the problem of antisatellite systems, exacerbated by the Soviet deployment over the past decade of a limited capability in this field. It would be a serious blunder if the West allowed the Russians to continue to enjoy their present monopoly.

The US intention to balance the established Soviet capability in this field is logical and prudent. On the other hand, we must recognise the heavy Western dependence upon the existing utilisation of space technology and particularly upon satellites for intelligence purposes. We must also recognise that the prospect, at a time of crisis, of either side being faced with the loss of its strategic eyes and ears would be gravely destabilising. It could provoke a new and even more threatening stage in any East-West confrontation.

The West must therefore strive to make its satellites less vulnerable. But there may equally be good grounds for negotiating some constraints upon elements of antisatellite activity. I welcome the readiness expressed by President Reagan at the UN last September to consider mutual restraint while the negotiations which have now been launched explore possibilities for concrete agreements.

One other factor must be recognised: the linkage between the development of antisatellite capabilities and the potential development of defences against ballistic missiles. It could be argued that by imposing constraints on antisatellite activity, decisions on the development of ballistic missile defences would be preempted before they could be properly considered. That however would be to ignore the important differences in the time-scales involved.

In the case of antisatellite systems, the future is now. The Soviet Union has already deployed such a system at low altitude, and the United States is in the middle of a successful testing programme. By contrast, any development beyond the research stage of defences against ballistic missiles, the most immediate nuclear threat, is in the prime minister's words many, many years away.

The Government takes the view that, if negotiations were to succeed in imposing mutual constraints on antisatellite systems, these could have a helpful impact over a period of years. We should take that opportunity now, if it is in the Western interest. Any such ASAT agreement could be limited if necessary to a fixed period, in order not to prejudge the future.

Against that background of present space activities, I should like now to consider the longer-term issue of active defences against ballistic missiles, and particularly defences which might be deployed in or into outer space. By active defences I mean systems specifically designed to prevent enemy missiles getting through, rather than deterrence based on the threat of retaliation.

Much has been said and written about President Reagan's Strategic Defense Initiative. The first point to make is that, as US spokesmen have made clear, this is a research programme, conducted in full conformity with the limits of the ABM Treaty. As a research programme, it is also full of questions. The answers may be clear or obscure. They may not even emerge at all. As the US Administration themselves recognise, the programme is geared to a concept which may in the end prove elusive.

The second point is that treaty obligations specifically allow for research to continue into defensive systems. Evidently, it is pointless to try to impose constraints which cannot be verified. Most activities in laboratories or research institutes come into that category. The ABM Treaty recognised this when it drew a distinction between research on the one hand and development, testing and deployment on the other.

The third, equally important point is that a balance must always be maintained between US and Soviet capabilities, in research as in other aspects. Given what we know of Soviet activities in the research field over a number of years, there is a clear need for the United States to match the present stage in Soviet programmes. It is for this reason that the prime minister has repeatedly expressed our firm conviction that US research should go ahead.

SDI: Implications of Deployment

But what should happen if and when decisions are required on moving from the research to the development stage? In evaluating the results of research, and in taking any such decisions, we shall need to ask ourselves some very basic questions about the future nature of Western strategy. In particular, we shall have to consider how best to enhance deterrence, how best to curb rather than stimulate a new arms race. At that stage, the judgements to be made will only partly depend upon technical assessments about the feasibility of defences. Even if the research shows promise, the case for proceeding will have to be weighed in the light of the wider strategic implications of moving down the defensive road.

But can we afford even now simply to wait for the scientists and military experts to deliver their results at some later stage? Have we a breathing space of five, ten, fifteen years before we need to address strategic concerns? I do not believe so. The history of weapons development and the strategic balance shows only too clearly that research into new weapons and study of their strategic implications must go hand in hand. Otherwise, research may acquire an unstoppable momentum of its own, even though the case for stopping may strengthen with the passage of years. Prevention may be better than later attempts at a cure. We must take care that political decisions are not preempted by the march of technology, still less by premature attempts to predict the route of that march.

The questions to be faced are complex and difficult. I want to emphasise that they are questions and not prior judgements. The answers will only emerge in time. There would inevitably be risks in a radical alteration of the present basis for Western security. How far would these risks be offset by the attractions of adopting a more defensive posture: that is to say, of developing what might prove to be only a limited defence against weapons of devastating destructive force. Could the process of moving towards a greater emphasis on active defences be managed without generating dangerous uncertainty?

Let us assume that limited defences began to prove possible, and key installations began to be protected by active defences. In his 1983 address President Reagan himself acknowledged that a mix of offensive and defensive systems could be viewed as fostering an aggressive policy. Uncertainty apart, would the establishment of limited defences increase the threat to civilian populations by stimulating a return to the targeting policies of the 1950s?

Most fundamental of all, would the supposed technology actually work? And would it, as Mr. Paul Nitze has noted, provide defences that not only worked but were survivable and cost-effective? These are the key questions to be answered by the research that is being undertaken on both sides.

It would be wrong to underestimate the enormous technological expertise and potential of the United States; but, as we all recognise, there would be no advantage in creating a new Maginot Line of the twenty-first century, liable to be outflanked by relatively simpler and demonstrably cheaper counter-measures. If the technology does work, what will be its psychological impact on the other side? President Reagan has repeatedly made it clear that he does not seek superiority. But we would have to ensure that the perceptions of others were not different.

What are the chances that there would be no outright winner in the everlasting marathon of the arms race? And if the ballistic missile indeed showed signs of becoming, in President Reagan's words, impotent and obsolete, how would protection be extended against the nonballistic nuclear threat, the threat posed by aircraft or cruise missiles, battlefield nuclear weapons or, in the last resort, by covert action. What other defences in addition to space-based systems would need to be developed, and at what cost, to meet these continuing threats?

If it initially proved feasible to construct only limited defences, these would be bound to be more vulnerable than comprehensive systems to counter-measures. Would these holes in the dyke produce and even encourage a nuclear flood? Leaving aside the threat to civilian populations, would active defences provide the only feasible way of protecting key military installations? Might we be better advised to employ other methods of protection, such as more mobile and under-sea forces?

Finally on the technology side, could we be certain that the new systems would permit adequate political control over both nuclear weapons and defensive systems; or might we find ourselves in a situation where the peace of the world rested solely upon computers and automatic decision-making?

Then there is the question of cost. The financial burden of developing and deploying defences goes far beyond the additional cost of providing defences against the nonballistic missile threat. No one at present can provide even a guesstimate of the total sums involved. But it is fair to assume that these will run into many hundreds of billions of dollars.

We know only too well that our defences must be cut to the cloth of our financial resources. We shall have to ask ourselves not only whether the West can afford active defences against nuclear missiles. We must also ask whether the enormous funds to be devoted to such systems might be better employed.

Deterrence

Are there more cost-effective and affordable ways of enhancing deterrence? Might it be better to use the available funds to improve our capability to oppose a potential aggressor at a time of crisis with a credible, sustainable and controllable mix of conventional and nuclear forces? In short, how far will we be able to impose new burdens on defence budgets already under strain? And what would be the effect on all the other elements of our defences, on which Western security will continue in large part to depend?

The implications for arms control must also be carefully considered. Would the prospect of new defences being deployed inexorably crank up the levels of offensive nuclear systems designed to overwhelm them? History and the present state of technology suggest that this risk cannot be ignored. Or could the same prospect—the vision of effective defences over the horizon—provide new incentives to both sides to start at once on reducing their present levels? This explains the importance of the second point agreed at Camp David last December.

In his statement to Congress last month President Reagan spoke of the need to reverse the erosion of the ABM Treaty. It represents a political and military keystone in the still shaky arch of security we have constructed with the East over the past decade and a half. But to go beyond research into defensive systems would be inconsistent with the terms of the ABM Treaty as it stands. It was agreed at Camp David last December that any deployment beyond those limits would have to be a matter for negotiation. We would have to be confident that that formidable task could actually be managed on a mutually acceptable basis.

We have heard recently from Moscow a lot of dogmatic statements and preconditions for the success of the new talks. I discount much of these. But I do attach importance to convincing the Soviet leadership that we in the West are indeed serious in our aim of maintaining strategic stability at significantly lower levels of nuclear weapons. We do not want to give them the impression that we have something else in mind. We are serious about arms control. And we must be seen and heard to be so.

Finally, as members of the Atlantic Alliance, we must consider the potential consequences for this unique relationship. We must be sure that the US nuclear guarantee to Europe would indeed be enhanced as a result of defensive deployments. Not only enhanced at the end of the process, but from its very inception.

Many years of deployments may be involved. Many years of insecurity and instability cannot be our objective. All the allies must continue at every stage to share the same sense that the security of NATO territory is indivisible. Otherwise the twin pillars of the alliance might begin to fall apart.

Other things being equal, we welcome any cost-effective enhancement of deterrence to meet palpable weaknesses on the Western side. But we also have to consider what might be the offsetting developments on the Soviet side, if unconstrained competition in ballistic missile defences beyond the ABM Treaty limits were to be provoked. In terms of NATO's policy of forward defence and flexible response, would we lose on the swings whatever might be gained on the roundabouts?

I have posed a lengthy list of questions, to which the answers cannot be simple. Some do not admit of answers now. But that does not acquit us of the duty to pose them. They are questions so vital to our future that we cannot afford to shrug them off. It is right to ponder and debate them as research continues. In this way we stand the best chance of reaching the right policies. The attractions of moving towards a more defensive strategy for the prevention of war is as apparent as are the risks. It would be wrong to rule out the possibility on the grounds that the questions it raises are too difficult.

But the fact that there are no easy answers, that the risks may outweigh the benefits, that science may not be able to provide a safer solution to the nuclear dilemma of the past forty years than we have found already—all these points underline the importance of proceeding with the utmost deliberation. Recent testimony to the US Congress by Secretary Shultz and Mr. Nitze, and the continual process of alliance and consultation, confirm the prime minister's statement last December that the US Administration sees matters in very much the same light.

Deterrence has worked; and it will continue to work. It may be enhanced by active defences. Or their development may set us on a road that diminishes security. We do not know the answer to that question. Meanwhile, four clear points emerge.

First, as the prime minister reminded the United States Congress last month, in the words of Sir Winston Churchill: Be careful above all things not to let go of the atomic weapon until you are sure, and more than sure, that other means of preserving peace are in your hands.

Secondly, impressions can be created by words as well as deeds. Policies, aims, visions—all these can and must be clearly stated. Without the approval of an informed public, the governments of the West are wasting their breath. But we must be especially on our guard against raising hopes that it may be impossible to fulfil. We would all like to think of nuclear deterrence as a distasteful but temporary expedient. Unfortunately we have to face the harsh realities of a world in which nuclear weapons exist and cannot be disinvented. Words and dreams cannot by themselves justify what the prime minister described to the United Nations as the perilous pretence that a better system than nuclear deterrence is within reach at the present time.

Thirdly, any deployments of space-based or other defences must be a matter for negotiation. The prime minister agreed this with President Reagan at Camp David, and they reaffirmed it in Washington. In the words of the White House statement of January 3, 1985, deployments of defensive systems would most usefully be done in the context of a cooperative, equitable and verifiable arms control environment. A unilateral Soviet deployment of such defences would destroy the foundation on which deterrence has rested for twenty years. I warmly welcome the clear statements of the US administration's view that deployments would need to be a cooperative endeavour, to be embarked upon in an arms control context.

Fourthly, the linkage between offensive and defensive systems. The White House statement of January 3 recognised the merit of controlling both the offensive and defensive developments and deployments on both sides. If defensive systems are to be deployed, they will be directed against the then levels of offensive forces. If the latter can be lowered dramatically, then the case for active defences may be correspondingly strengthened. Conversely, radical cuts in offensive missiles might make the need for active defences superfluous. Equally, the effectiveness of defences will be directly governed by the numbers of missiles and warheads which they are intended to destroy. If the levels rise dramatically, then the effectiveness of defences may not be adequate.

It is therefore clear that there is and will continue to be an integral relationship between measures to control offensive forces and any decisions to move to the development of active defences. The US administration have always recognised such a linkage. Belatedly, the Russians now seem to have reached the same conclusion. As the new negotiations, finally launched this week, get underway, a key question for all our futures will be the extent to which reductions in offensive forces prove possible and the impact this will have upon the incentive to develop defences.

US-Soviet Talks

We warmly welcome the renewal of US-Soviet talks. The joint communique agreed between them as the basis for the talks describes their aim as the prevention of an arms race in space and its termination on earth. We applaud and endorse that aim. There is no doubt however that progress will be painstaking and slow. We cannot expect sudden breakthroughs. In the words of Tolstoy's General Kutuzov, patience and time will have to be our watch words. There is no quick or easy route to success.

25

Britain and the Strategic Defense Intitiative

Michael Heseltine

THE British government's policy towards the Strategic Defense Initiative remains firmly based on the four points agreed between the prime minister and President Reagan in December 1984: that the Western aim is not to achieve superiority, but to maintain balance taking account of Soviet developments; that SDI-related deployment would, in view of treaty obligations, have to be a matter for negotiation; that the aim is to enhance, and not to undermine deterrence; and that East-West negotiation should aim to achieve security with reduced levels of offensive weapons on both sides. It was in that context that, at Camp David, the prime minister told President Reagan of her firm conviction that the SDI research program should go ahead as a prudent hedge against Soviet activities in the same field.

Earlier this year [1985], the United States invited her NATO and certain other allies to participate in the SDI research program. Following that invitation, we have engaged in detailed discussions with the United States Government on the nature and scope of the research which could sensibly be undertaken by United Kingdom firms and institutions. Those complex discussions have now been completed and agreement has been reached on an information exchange program, on the areas where British companies and institutions have enterprise which might form part of the United States-funded SDI research program, and on the mechanisms to facilitate that cooperation.

The confidential memorandum of understanding reached between the two governments safeguards British interests in relation to the ownership of intellectual property rights and technology transfer, and provides for consultative and review mechanisms in support of the aims of the memorandum.

Mr. Heseltine made this statement to the House of Commons on December 10, 1985.

The SDI research program goes to the heart of the future defence technologies. Participation will enhance our ability to sustain an effective British research capability in areas of high technology relevant to both defence and civil programs.

Now that agreement has been reached on the memorandum, British companies, universities and research institutions have the opportunity to compete on a clearly defined basis for the research contracts which are on offer from the United States Government, as well as to participate in an information exchange program on a fully reciprocal basis for the mutual benefit of the United Kingdom and the United States.

To act as a focal point for British participation, and to liaise with the United States SDI participation office, I am establishing immediately within the Ministry of Defence an SDI participation office with representation from other interested departments. That office will work in the closest concert with British firms and institutions interested in such participation.

This agreement opens for Britain research possibilities which we could not afford on our own in technologies that will be at the center of tomorrow's world. It will bring jobs that would otherwise be created abroad, and I commend it to the House.

26

Press Conference on the Strategic Defense Initiative

Brian Mulroney

MULRONEY: On March 26 [1985] the United States invited Canada and other friendly countries to participate directly in research under the Strategic Defense Initiative. After careful and detailed consideration the government of Canada has concluded that Canada's own policies and priorities do not warrant a government to government effort in support of SDI research. Although Canada does not intend to participate on a government to government basis on the SDI research program, private companies and institutions interested in participating in the program will continue to be free to do so.

As stated in the House of Commons on January 21, 1985, by the secretary of state for external affairs, this government believes that SDI research by the United States is both consistent with the ABM Treaty and prudent in light of significant advances in Soviet research and deployment of the world's only existing ballistic missile defense system.

I conveyed this decision today to the president of the United States and informed him of this. I had discussed it, as you might imagine, with my caucus and cabinet. And that is our position with regard to this particular item.

QUESTION: *Is there a danger, a possibility, that this decision could weaken Canada's position when the time comes to negotiate other agreements, for example dealing with the expansion of trade or with acid rain, et cetera?*

Canadian Prime Minister Mulroney held this press conference on September 7, 1985. All questions and answers in the French language have been translated by Wayne C. Thompson.

MULRONEY: No. The decision was taken by the Canadian national government in the interests of Canada. We are friends and loyal allies of the United States and are members of NATO, and we have proven ourselves in all these domains. But the government of Canada is also bound by the obligation always to act in accordance with Canada's interests. We recognize the fact that, under the circumstances, such research by the United States is prudent, and we of course remain committed to our roles as ally and member of NATO. We have nevertheless decided that the interests of Canada in maintaining a foreign policy which is at all times independent are the sole criteria which have motivated our decisions. This is the sole objective criterion which motivated today's decision. And I can tell you that I transmitted this as such to President Reagan, as a decision in the interest of Canada, and he understood.

QUESTION: *How much disappointment do you think there will be in the White House by Canada's decision not to embrace this invitation?*

MULRONEY: I don't think there will be any disappointment in the White House. This is a decision made by a sovereign state, an independent foreign policy which we have always had. The government of Canada was the first to provide support for the United States of America on January 21st with regard to the research initiatives it had undertaken on SDI. We stated publicly in the House of Commons that such research was prudent. Lord Carrington came to Ottawa and indicated that that degree of prudence was required, in his judgment, for the alliance. We have supported the concept of the research, as I indicated in an interview with Global Television. Only a naive six-year-old would fail to understand that the Americans are involved in this research because the Soviets have been doing it for a long period of time, expended billions of dollars, and committed thousands of personnel to it. And so the government of Canada felt that it was appropriate as an ally and a member of NATO that we convey that support on the 21st of January. Since that time we were handed an invitation to participate on a government to government basis by Secretary Weinberger. And I have instructed the minister of defence today to advise Secretary Weinberger that that is not in our national interest and that we will not be accepting the invitation to participate on a government to government basis.

QUESTION: *Why isn't it in Canadian national interest to participate on a government to government basis?*

MULRONEY: For the reasons that we have given, the reasons that I have given. It does not diminish our commitment to the Alliance or our belief in the prudence. But some of you may remember my statement in Baie Comeau when I was asked some time ago about getting involved in a situation where the parameters are beyond our control and where the government of Canada does not call the shots. Our national integrity and our national commitment is to the welfare of Canada, and the conduct of the foreign policy of Canada will always be in the interests of this nation. And that decision was taken on that basis.

QUESTION: *Mr. Prime Minister, does the decision which you are announcing today change anything in your political position announced in January, namely that you support the concept of the Strategic Defense Initiative?*

MULRONEY: As Mr. [Joe] Clark [secretary of state for external affairs] said, we support the research which the United States has undertaken. Isn't that right? That with which we are dealing today is a direct question. Mr. [Caspar] Weinberger has sent Mr. [Erik] Nielsen [then minister of defence] a letter inviting us to participate in a research project on a government to government basis. We considered that. I ordered the formation of a parliamentary commission in order that we could learn about the opinion of Canadians. That opinion seems to be divided. More and more we see it to be in favor of some kind of participation; nevertheless, opinion is essentially split. But, be that as it may, the government came to the conclusion that there can be no question of our participating on a government to government basis, and I gave Mr. Nielsen instructions today to inform Mr. Weinberger of this decision.

27

Reykjavik Summit: The Soviet View

Mikhail Gorbachev

GORBACHEV: . . . One . . . problem in view of our setting about the practical elimination of nuclear weapons is this: Each side should have a guarantee that during that time the other side will not be seeking military superiority. I think that that is a perfectly fair and legitimate requirement, both politically and militarily.

Politically, if we begin reductions, we should take care that all existing brakes on the development of new types of weapons be not only preserved but also strengthened. Militarily, indeed, care should be taken to preclude the following situation: As both sides are reducing their nuclear potentials and while the reduction process is underway, one of the sides secretly contemplates and captures the initiative and attains military superiority. This is inadmissible. I apply this to the Soviet Union. And we have all rights to lay similar demands on the American side.

In this connection, we raised the question in the following way: When we embark on the stage of a real, deep reduction and, after ten years, the elimination of the nuclear potential of the Soviet Union and the United States, it is necessary that during this period the mechanisms restraining the arms race, especially those like the ABM Treaty, should not be disturbed. These mechanisms should be consolidated.

Our proposal was reduced to the following: The sides consolidate the ABM Treaty of unlimited duration by assuming equal pledges that they shall not use the right to withdraw from the treaty within the next ten years.

Printed here are extracts from General Secretary Gorbachev's October 12, 1986, press conference in Reykjavik, Iceland, and from his October 14 television speech in Moscow. Transcripts were released by the Information Department of the Soviet Embassy, Washington, D.C.

Is this proposition correct and logical? It is logical. Is it serious? It is serious. Does it meet the interests of both sides? It does meet the interests of both sides.

Simultaneously, we suggested that all ABM requirements be strictly observed within these ten years, that the development and testing of space weapons be banned and only research and testing in laboratories be allowed. What did we mean by this? We are aware of the commitment of the US administration and the president to SDI. Apparently, our consent to its continuation and to laboratory tests offers the president an opportunity to go through with research and eventually to make clear what SDI is, what it is about, although it is already clear to many people, ourselves included.

And it was at that point that a true battle of two approaches to world politics, including such questions as the termination of the arms race and a ban on nuclear weapons, began. The US administration and the president insisted to the end that America should have the right to test and study everything involved in SDI not only in laboratories but elsewhere, including outer space. But who will agree to this? A madman? But madmen, as a rule, are kept where they should be, where they are given medical care. Anyway, I do not see any in leadership positions, especially at the helm of states.

We were on the brink of taking major, historic decisions because up to then the point had always been merely arms reductions. We took decisions on ABM, SALT I, SALT II, etc. And since the US administration, as we understand now, is confident of US technological superiority and is hoping to achieve through SDI military superiority, it has even gone as far as burying the accords already achieved. We suggested that instructions be given for drawing up treaties with a view to their practical fulfillment. They could be signed during a meeting in Washington, but the American side torpedoed all this. I told the president that we were missing a historic chance. Our positions had never been so close.

Bidding me good-bye, the president said that he was disappointed that I had from the outset come unwilling to look for agreements and accords. Why do you display such firmness on SDI and the problem of testing, all this range of problems, because of one word? But I think that the matter is not words but substance. Herein lies the key to understanding what the US administration has on its mind. And I think that it has on its mind what, as I now see, is on the mind of the military-industrial complex. The administration is captive to the complex and the president is not free to take such a decision. We had breaks and held debates, and I saw that the president was not given support. And that was why our meeting failed when we were already close to producing historic results. . . .

QUESTION (*Pravda*): *Mikhail Sergeyevich, what's your opinion? Why did the US administration decide to wreck the negotiations, taking such an irresponsible decision and ignoring world public opinion?*

GORBACHEV: I think that America has yet to make up its mind. I think it has not done this yet. This, as we felt, was evident in the president's stand. . . .

QUESTION (Icelandic Radio and Television): *After the negative result of the summit, will the Soviet Union counter the American SDI program with something else and will it not launch its space arms program full blast?*

GORBACHEV: I think that you have understood the essence of the Soviet position. If we have now approached a stage at which we start a drastic cut in nuclear weapons, both strategic and medium-range missiles—we have already approached an understanding with the Americans to do this in the next decade—we have the right to demand that we should have a guarantee that during this period nothing surprising and unforeseen will take place. This also includes such spheres as space and the deployment of space-based ABM systems.

I told the president—perhaps I would open the curtain slightly on our exchange of opinions—that SDI does not bother us militarily. In my opinion, nobody in America either believes that such a system can be created. Moreover, if America eventually decides to do this, our response would not be symmetrical. True, I told him, Mr. President, you know that I have already been turned into your ally on the issue of SDI. He was surprised by that. It turns out, I told him, that since I so sharply criticize SDI, it offers him a convincing argument that SDI is needed. You just say: If Gorbachev is against it, it means that it's a good thing, and you will win applause and financing. True, cynics and skeptics have appeared who say: What if this is Gorbachev's crafty design—to stay out of SDI and to ruin America. So you figure it out for yourself. But, in any case, we are not scared by SDI. I say this with confidence, since it is irresponsible to bluff in such matters. There will be a reply to SDI. An asymmetrical reply, but there will be a reply. And we shall not sacrifice much at that.

But what is its danger? For one thing, there is a political danger. A situation is created right away that brings uncertainty and stirs up mistrust and suspicion of each other. Then the reduction of nuclear weapons will be put aside. In short, quite another situation is needed for us to take up thoroughly the question of reducing nuclear weapons.

Second, there is a military aspect after all. The SDI can lead to new types of weapons. We also can say this with competence. It can lead to an entirely new stage of the arms race with serious, unpredictable consequences.

It appears that, on the one hand, we are supposed to agree to begin a reduction of nuclear weapons—at present the most dangerous and dreadful—and, on the other hand, we should give our blessing to research, and even to its conduct in space, under natural conditions, that would result in the creation of the latest weapons. This does not make sense. . . .

QUESTION (ABC television network): *Mr. General Secretary, I don't understand why, when you had an opportunity to achieve agreement with President Reagan on cuts in nuclear weapons, the Soviet side did not agree to SDI research. You yourself said in Geneva that you were ready to pay a high price for nuclear arms cuts, and now, when you had such an opportunity, you missed it.*

GORBACHEV: Your question contains an element of criticism, so I will answer it in some detail.

First, the US president came to Reykjavik with empty hands and empty pockets. The American delegation, I would say, brought us trash from the Geneva talks. Only thanks to the far-reaching proposals of the Soviet side were we about to reach most major agreements—they were not formalized, mind you—on cuts in strategic offensive weapons and on medium-range missiles. Naturally, we hoped in that situation—and I think it is perfectly clear to a politician, to a military man and to any normal person in general—that if we were to sign such agreements on major cuts in nuclear weapons, we would take care to ensure that nothing could thwart that difficult process, toward which we have been moving for decades.

And then we raised the question that we stood for strengthening the ABM Treaty. The American side is constantly burrowing under the ABM Treaty. It has already called into question SALT II. Now it would like to stage a funeral for the ABM Treaty in Reykjavik—and, moreover, with the participation of the Soviet Union and of Gorbachev. That will not do. The world as a whole would not understand us, it is my conviction.

All of you who are sitting here are convinced that we are worthless politicians if we begin to attack the ABM Treaty, the last mechanism that has contributed so much to constraining, in spite of everything, the arms race. But it is not enough to preserve the terms of that treaty at a time when deep cuts in nuclear weapons are initiated. We think that the treaty must be strengthened. And we proposed a mechanism for strengthening it—not to use the right to pull out of the ABM Treaty during the ten years in which we will totally reduce and destroy the nuclear potential in our countries.

At the same time, to ensure that the Soviet Union does not seek to overtake America in space research and to achieve military superiority and that America does not seek to overtake the Soviet Union, we said that we agreed to laboratory research and testing but opposed the development with that research and testing of components of space-based ABM defense. This is our demand. Our demand was constructive, reckoning with America's stand. If the United States agreed, it would have an opportunity to resolve its problems within the framework of continued laboratory research but without attempts to develop space ABM defenses. I think there is iron logic here, as the children say, and sometimes we should learn even from children. . . .

The Americans think that they will achieve military superiority over us through outer space. One of their presidents said: "He who dominates outer space will dominate Earth." This shows that what we have to deal with is imperial ambitions. But the world today is not what it once was. It does not want to be and will not be the happy hunting grounds of either the United States or the Soviet Union. . . .

America must be very nostalgic about olden times, when it was strong and militarily superior to us, as we all had emerged from the war economically weakened. There must be nostalgia in America. Yet we should wish our American partners to come to grips with today's realities. They ought to do so, too. If the Americans do not start thinking in today's terms and proceeding from today's realities, we will not make progress in our search for correct solutions. . . .

QUESTION (NBC television network): *As I understand it, you are directly calling on other members of the world community to act as a kind of lever in order to influence the United States and make it change its mind.*

GORBACHEV: We know how developed lobbying is in your country, how the political process goes on in America. Perhaps that is why it was difficult for the president to make a decision at this meeting.

But when the matter at hand is related to consolidating peace and undertaking real steps to this end, when concerted efforts are needed—this concerns all, not only the United States and the Soviet Union—then, I think, one should speak not about lobbying, but about the sense of responsibility, the common sense of peoples, about the appreciation of today's peace and the need to protect it.

It is, therefore, insulting to accuse peoples or movements campaigning for peace of being lobbyists for the Soviet Union. The point at issue is that people uphold their political and civic stance. . . .

Two days later, on October 14, 1986, General Secretary Gorbachev addressed the Soviet people in a televised speech concerning the Reykjavik summit. Excerpts from that speech follow.

We strove for the main questions of world politics—ending the arms race and nuclear disarmament—to be given top priority at the meeting in Reykjavik. And that is how it was. . . .

A whole package of major measures was submitted at the talks. These measures, once accepted, would usher in a new era in the life of mankind—the nuclear-free era. This makes up the essence of the radical turn in the world situation, the possibility of which was obvious and realistic. The point at issue was no longer the limitation of nuclear arms, as in the SALT I and SALT II and other treaties, but the elimination of nuclear weapons within comparatively brief periods. . . .

The third question that I raised during my first conversation with the president and which formed an integral part of the package of our proposals, was the existing Anti-Ballistic Missile (ABM) Treaty and the Nuclear Test Ban Treaty. Our approach is as follows: Since we are entering an entirely new situation, when a substantial reduction of nuclear weapons and their elimination in the foreseeable future will be started, it is necessary to protect ourselves from any unexpected developments. We are speaking about weapons which to this day make up the core of this country's defenses.

Therefore, it is necessary to exclude everything that could undermine equality in the process of disarmament, to preclude any possibility of developing weapons of a new type ensuring military superiority. We regard this stance as perfectly legitimate and logical.

And since that is so, we firmly stated the need for strict observance of the 1972 ABM Treaty of unlimited duration. Moreover, in order to consolidate its regime, we proposed to the president adopting a mutual pledge by the US and the Soviet Union not to use the right to pull out of the treaty for at least ten years while abolishing strategic weapons within this period.

Taking into account the particular difficulties which the administration created for itself on this problem, when the president personally committed himself to space weapons, to the so-called SDI, we did not demand the termination of work in this area: The implication was, however, that all provisions of the ABM Treaty will be fully honored—that is, research and testing in this sphere will not go beyond laboratories. This restriction applies equally to the US and to the USSR. . . .

As for the ABM, we propose to preserve and strengthen this fundamental, important agreement, and you want to give it up and even propose to have it replaced with some new treaty, and thereby—following the departure from SALT II—also to wreck this mechanism standing guard over strategic stability. This is incomprehensible too.

We grasped the essence of the SDI plans as well, I said. If the United States creates a three-tiered ABM system in outer space, we shall respond to it. However, we are concerned over another problem: The SDI would mean the transfer of weapons to a new medium, which would destabilize the strategic situation and make it even worse than today. If this is the purpose of the US, then this should be said plainly. But if you really want to have reliable security for your people and for the world in general, then the American stand is absolutely unsupportable. . . .

The ABM Treaty in this situation acquired key significance. Its role was becoming even more important. Could one wreck, I said, what has so far made it possible to somehow restrain the arms race? If we now start reducing strategic and medium-range nuclear weapons, both sides should be confident that over that time nobody will develop new systems which would undermine stability and parity. Therefore, in my view, it is absolutely logical to fix the time frame—the Americans mentioned seven years, and we proposed ten years during which nuclear weapons ought to be eliminated. We proposed ten years during which neither the Soviet side nor the American side will avail itself of the right—and they have such a right—to withdraw from the ABM Treaty, and to conduct research and tests only in laboratories. Thus, I think you understood why exactly ten years. This is not casual. The logic is simple and honest. Fifty percent of the strategic armaments are to be reduced in the first five years and the other half in the next five years. This makes it ten years.

In this connection I proposed to instruct our high-ranking representatives to start full-scale talks on the discontinuation of nuclear explosions in order to ultimately work out an agreement on banning them once and for all. In the course of the preparation of the agreement—and here we again displayed flexibility and assumed a constructive stand—specific problems connected with nuclear explosions could be resolved simultaneously.

In answer we again heard from President Reagan the reasoning which had been familiar to us since Geneva and from his public statements: that the SDI was a defense system, that if we started to eliminate nuclear weapons, how could we protect ourselves from some madman who might get hold of them, that he was ready to share with us the results obtained within the framework of the SDI. Answering this last remark, I said: Mr. President, I do not take seriously your idea of sharing the SDI developments with us. You do not want to share even oil equipment and equipment for dairy factories with us, and still you expect us to believe your promise to share SDI developments with us. It would be a kind of "second American revolution," and revolutions do not happen too often. I said to President Reagan: Let us be realists and pragmatists. It is more reliable this way. The things we are talking about are too serious.

By the way, yesterday, trying to justify his stand on the SDI, the president said he needed the program for America and its allies to remain invulnerable against a Soviet missile attack. As you see, in this case he already made no mention of madmen. The "Soviet threat" was again brought to light.

But this is nothing but a trick. We suggested that not only strategic armaments but also all the nuclear armaments in the possession of the US and the USSR be eliminated under strict control. Why the need to protect the "freedom of Americans" and their friends from Soviet nuclear missiles since these missiles will no longer exist? If there are no nuclear weapons, why have protection against them? It means that the entire Star Wars undertaking is of a purely military nature and is directed at gaining military superiority over the Soviet Union. . . .

In these discussions, the president sought to handle ideological problems as well, demonstrating, to put it mildly, total ignorance and inability to understand both the socialist world and what is taking place in it. I rejected the attempts to link ideological differences with questions of ending the arms race. I persistently drew the president and the secretary of state back to the subject that brought us to Reykjavik. It was necessary to remind our interlocutors again and again about the third element of the package of our proposals, without which it was impossible to reach accord on the whole. I mean the need for strict compliance with the ABM Treaty, consolidating the regime of this major treaty and banning nuclear tests.

We had to draw attention again and again to what seemed to be perfectly clear: Since we agreed to effect deep reductions in nuclear arms, we needed to create a situation that would preclude attempts—both in deeds and in thoughts—to shake strategic stability or circumvent the agreements. That is why we should be confident about the preservation of the ABM Treaty of unlimited duration. You, Mr. President, I said, ought to agree that once we start reducing nuclear weapons, there should be firm confidence that the US will do nothing behind the back of the USSR, while the Soviet Union will not do anything that would jeopardize US security, that would depreciate the agreement and create difficulties behind your back.

Hence the key task is to strengthen the ABM regime and not to go into outer space with the results of work under this program, which should remain within laboratories. Ten years of not using the right to pull out of the ABM Treaty are necessary to create the confidence that, while resolving the problem of reducing arms, we ensure security for both sides and ensure worldwide security. But the Americans obviously had other intentions. We saw that the US actually wants to weaken the ABM Treaty, to review it so as to develop a large-scale space-based ABM system for its own egoistic ends. To agree with this would be simply irresponsible on my part.

As for nuclear tests, here too it was totally clear why the American side did not want to conduct talks on this issue in earnest. It would have preferred to make them timeless, to put off the solution of the problem of banning nuclear tests for decades. And once again we had to reject attempts to use talks as a screen for a free hand in the field of nuclear explosions. I stated frankly: I am having doubts about the honesty of the US position [and am wondering whether] there is anything in it that might inflict damage to the Soviet Union. How can one reach agreement on the elimination of nuclear arms if the United States continues perfecting them? Still we had the impression that the SDI was the main hitch. If it had been removed, we would have had an opportunity to reach an accord on the banning of nuclear explosions as well.

At a certain stage of the talks, when it became absolutely clear that to continue the discussion would be a waste of time, I reminded the other side: We have proposed a definite package of measures, and I ask you to consider it as such. If we have worked out a common position on the possibility of a major reduction of nuclear arms and at the same time failed to reach agreement on the matter of the SDI and nuclear tests, then everything that we have tried to create here falls apart.

The president and the secretary of state took our firmness badly. But I could not pose the question in a different way. This concerned the security of our country, the security of the whole world, all peoples and continents. We proposed major, really scopeful things clearly in the nature of a compromise. We made concessions, but on the American side we did not see even the slightest desire to respond in kind, to meet us halfway. We were deadlocked, and we began thinking about what to conclude the meeting with. Yet we continued the efforts to make our partners engage in a constructive dialogue. . . .

But this time, too, our attempts to come to terms were to no avail. For four hours we were again trying to persuade the interlocutors that our approach was well-founded, that it threatened them with nothing and did not affect the interests of the genuine security of the United States. But the further we went, the clearer it became that the Americans would not agree to limit SDI research, development or tests to laboratories. They are bent on going to outer space with weapons.

I said firmly that we would never agree to help undermine the ABM Treaty with our own hands. This is, to us, a question of principle, a question of our national security. Thus, being virtually one or two to three steps from taking decisions which could become historic for the whole nuclear-space era, we were unable to make that step or those steps. No turning point in the world's history occurred though, I say it again with full confidence, it was possible. However, our conscience is clear, and no one can reproach us with anything. We did all we could. . . .

Foreigners asked me there in Iceland and my comrades ask me here: What, in my opinion, are the root causes of this attitude of the American delegation at the Reykjavik meeting? There are a number of causes, both subjective and objective ones. However, the main cause is that the leadership of this great country excessively depends on the military-industrial complex, on the monopolistic groups which turned the nuclear and other arms race into business, into a way of making profits, into the objective of their existence and the meaning of their activities.

In my opinion, the Americans are making two serious mistakes in their assessment of the situation.

The first one is a tactical mistake. They believe that the Soviet Union will sooner or later reconcile itself to the attempts to revive American strategic diktat and will agree to the limitation of only Soviet weapons, to the reduction of only Soviet weapons. It will do so because they think it is interested in disarmament agreements more than the US. But this is profound delusion. The quicker the US administration overcomes it—I am repeating this perhaps for the hundredth time—the better it will be for them, for our relations, for the world situation in general.

The other mistake is a strategic one. The United States wants to exhaust the Soviet Union economically through the buildup of sophisticated and costly space weapons. It wants to impose hardships of all kinds on the Soviet leadership, to foil its plans, including in the social sphere and the sphere of improving our people's living standards, and thus foment discontent among the people with their leaders, with the country's leadership. . . . It is not, of course, difficult to predict all the long-term consequences of such a policy. One thing is already clear to us: It will not bring, it cannot bring anything that is positive to anyone, including the United States.

Before addressing you, I read the US president's statement on Reykjavik. It is noteworthy that the president ascribes all the proposals discussed to himself. Well, these proposals are probably so attractive to the Americans and the peoples around the world that one resorts to such a ruse. We are not consumed with vanity. But it is important that people get the truthful picture of what happened in Reykjavik. . . .

We are realists and we clearly understand that the questions which had remained unresolved for many years and even decades can hardly be resolved at a single sitting. We have quite a lot of experience in doing business with the US. We know how quickly the domestic political climate can change there, how strong and influential opponents of peace across the ocean are. There is nothing new in it for us.

And if we are not despairing, if we do not slam the door and give vent to our emotions, although there is more than enough reason to do so, this is because we are sincerely convinced of the need for new efforts aimed at building normal relations between states in the nuclear epoch. Any other way just does not exist.

Another thing: After Reykjavik the infamous SDI became more conspicuous as a symbol of obstruction in the way of peace, as a concentrated expression of militaristic designs and the unwillingness to avert the nuclear threat looming large over mankind. It is impossible to perceive it otherwise. This is the most important lesson of the Reykjavik meeting. . . .

28

Space-Strike Arms and International Security

Committee of Soviet Scientists for Peace, Against the Nuclear Threat

MODERN military-strategic balance reflects the qualitative and quantitative correlation of the sides' forces and the factors determining the strategic situation. Their sum total may be regarded as a complex dynamic macrosystem, the major element of which consists of nuclear, primarily strategic weapons. The state of this macrosystem depends not only on nuclear weapons, but also on many other elements and factors. There exists a well-known close dialectical connection between the sides' offensive weapons taken separately, as well as between the offensive and defensive systems both of the confronting sides and within each of them.

The dialectics of the development of the strategic balance is such that the appearance (even R&D and testing, not to mention deployment) of a presumably effective new "defensive" weapon may disturb the balance as much as, if not more, than the emergence of a new offensive weapon.

In the late 1960s and early 1970s, when the USSR and the USA began to discuss the problem of strategic weapons, they both recognized the existence of a military-strategic parity between them and of an inseparable link between the sides' offensive and defensive systems. It is not accidental that in 1972 the two countries concluded simultaneously a permanent Treaty on the Limitation of Anti-Ballistic Missile Systems and the first Interim Agreement on the Limitation of Strategic Offensive Armaments (SALT I).

In order to maintain the military-strategic balance achieved at that period, in conditions when almost every five years the United States initiated another spiral of the arms race and adopted new programs for the building up of its military power, the Soviet Union was forced to take appropriate

---

Printed here is an extract from the English-language version of the Report of the Committee of Soviet Scientists published in Moscow, October 1985.

measures aimed at ensuring its security and preventing anyone from attaining military superiority over it. Thus, a series of such measures for preventing the disruption of the military balance was launched by the USSR together with its Warsaw Treaty allies in response to the start of deployment in Western Europe of Pershing-2 ballistic missiles and ground-based long-range cruise missiles. It should be noted that the balance was restored, though at a higher level (through no fault of ours): the sides increased the number of warheads aimed at each other's targets, and the time for decision-making in the event of a nuclear attack or the accidental emergence of a nuclear situation became considerably shorter. Besides, confidence between states was seriously damaged.

In this context one can probably speak about a law of "decreasing effectiveness" being at work in the field of strategic balance. Growing investment of resources in nuclear weapons (considering their tremendous stockpiles and the ability of the other side to take countermeasures) is becoming less and less effective in bringing about a significant change in the correlation of forces. It merely multiplies the already tremendous overkill potential.

Apparently, the world has gone beyond the point where further accumulation and improvement of nuclear weapons are not only dangerous but also senseless.

There is yet another aspect of the problem of unrestrained nuclear arms build-up. The existence of nuclear stockpiles as such gives rise to well-founded anxiety, increases nervous tension and the danger of nuclear war breaking out as a result, for example, of a technical error or human miscalculation. Such possibilities grow with the growth in the number of weapons.

Many scientists, statesmen, military, and political leaders of the West acknowledge that the alignment of the sides' nuclear forces that has existed for the past twelve to fifteen years does not allow either of them to win a nuclear war even by striking first. American military theorists have called this strategic situation "mutual assured destruction." Different Western leaders sometimes say that it is the "balance of terror." The fact that it is understood and accepted by both sides has been the main guarantor of peace which has so far prevented a new world war.

There is no doubt that peace would be much more durable if neither the USSR nor the USA, nor any other power, had nuclear weapons. A much more reliable way of ensuring security would be, instead of maintaining the "balance of terror" (especially considering the high military-political tensions and the enormous number of nuclear weapons of the confronting sides), to normalize the political situation, develop and deepen economic, scientific, technological and cultural cooperation and broad contacts among nations, and settle all disputed issues peacefully, through diplomatic channels.

These objectives are promoted by the Soviet proposal on a 50 percent reduction by the USSR and the USA of their nuclear weapons capable of reaching each other's territory.

Paradoxical as it may seem, the situation of "mutual assured destruction" is being criticized in the West not only by pacifists, but also by politicians and scientists, and even by some military men, whose motives here have little to do with a desire for peace. They do not accept arms reductions and disarmament on the principle of parity and equal security. But they also oppose "mutual assured destruction," for it implies a recognition of "nuclear stalemate" characteristic of parity which calls into question the use of military force as an active instrument of policy, thereby rendering invalid the idea that further development of the means of warfare is necessary.

The plans to create a large-scale antimissile system with space-based elements under the program of so-called "strategic defense initiative" (SDI) may be regarded as a major US attempt to break the "nuclear stalemate."

Approximately from the late 1950s one of the main principles of US strategy was that it is impossible to reduce the destructive effects of an all-out nuclear war on the USA to an acceptable level. This was due to the fact that the USSR had developed its own ICBMs and that both sides had increased their stocks of nuclear weapons in absolute figures. After toying with short-lived strategic experiments based on "counterforce" and "limited damage" concepts in the early 1960s (which provided for reducing US losses by means of strikes at part of Soviet strategic missiles on launching pads), US strategists came to recognize in varying degree the principle of "mutual assured destruction." Back in the 1960s Robert McNamara, defense secretary in the Kennedy administration, came to the conclusion that "unacceptable damage" for the sides is the destruction of from one-fourth to one-third of their population and from one-half to two-thirds of their industrial potential. This, according to his estimates, could be effected by a nuclear strike with a total yield of four hundred megatons. This principle, with some modifications (in the form of such concepts as "selective strikes," "limited nuclear war," etc.) remained the basis of the United States' declared nuclear strategy from the late 1960s to the early 1980s. This strategy envisaged that US security, with both sides having stockpiled huge thermonuclear potentials, was assured not by the possibility of reducing damage to the United States in the event of an all-out nuclear war to an acceptable level but by the possibility of deterring the would-be enemy from using nuclear weapons by threatening to inflict equal or even greater damage on him. Concepts and technical systems relating to the direct defense of US territory disappeared from official papers and US government statements, including those of the Defense Department. Antiballistic missile systems (ABM) were discussed almost exclusively in connection with ensuring the survivability of the US strategic force.

President Reagan's statement of March 23, 1983, and the putting forward of the SDI program meant a radical departure from the fundamental concepts of the United States' declared military-political strategy. Deterrence linked with the ability of the two great powers to destroy each other by a retaliatory strike was declared "evil." As an alternative, the idea of defending US territory from nuclear attack by every possible means, including space-based ABM systems, was advanced. The majority of US specialists consider these systems to be decisive for creating an "all-embracing antimissile shield." This change in official US policy, if it becomes consolidated, may create a fundamentally new scientific, technological, strategic, political and psychological context in which decisions on both offensive and defensive military programs would have to be made. But before discussing these consequences, it would be pertinent to analyse the scientific and technological aspects of a large-scale ABM system on which these ideas rest.

Measures and Means of Counteracting Strike Outer Space Weapons

The main objective of such counter-measures is to maintain the ability to inflict a retaliatory, unacceptable for the aggressor strike, roughly equal to the strike which the attacking side theoretically relies on, under any scenario of nuclear attack.

Such counter-measures can be active and passive. They may include both the development of special means of neutralization and of destruction of different elements of antimissile system echelons, and the build-up, modification and diversification of strategic nuclear weapons.

According to the time of their activation counter-measures can be divided into quick-reaction measures whose activation is directly timed to the moment of a retaliatory strike, and long-term measures embracing an advance preparation of the retaliatory-strike potential, including its structural changes (both quantitative and qualitative).

It is obvious that the picture of potential counter-measures can be made complete when the concept of a large-scale ABM is finally formed; but today we can already include in them some local measures which can be used to destroy such vital but quite vulnerable of its elements as:

—outer space communication which can be disrupted;

—the management system the most vulnerable link of which is the central control computers which, even if duplicated, will be deployed in limited numbers because of their complexity and high cost;

—various energy sources and energy systems (nuclear power plants, explosives, combustibles, etc.).

1. Special means of destruction and neutralization of a large-scale ABM system with space-based elements

Active measures of this kind include various ground-, sea-, air- and space-based means using for destructive effect kinetic energy (of missiles and shells), laser and other kinds of high-energy radiation. Active counter-measures are especially effective against elements of the outer space echelons of antimissile defense, that are in the orbits with known characteristics for a long time, which greatly simplifies the task of their neutralization, suppression and even complete physical elimination.

For example, a system of outer space battle stations seems very vulnerable to a wide range of active counter-measures. Since outer space stations, in accordance with their primary goal, will be designed to destroy strategic ballistic missiles, specially designed small missiles of different types of basing, the use of which could be combined with diverse means of disguise, may prove an effective means of their destruction. Such missiles must obviously have powerful propulsion for the rapid passage through the atmosphere and for shortening to the minimum the boost-phase. They must also be protected against laser radiation. Analogous types of such means already exist. Similar characteristics are to be found, for example, in the American antimissile "Sprint" capable of withstanding high aerodynamic pressure and temperature while moving through dense layers of the atmosphere.

So-called "space mines"—satellites launched into orbits close to the orbits of the other side's battle-stations and carrying powerful explosive charges activated by command from the Earth, could be an effective means of active counteraction to destroy simultaneously a great number of space battle-stations. Such "mines" can be equipped with detonators of different types, e.g. reaction to heat or mechanical action.

Ground-based lasers of great power can be used as an active counter-measure. The creation of such lasers is much more simple than those designed for outer space battle-stations and aimed at destroying ballistic missiles in flight. In the contest of "laser versus missile" and "laser versus outer space platform" the advantage may be with the latter. It can be explained by a number of factors. First, space battle-stations are larger objects for laser destruction than ICBMs, which simplifies the task of targeting the laser beam on them and their destruction. Second, the number of such battle-stations will be much smaller than the number of objects—ICBMs or their warheads—which are to be destroyed at the time of their mass launching or in flight. This practically removes the problem of super-fast retargeting of laser beams. Third, space battle-stations are in the visual field of a ground-based laser installation for a long period of time, which makes it possible considerably to increase the time of exposure (up to one thousand seconds) and, consequently, to lower requirements for its power. Besides, ground-based installations are much less limited, in comparison with outer space systems, by their weight, size, energy-consumption, efficiency, etc.

Obstacles in the orbits of battle-stations, created by clouds of small objects—"shrapnel"—moving in such a way that their speed relative to that of the station is high enough, can be a very effective means of counteraction. For example, the relative speed of a "shrapnel" cloud moving head-on is 15 kilometers per second. At such a speed a particle of 30 grams is capable of piercing through a steel screen (or the station's "skin") 15 centimeters thick. Such vulnerable elements of laser battle-stations as fuel tanks, energy systems and reflecting mirrors may be the most attractive targets in such a scheme of counteraction. Spraying a small cloud of even microscopic particles in the orbit can cause defects on the surface of a reflecting mirror, which will prevent the focusing of a laser beam.

In the case of weapons employing ground-based excimer lasers with mirrors in geo-stationary and low orbits, the spraying of light substances with large absorption of laser radiation directly in the base area of the mirror or laser, can also be an effective counter-measure, besides destroying a ground-based laser.

As for the choice of a possible counteraction measure against the deployment in outer space of X-ray lasers pumped by a nuclear explosion, the following should be noted. In accordance with one of the SDI concepts ("pop-up-defense") they are intended to be put in orbit by missiles launched by submarines (SSBNs) at the latest moment. Such SSBNs must be deployed in the World Ocean close to the Soviet borders (under the concept we are examining the launching of lasers into orbit from US territory is out of the question because of the long time needed for bringing them to altitudes most effective for destroying flying ICBMs with a laser beam). Calculations show that even most powerful carrier rockets, if launched from US territory, cannot bring the laser up to the required altitude (up to three thousand kilometers) before the end of the boost-phase of ICBM trajectories. That is why, in particular, plans of deploying X-ray lasers on missiles launched by submarines patrolling in the northern part of the Indian Ocean or in the Norwegian Sea are examined within the framework of the SDI program. Such a plan can obviously be vulnerable to antisubmarine warfare which the opposite side will try properly to develop.

The systems of acquisition and targeting of outer space weapons will be quite vulnerable. The task of "blinding" them can be achieved by means of a nuclear explosion in the upper layers of the atmosphere. Finally, the traditional measures of radio-electronic warfare, used against space echelons of a large-scale ballistic missile defense (BMD), can significantly affect its effectiveness.

A brief review of the possible means of neutralization and suppression of a large-scale BMD with space-deployed echelons of strike weapons shows that it is not at all necessary to set the task of completely destroying it. It is enough to weaken such a BMD system by crippling its most vulnerable elements, making a "breach" in the so-called defense, in order to maintain the power of a retaliatory strike unacceptable for the aggressor.

2. The development of strategic nuclear weapons as a means of maintaining the capability of an adequate retaliatory strike

Among the hypothetical counteraction measures we should single out the build-up of a "retaliatory" arsenal of strategic nuclear weapons, first of all, of ICBMs and so-called dummy missiles. The deployment of a large-scale antimissile system, or its separate battle subsystems by the United States will directly violate the 1972 ABM Treaty. In such a situation it would be quite natural that the Soviet Union may have to free itself, in the interests of its security, from the observance of both Article XII of the treaty, prohibiting deliberate concealment measures which impede verification by national technical means, and the SALT 2 Treaty, unratified by the United States, which limits the number of ICBMs and the construction of additional launchers for them. The quantitative build-up of ICBMs, and, consequently, the appearance of greater capabilities of the other side of using its ICBMs massively in a retaliatory strike will create additional difficulties for the acquisition systems of the space-based antimissile defense and will cause a drastic decline in the effectiveness of its systems of interception and targeting strike weapons. All this will enhance the "penetrating" capacity of ICBMs and reduce the reliability of an "outer space shield."

The increased number of warheads on ballistic missiles will lead to similar results. This measure can compensate to a great extent missile losses during boost-phase of their flight because of greater difficulties of intercepting them at next phases.

Further "saturation" of the antimissile system can be achieved by additional deployment of relatively inexpensive "dummy missiles," equipped with a simplified guidance system and without warheads. The deployment of such dummies, which cannot be reliably identified by the existing technical means, will be a simple and economically effective measure (if we compare their costs with the cost of the ABM system), which will complicate the operation of the ABM system, and during an exchange of blows it will cause it to fire uselessly.

The tactics of launching ICBMs, which is aimed at "exhausting" the outer space antimissile defense by its early activation by means of a specially arranged order of a retaliatory strike, can also be an effective countermeasure. It may include combined launches of ICBMs and "dummy missiles," ICBM launches with a wide range of depressed and steep trajectories, ICBM launches in different azimuth directions, etc. All this will lead to a great expenditure of the energy resources of the ABM outer space echelons, the discharge of X-ray lasers and electromagnetic guns, and other premature losses of the ABM system's fire-power (e.g., as a result of rapid and orderless retargeting of strike-space weapons). It can sharply increase the "penetrability" of such a system as a whole.

A possible build-up of an arsenal of weapons, for which no satisfactory means of interception have been devised, should be mentioned as a measure for maintaining the capability to effect an adequate retaliatory strike. These may include submarine-launched ballistic missiles (SLBMs) launched on depressed trajectories. A great part of the trajectory of these missiles is within the limits of stratospheric altitudes, where the effectiveness of some ABM systems is sharply reduced. The massive deployment of cruise missiles of different basing modes can be another such measure. None of the proposed today versions of outer space weapons is capable of the reliable acquisition and interception of small cruise missiles with an extremely low radar cross-section, flying at low altitudes. The organization of the interception of thousands of long-range cruise missiles of different basing modes is a complicated and costly task.

A shortened boost-phase is an effective means of passive counteraction against the enemy's ABM system, increasing the survivability of ICBMs in the process of its penetration. The parameters of the boost-phase of ballistic missiles are mainly determined by considerations of reducing the overloading of the missile body and of using trajectories which are most suitable from the energy-consumption point of view. Specialists speaking before the Fletcher Commission pointed out that it is potentially possible to shorten the boost-phase to forty seconds and finish it at altitudes of no more than eighty kilometers. According to their estimates, such characteristics can be achieved at a relatively low cost connected with increasing the weight of the missile by roughly 15 percent while preserving the initial payload and range. A shortened boost-phase will create additional difficulties for the systems of acquisition, tracking and targeting, which in its turn will reduce the effectiveness of antimissile weapons.

All other counter-measures at boost-phase can be divided into two main groups: measures complicating the targeting of antimissile weapons, and measures strengthening the protection of the missile shell. The first group of such measures includes changing the brightness and shape of the missile exhaust flame. The target is not the exhaust flame itself, but the missile, which is a certain distance away from it, and any infra-red guidance system must use an algorithm, calculating the position of the missile in respect to its exhaust flame. Besides, a laser beam must be fixed for several seconds on a certain portion of the missile shell. This makes it possible to complicate the problem of guiding and fixing the beam by changing the brightness of the exhaust flame and its shape, because the changes of the flame, registered by infra-red sensors, will cause, in accordance with the standard algorithm used, the shift of the laser beam itself. Such an unstable burning of the exhaust flame can be achieved mainly by introducing different additives to the missile fuel.

Concealment of missile launches can also be included in this group of counter-measures. It can be achieved by means of smoke screens over the launching areas or by different ways of camouflaging the missile in flight, e.g. fitting it with disguising screens.

There are also diverse means of protecting missiles against laser radiation. They may include the protection of the missile shell with the help of reflecting or absorbing coatings, or rotating it round its axis, which will prevent the fixing of a laser beam on a certain portion of the shell. The provision of the missile shell with an additional cooling system or the installation of a movable absorbing screen on it, which can be lowered into the heating zone, may be an effective measure. For example, a screen with a graphite coating one centimeter thick is enough to absorb a heat energy of 200 MJ/m^2. The dispersion of various substances in the atmosphere to produce smoke or aerosols, i.e. screens absorbing laser radiation, can be a promising counter-measure. It may be worth recalling the design of the very first missiles. In the case of the German ballistic missile "V-2," for instance, fuel and oxydizer tanks were inside the load-carrying shell of the missile body. The rejection of the load-carrying structure of tanks and return to a two-layered structure with additional light heat-protecting interlayers between them can considerably increase the survivability of ICBMs.

The use of the above-mentioned and other means in different combinations will help considerably to decrease the vulnerability of ballistic missiles at boost-phase, while their higher survivability at this phase will, in its turn, substantially complicate the task of intercepting them at all subsequent phases of their flight. A complex of passive measures can be also applied to the intermediate phase and the mid-course of ICBMs. The intermediate phase of the trajectory, i.e. the flight along a ballistic curve from the moment of the engine cut-off of the last stage of the missile and the separation of the bus until the warheads' reentry into the atmosphere, is usually divided into two stages. The first is the flight of the bus as a whole until the deployment of warheads and the release of decoys. The second is the independent flight of warheads and decoys until their reentry into the atmosphere.

The first stage of this phase, due to fewer objects and the absence of decoys, which complicate the identification of buses, naturally seems more suitable for interception. But missiles may complete their boost-phase within the atmosphere with an earlier separation of busses and their deployment of warheads. That is why most of the researchers emphasize the intermediate phase as the stage of flight of deployed warheads.

The considerable duration of this phase (twenty minutes for ICBMs and about ten minutes for SLBMs) increases the possibilities for interception. Besides, at this phase the strength of warheads is subjected to a crucial test by the prolonged effect both of separate means of destruction and of their different combinations.

On the other hand, during this phase of missiles' trajectory the antimissile system will have to deal with many more objects to be identified and intercepted, the number of which can reach several tens of thousands in a massive strike. All these objects, both warheads and decoys, have practically the same velocity and similar trajectories. As a result interception during this phase creates formidable difficulties for the systems of acquisition, tracking and battle management. These difficulties are further aggravated if the massive strike is not sufficiently weakened during the previous phases of missiles' flight.

The two principal considerations mentioned above lead to the conclusion that from the point of view of penetrating the air defense system at this phase the emphasis should be on passive counter-measures against the tracking and targeting facilities of the ABM system. The acquisition and tracking of targets, i.e. warheads, during this phase are much more complicated by the comparatively small size of a great number of moving objects and the absence of exhaust flames. The acquisition, identification and targeting functions in the projects of space-based echelons of a wide-scale antimissile system now under discussion in the United States are to be implemented by a large set of active and passive means, including optical, infra-red, radar and other facilities which can be land-, air- or space-based. Not only are all these means vulnerable to the above-mentioned counteraction measures but it is also possible to develop specific counter-measures against them.

As it has been already confirmed more than once, one of the most effective means of such counteraction is different decoys. For example, during the deployment of warheads a cloud of very small and light metal objects can be created around them, which would not only absorb and reflect radiowaves but also disperse the radar radiation reflected from the warheads. Spraying of an aerosol cloud around warheads, which is a source of infra-red radiation, can also be a counter-measure against infra-red detection facilities. The infra-red radiation of such an aerosol cloud can be used to camouflage warheads' own infra-red radiation. All these counter-measures may be quite effective and, most important, available for wide application.

The servicability of space ABM sensors can be significantly decreased by the use of different kinds of jamming, suppression or distortion of signals by the other side, and equipping decoys with devices which imitate the reflection of laser, radar and visual signals from warheads. Several studies have described the concealment of warheads inside light, multilayer metallized reflecting balloons. For each warhead inside such a balloon ten empty balloons can be deployed. It is important here that, besides the unidentifiability of "filled" and "empty" balloons by their reflected signals, it is also possible to achieve their identity by ballistic parameters.

The list of counter-measures given above evidently does not exhaust the possibilities of countering the facilities for acquisition and targeting of the strike weapons of the space defense system at flying ICBMs.

During the terminal phase (after reentry) warheads and decoys can be selected by the acquisition sensors of the antimissile system due to the difference in weight and aerodynamics. But this phase of trajectory does not exceed sixty seconds, and this requires very fast interceptors. One way of countering such interceptors is the use of maneuvering and high-speed warheads. It is also possible to increase the yield of warheads and to use detonators which will fire the warhead before it would be destroyed by interceptor. Calculations show that in this case, even if warheads explode at an altitude of more than ten kilometers above the Earth, the destructive effect would be significant. The use of these counter-measures will certainly create additional problems for the retaliatory forces, such as increasing the weight of missiles and decreasing their payload. But the quantitative build-up of ICBMs can to some extent compensate such losses.

In concluding the analysis of possible counter-measures available for the other side in case of the deployment of strike space weapons by the United States in keeping with the SDI program it should be noted that some SDI proponents claim that the echeloned (multilayered) structure of antimissile defense in outer space is little affected by a decrease of its individual echelons' effectiveness. To prove this they usually use simplified calculations of "penetration" probability for the entire ABM system. These calculations are based on the false premise about the independence of functioning of the echelons ("layers") and ignore the great variety of possible counter-measures. The case when the battle-management link of the ABM system (acquisition, tracking, selection and targeting) is damaged clearly illustrates the fallacy of such an approach. Since the different echelons of the antimissile system are interdependent and based on a common battle-management system it is evident that effective counteraction against this most important element of the ABM system can sharply reduce the entire system's efficiency.

Thus if we assess the overall effectiveness of the possible counter-measures against a wide-scale ABM system with space echelons it can be predicted with sufficient certainty that what may be claimed to be an "antimissile shield" is very far from an ideal "leakproof" defense. There is a broad range of effective, available and much less costly facilities which can be freely employed by the side against which the antimissile system is planned to maintain the potential of retaliation. In any case this system is an offensive one, and it functions effectively only when the side possessing it strikes first.

Studies conducted by the Committee of Soviet Scientists with the help of general and special methods of system analysis also lead to the following general conclusion concerning the strategic balance. The point is that several combinations of the above-mentioned counter-measures would actually neutralize the danger of unilateral disruption through the SDI deployment of the military-strategic parity by relatively cheaper means than the reciprocal build-up of an antimissile arsenal of outer space strike weapons. In one of the combinations analyzed in the studies the estimated cost of a counter-measure complex was only a few percent of the cost of a large-scale ABM system with space-based echelons.

Military and Political Consequences of Creating a Large-scale Antimissile System with Space-based Echelons

1. General military and political consequences of creating a large-scale antimissile system to cover a country's territory

An antimissile system, even an ideal one from a scientific and technological point of view, will not bring about a fundamental change in strategic thinking from "mutual assured destruction" to "mutual assured survival," as it has been claimed by some US state leaders, as it does not guarantee complete protection against ballistic missiles and strike space-based weapons. Therefore, all the arguments in favor of the stabilizing role of a large-scale antimissile system lack physical sense. They might have some weight only if the Reagan administration, simultaneously with taking a decision to launch the SDI program, renounced the buildup and modernization of offensive nuclear forces. What is happening, however, is just the opposite. The United States is stepping up the development of its strategic offensive weapons, intermediate-range nuclear systems and tactical nuclear weapons. Therefore, the building of an antimissile system with space-based echelons will only substantially complicate the task of mutual deterrence and make it more uncertain. The framework of strategic balance will become less steady and more frail.

The assessment of the prospective US antimissile system as one of the means of providing first-strike capabilities is determined by the fact that the USA is refusing to make the no-first-use of nuclear weapons commitment and is building-up its first-strike potential. The deployment of American intermediate-range nuclear systems, above all Pershing-2 missiles in Europe, is an important element of this policy.

The Soviet Union, taking into account the special significance of the efforts to strengthen strategic stability in conditions of growing military and political tensions, unilaterally pledged in July 1982 not to be the first to use nuclear weapons.

In addition to what has been said about counter-measures against antimissile systems with space-based echelons there is a good reason to believe that in response to their development there will appear weapons to counter such systems. One can also agree with those specialists who believe that while an antimissile system is being created and deployed, the means of penetrating it by strategic offensive forces will be improved at a high pace. The creation of space-based antimissile systems may also heavily stimulate the quantitative buildup of strategic delivery systems and nuclear warheads, in particular, long-range cruise missiles, including sea- and ground-based ones, the deployment of which is extremely difficult to verify by national technical means.

Space-based antimissile systems are also justifiably regarded as anti-satellite weapons. This factor will also destabilize the strategic situation, even under conditions of a limited deployment of such systems. The strategic balance is sure to lose its solid basis because it largely depends on the degree of both sides' confidence in the reliability and security of the warning, control and monitoring systems based on different types of satellites.

System-analytical studies of a number of aspects of the strategic balance, dealing with the examined problem, show that, contrary to the statements of SDI advocates, if the two sides possess large-scale antimissile systems with space-based echelons instability in the strategic balance will increase significantly, especially if we take into account the above-mentioned wide range of available counter-measures and the vulnerability of such systems.

It is also worth mentioning that space-based strike weapons which are being developed in the USA appear to be intended not only for knocking out the other side's satellites and strategic missiles after their launch but also as preemptive weapons to be used against ground targets, for performing the first strike. Accurate and powerful enough to destroy strategic ballistic missiles in flight, space-based weapons can also be used to hit other types of strategic weapons, e.g. planes on air fields before their takeoff. Besides, it has been reported in the American press that space-based strike weapons could also hit other ground and sea targets, including command posts, communication and control networks, large floating targets and key economic facilities (oil and gas refineries, chemical plants, power stations, etc.).

Modifications of space-based antimissile systems may well be used for delivering strikes at ground targets from outer space, in which different types of missiles can be used.

The danger of the above-mentioned US projects is aggravated by the fact that they appeal to human beings' inborn natural desire to find a shield against the all-destructive power of nuclear weaponry. SDI peddlers unrestrainedly exploit these feelings.

A number of Western experts also claim that, in accordance with the dialectics of war, the prevalance of offense should give way to the superiority of defense, as it occurred many times in the past, and nuclear weapons, which have dominated the scene for several decades, should in their turn be replaced by fundamentally new types of weapons—in this case, directed-energy weapons.

However, with respect to antimissile systems, this reference to the dialectics of the development of means of warfare which was, incidentally, best of all elucidated by the classics of Marxism (e.g., Friedrich Engels in his work "Anti-Dühring"), simply ignores the core of the issue. Indeed, competition between the offensive and defensive means has been going on with varying success in the course of history. Yet, it should not be forgotten that parallel to that a general trend towards increasing the destructive consequences of wars, above all, for a civilian population, has also been developing. Suffice it to recall World War I, a classical example of the preponderance of defense, which accounted for its largely positional style of hostilities. However, they were accompanied by the hitherto unheard-of destructions in battlefield areas (the Marne, Verdun, Galicia, etc.). In this respect nuclear armaments occupy quite a special place as weapons specially made and used for the first time by the United States for the mass destruction of civilians and material values. The prospect of the total annihilation of civilians and the devastation of large areas has always accompanied all attempts by Western strategists to invent some ways of using these weapons for carrying out more traditional combat missions by launching "limited" or "selective" strikes.

Supporters of the Strategic Defense Initiative in the United States actively play around with the assertion that in comparison with the period between the late 1960s and early 1970s when Soviet-American ABM Treaty of unlimited duration was being developed, there emerged new scientific and technological capabilities which drastically changed the correlation between defense and offense in the nuclear age. Far-reaching conclusions of scientific, technological and political nature are made on this basis.

Speaking about recent technological achievements which help create defense capabilities that have not yet existed, SDI proponents mention progress in computer technology and its software (higher speed, microminiaturization, the use of artificial intellect elements, etc.), in sensor technology (in infra-red, optical range, etc.), in the development of various types of lasers, neutral-particle accelerators, electrodynamic mass accelerators and in the development of new types of increased-payload space delivery vehicles. It has already been mentioned in other parts of this report that progress has been made in some of these areas as compared with the period between the late 1960s and early 1970s. It should be pointed out, however, that if we examine all the major areas of science and technology connected with the development of various potential types of antimissile weapons, we shall see that progress in these areas has been far from even in recent years.

Many American sources, including governmental ones, admit that these achievements now and for a foreseeable future are clearly insufficient to ensure the creation of a large-scale antimissile system with space-based echelons. Besides, the composition and integration of all different elements and subsystems (those of detection, acquisition and targeting, subsystems of battle management and compound striking elements) into a united macrosystem with extremely complex intersystem relations are a fundamentally new and exceptionally complicated task.

Putting forward the thesis of "significant development of technology, suitable for means of defense," its advocates fail to mention, first, that many of its elements, which, they assert, are intended for the development of a large-scale antimissile system, can with the same or even better results be used as part of means of destruction and neutralization of this system (which are analyzed in this report). Secondly, offensive means have also been considerably modernized in the past ten to fifteen years thanks to US initiative. Mention should be made here above all of the creation and rapid improvement of MIRVed ballistic missiles, long-range cruise missiles, and the equipment of the delivery vehicles of nuclear weapons with improved guidance systems which considerably increase their accuracy compared to the level achieved in the late 1960s and early 1970s. We are witnessing the creation in the USA of a new generation of SLBMs with a destructive potential close to that of ICBMs.

It should also be pointed out that, speaking about the "brilliant prospects" of developing antimissile systems' technology, SDI advocates intentionally or unintentionally almost completely forget to mention the prospects of developing offensive weapons which can also be improved considerably within the same time limits. These improvements, as has already been mentioned above, can include the equipment of ballistic missiles and other types of strategic delivery vehicles with MARVs, the use of depressed trajectories of SLBMs, widening the inventory of dummies, shortened boost-phase of ballistic missiles, etc.

Thus, on the whole, if we compare the development of defensive technology with that of offensive technology, the advantage of "defensive systems" does not seem so convincing as SDI advocates try to prove. It should be added that, together with growing interaction between defensive and offensive systems, in case of SDI implementation there will emerge the necessity and therefore the opportunity, as it was mentioned in the previous section, of developing a whole class of new types of weapons, specially designed to destroy and neutralize elements of a large-scale antimissile system, above all its space-based echelons.

It should be stressed once again that many technological means which can be used for this purpose are now at such an advanced state in the US itself that, according to American experts, they can be realized much sooner and cheaper than a large-scale antimissile system with space-based echelons. Moreover, in conditions of growing international tensions, the SDI program, in the opinion of many Western specialists, will stimulate an accelerated development of those weapon systems of mass destruction against which a space-based antimissile system will be ineffective. And it does not look like simple coincidence that the United States, together with initiating the SDI program, began intensive discussions on the need to create alternative counter-measures against potential Soviet space-based antimissile systems (which are not being developed by the Soviet Union).

Should space-based weapon tests (let alone their deployment) begin, the permanent USSR-USA Treaty on the Limitation of Anti-Ballistic Missile

Systems, signed in Moscow on May 26, 1972, will be undermined. In accordance with Article I of the treaty, each side has undertaken "not to deploy ABM systems for a defense of the territory of its country and not to provide a base for such a defense." Besides, in accordance with Article V of the treaty, both sides have undertaken "not to develop, test, or deploy ABM systems or components which are sea-based, air-based, space-based, or mobile land-based."

It is quite obvious that the SDI program is a flagrant violation of the ABM Treaty. First, because the aim of the program is to create an antimissile defense of the entire American territory as well as the territories of US allies. Second, a space-based antimissile defense is prohibited by Article V of the treaty.

The SDI supporters, trying to distort an absolutely clear matter, refer to one of the Agreed Statements to the treaty, the so-called Statement D, which, they allege, allows the creation of ABM systems based upon other physical principles (lasers, particle beams, etc.). They "forget," however, that such means are allowed only in relation to limited ABM areas and only to fixed land-based systems. There can be no other interpretation of Statement D, which has also been confirmed by US participants in the ABM Treaty negotiations. Thus, the SDI program violates the ABM Treaty whose importance with regard to the present day and to the future is indisputable.

The above-mentioned limitations are not applied to basic research, i.e. research aimed at acquiring knowledge or eliciting the main aspects of phenomena or observed facts. But this kind of research is not in line with the solution of special applied problems, let alone such a specific program of creating new weapon systems for outer space and against it as the US SDI program. As far as real basic research is concerned, such research is under way and will continue. The USSR is carrying out this kind of research in outer space, but it is not aimed at developing components of space-based strike weapons. In its military aspect it is aimed at improving early warning, surveillance, communications, navigation and meterological space systems. It goes without saying that neither of these areas in research has anything to do with a program that could be regarded as analogous to the US SDI program.

Even if in the foreseeable future Soviet-American relations improve to such an extent that the American side will be politically ready to conclude mutually acceptable and equitable agreements on the limitation and reduction of strategic arms, the existence, even on a limited scale, of tested and deployed components of a space-based ABM system will tremendously complicate negotiations and reduce chances to reach such agreements. The experience of negotiating the SALT I and SALT II treaties proves this fact. It would have been absolutely impossible to negotiate them without the ABM Treaty.

If yet another, qualitatively new component (such as a large-scale ABM system with space-based echelons) is added to the strategic forces of one or both sides, this will complicate to a great extent and confuse the whole system of assessing the strategic balance, and create additional difficulties in estimating the balance of forces of partners in the talks. Furthermore, as was the case with strategic offensive arms, the development of antimissile systems by both leading nuclear powers will very likely go along different lines, and this will increase even more the asymmetry of their strategic forces and make them even more difficult to compare. This asymmetry may be even greater, if we take into account potential anti-SBAM systems and counter-anti-SBAM systems that can be built. Thus, a vicious circle will be created, which is well known to all scientists and military men: weapons—anti-weapons—counter-anti-weapons and so on and so forth. Because the starting point—a weapon as a material substance—contains an element of self-negation, i.e. an anti-weapon.

The Soviet Union will not allow the strategic and military balance to be upset, no matter how hard various bellicose groupings in the United States try to do it. It will invariably be restored, but it will be restored, through no fault of the Soviet Union, of course, at a higher level. The number of nuclear warheads will increase. The time available for making responsible decisions in the event of a nuclear attack or an accidental nuclear-pregnant situation will be shortened. A trend towards the increased danger of an accidental nuclear war will become stronger.

Speaking about international political consequences of the deployment of an antimissile space-based system by the United States, it should be mentioned that such deployment will practically block Soviet-American cooperation in the use of outer space for peaceful purposes. The potential value of such cooperation appears to be considerable from the economic, scientific and technological points of view because Soviet and US space programs complement each other in many ways. Such cooperation would also be of great political importance: it would help to improve the entire climate of Soviet-American relations and ensure trust between the peoples of the two great powers.

The deployment of strike weapons in Earth orbits and the creation of antisatellite weapons will adversely affect the prospects of broad international cooperation in peaceful exploration of outer space for the benefit of all mankind.

2. Limited variants of an anti-missile defense system and the stability of the strategic balance

The focus of attention of Western experts in 1985, unlike 1983, has shifted from debating plans and options of creating a comprehensive antimissile system to cover the territory of the United States and its allies to considering questions concerning antimissile systems of limited capability—zonal and point systems.

Experts cannot neglect enormous difficulties, huge expenses and potential counter-measures of the other side that make it unreal to accomplish the objective set by President Reagan in his well-known speech of March 23, 1983.

A series of reports by both Soviet and American scientists corroborate a conclusion made by Soviet academicians back in April 1983 in their "Appeal to the Scientists of the World." It is stated in the "Appeal" that antimissile weapons can do little to help a country which suffered an unexpected massive attack because they cannot protect the bulk of the population. It is pointed out with good reason that, given the counter-measures taken by the other side, this system also cannot fully disrupt the retaliatory strike by that side.

Despite these conclusions, SDI supporters try to speed up research on and the development of corresponding weapons, and make this process irreversible. In order to disguise their real intention to gain military superiority over the Soviet Union, they are putting forward various arguments in favor of different types of an antimissile system which are not supposed to be highly efficient. These arguments are aimed at misleading public opinion. The advocates of SDI allege that even a limited in capabilities, functions and scale antimissile system, which can be created before 2000, will have a "stabilizing influence" over the military-political and military-strategic situation in the world. At the same time they do not question the need to deploy in further perspective an antimissile system on a global scale. It is remarkable that not a word is being said about renouncing nuclear weapons. As for the limited variants of the system, they are looked upon by a number of American politicians as an intermediate step towards its full-fledged deployment.

The need to create interim variants (for which no high efficiency is claimed) of an antimissile defense is being explained, in particular, by the necessity to protect the USA from "third countries," i.e. states which can acquire nuclear weapons in the near future and which could, in the opinion of some Western experts, use their nuclear weapons to blackmail even great powers. It is also said that a limited antimissile defense, unable to ward off a more or less powerful nuclear strike (first or retaliatory) against industry and population, could nevertheless defend the country against a casual, unauthorized use of nuclear weapons.

The supporters of a partially effective ABM system in the USA also argue that such a system could allegedly strengthen the deterrence factor by raising the level of uncertainty in the enemy's strategic planning and creating "disproportionately high complications for first-strike planning by a potential adversary."

Some highly-placed members of the US administration think it desirable and technically feasible in the foreseeable future to deploy a large number of antimissile complexes (predominantly land- and air-based) for the point defense of intercontinental ballistic missile silos which, in their opinion, are becoming ever more vulnerable due to the increasing accuracy and destructive power of the other side's warheads.

These arguments put forward by the inventors of different types of an ABM defense do not hold water. In terms of robust political and strategic logic there is a far more effective way to protect states from nuclear blackmail, let alone a nuclear attack, which lies in strengthening the nuclear nonproliferation regime, lowering the level of international tension in general and particularly in those regional subsystems of international relations, whose subjects are potential possessors of nuclear weapons. Clearly, the United States and the other nuclear powers by following the Soviet example should in practice demonstrate to other states their desire to limit and reduce their nuclear armaments in compliance with Part VI of the Non-Proliferation Treaty.

As for the argument about the need to create large-scale ABM systems with space-based components to defend against casual, unauthorized launches of nuclear-armed missiles, it may seem attractive to someone at first sight. But those who put it forward deliberately keep silence about the fact that the risk of such launches can be reduced by carrying out measures technically far more simple and less destabilizing in military-strategic and political sense, for example, by enhancing the reliability of self-detonating devices in strategic carrier systems with autonomous homing. This could enable the detonation of the carrier (without setting off the nuclear warhead) on the order from the command post with minimal damage to population and environment.

It is necessary to set the risk of a casual, unauthorized missile launch not only against military-political and economic costs of creating a large-scale ABM system, but also against the danger of self-activation of such a system as a result of error in a subsystem of detection, identification of the target or in a chain of combat control. Some calculations show that the probability of error, malfunction in a combat control subsystem operating an American antimissile system will be much higher than the probability of an unauthorized launch, particularly if the nuclear arsenals of both sides are substantially cut as a result of reaching mutually acceptable agreements (under their terms in particular the ratio of warheads on one side to the strategic force targets on the other would not be increased) and the reliability of control systems operating the corresponding forces is purposefully increased.

The assertions of some members of the present US administration that a limited ABM system will have a stabilizing effect due to a rise in the level of uncertainty in the strategic planning of a nuclear strike by the other side, run counter to the dialectic of the present-day strategic balance as a complex dynamic macrosystem.

First, the authors of this argument deliberately ignore the Soviet Union's unilateral commitment not to be the first to use nuclear weapons. Under it, incidentally, a still tougher framework is set to organize strict control which ensures the exclusion of an unauthorized nuclear weapons launch. If the USA and its nuclear allies assumed such an obligation the situation would become much more stable and secure both in terms of reducing the probability of delivering the first strike and in terms of casual and unauthorized missile launches.

Second, a significant measure of uncertainty is inherent in the present-day strategic situation because of the very nature of nuclear weapons and their stocks, the complexity of control and communications systems, and some other factors. So why increase this uncertainty? This additional measure of uncertainty added to the strategic and operational planning of one side will inevitably affect the level of uncertainty for the other side, which will reduce the stability of the existing strategic balance and increase the danger of an outbreak of nuclear war. In the opinion of many authoritative specialists, uncertainty in planning the "first disarming strike" is today still very high. Both sides have capabilities for a secure retaliatory strike because there exist adequate, multiple-duplicated early-warning command control and communications systems; the strategic forces of both sides are kept in a high state of alert. Needless to say that in a further-off perspective the danger of this kind may grow because the United States continues to impose the drive to increase the number of warheads and improve their accuracy and destructive capabilities, and because of possible "breakthroughs" in antisubmarine warfare. All this speaks of an urgent need to take drastic steps to prevent a situation where the danger of the "first strike" may theoretically grow higher and higher.

One of the truly effective measures to put an end to the creation of first strike potentials would be to stop the qualitative and quantitative buildup of nuclear weapons as a first step towards their substantial reduction. Such actions taken by the sides would prevent, among other things, the buildup of highly accurate nuclear weapons in the arsenals of both sides, weapons capable of destroying highly protected targets. It is also necessary to take mutually coordinated steps to limit antisubmarine warfare activity which increases strategic instability, as well as other measures reducing the probability and possibility of the first disarming strike. Proposals to this effect have been repeatedly put forward by the Soviet Union.

Finally, a US multicomplex antimissile system, designed to cover ICBMs outside the limits set by the 1972 ABM Treaty and the 1974 Protocol to it, would in itself be a destabilizing system representing one of the most important means of ensuring material support of the concepts of a "protracted" and "limited" nuclear war, which are now popular in the United States.

Influential Washington strategists speak of a "limited," "controlled" exchange of strikes against ICBM silos without damaging industrial enterprises and administrative centers, and without inflicting catastrophic losses on the civilian population, an exchange that would result in the cessation of hostilities "on terms more favorable to the USA." The Soviet military doctrine, based on realistic concepts of the nature and character of nuclear war, rejects the idea of "limitation" as an unsound, illusory and extremely dangerous one.

At the same time the USSR and its allies have to take into account these trends in American military-political and military-stratetgic thinking, however unrealistic these trends are.

Before the nuclear era set in, the adoption by a state of an unrealistic scheme of conducting a war meant, in the first place, the danger of that state suffering a landslide defeat in the war. In military terms, it played into its adversaries' hands. Today one has to look upon the matter in a different way. State leadership that adopts concepts and doctrines which do not take into account the real nature of war and the system character of the strategic balance, and which are based on the assumption of the "controllability" and "limitability" of a military conflict in which weapons of mass destruction are used, condemns its own country and its allies to unavoidable annihilation in the event of war and at the same time may plunge the whole of humanity into non-existence.

3. *Antimissile defense and European security*

The military-strategic and international political consequences of creating a large-scale antimissile system with space-based components, which were considered above, do relate in general to the situation in Europe because of their global nature. At the same time it seems necessary to consider more single-mindedly and at least from several principal angles some aspects of the given problem in terms of the stability of the strategic equilibrium and international security in Europe.

The Reagan administration counts mainly on convincing the West European members of NATO that by creating an antimissile shield the United States will allegedly be able to cover not only itself but also the appropriate states in Western Europe. It goes without saying that West European countries will have to contribute to financing the military and technological development of the program.

Considering the possibility of creating an antimissile defense with some reasonable level of efficiency, one should first of all bear in mind those concrete nuclear weapon delivery vehicles which determine the balance of forces in the given region. At present the NATO countries have in Europe several hundred intermediate-range missiles and a larger number of tactical nuclear weapons delivery vehicles. They are opposed by respective Soviet armaments.

The flight time of intermediate-range ballistic missiles (IRBMs) is substantially lower (by the order of two to three) than that of ICBMs or even a considerable part of SLBMs. IRBMs are lighter than ICBMs, and, consequently, have a shorter acceleration path, on which, as it has been noted in the previous parts of the report, ballistic missiles are more vulnerable to a space-based ABM system. The acceleration time of tactical ballistic missiles is still shorter.

A major part in the balance of intermediate-range weapons in Europe is played by aircraft which are not subject to interception by ABM space-based echelons. And it goes without saying that the delivery vehicles of various battlefield nuclear weapons are completely outside the scope of ABM defense. This also fully refers to long-range cruise missiles with various basing modes which can be used to hit targets on European territory.

In view of what has been said, as well as a particularly high population density in Western Europe and the proximity of military installations to populated areas, the deployment of point ABM complexes in this region seems even less sensible than their deployment on US territory. On the whole, there is every reason to believe that arguments in favor of an American (or NATO) shield covering Western Europe are baseless, and in reality American strategists intend to use this shield to protect the United States from a retaliatory strike in a crisis situation, while using Europe as a theater of military operations. This seems to be linked with the deployment by the United States of its intermediate-range missiles in Europe and the adoption of the "Airland Battle Doctrine" by the US Army and the "Deep Strike Doctrine" by the NATO Defense Planning Committee.

The renouncement by the United States of the principle of no-first-use of nuclear weapons, which in the context of the military-political and military-strategic situation in Europe has a special meaning, also shows that the above interpretation of American plans to create an antimissile shield is correct.

Some American supporters of a large-scale antimissile system say that the deployment of such systems by the United States and the Soviet Union would allegedly solve many problems of the West-East strategic balance by devaluing British and French nuclear forces.

The authors of such ideas deliberately ignore the fact that a more or less system approach to this question clearly shows that in terms of strengthening mutual security and strategic balance stability there are no serious grounds for supporting their conclusion. Suffice it to say that France and Britain can hardly be expected to remain passive in such a situation. Evidently, many of the above-mentioned measures to counterbalance antimissile weapons (both through modernizing and building up offensive armaments and through employing special means to neutralize and destroy the space-based echelons of antimissile systems) are within the reach of the two states, let alone a larger West European conglomerate.

The most sensible alternative to the destabilizing introduction of anti-missile weapons into the military-strategic balance equation in Europe is the complete removal of nuclear weapons, both intermediate-range and tactical, from the region. The creation of a "nuclear-free corridor" along the line between the NATO and the Warsaw Pact, proposed by the Palme Commission, could be the first step in this direction. The establishment of a nuclear-free zone in Northern Europe and other parts of the continent could serve the same purpose.

Conclusion

The comprehensive analysis of the scientific, technical, and military-strategic aspects of the development of a large-scale ABM system with space-based elements by the USA, and of the potential influence of its deployment on strategic stability, parity and international security, carried out by the Working Group of the Committee of Soviet Scientists for Peace, Against the Nuclear Threat, makes it possible to draw quite definite conclusions. Contrary to what its supporters say, such a system is obviously incapable of rendering nuclear weapons "impotent and obsolete." Nor can it reliably protect the territory of the USA, let alone that of its allies in Western Europe or in other regions of the world.

Washington's hopes, which are clear to all, if not much publicized, to gain significant political, not to mention military-strategic, advantages over the USSR and its allies by creating and deploying such a system are groundless, considering the tremendous economic, scientific and technical potential of the USSR and its rich experience in finding optimal ways of maintaining rough military-strategic parity.

Also unrealistic are the hopes of the "hawks" in the United States to "exhaust the Soviet Union economically" by imposing on it an arms race in space together with an escalation of the race in nuclear weapons.

One of the major conclusions of the present study is that the Soviet Union has a wide range of possibilities and means for taking relatively uncostly measures to counter the new threat from space due to the development by the USA of a large-scale ABM system as a means of delivering a first strike with impunity. A thorough analysis of this problem by the Committee of Soviet Scientists shows in particular that in one of the combinations studied the cost of a comprehensive system of measures countering a large-scale ABM system with space-based elements will amount to just a few percent of the latter's cost.

If the bellicose forces in the US ruling political, military and industrial quarters follow the road of developing and deploying "star wars" weapons, contrary to the ABM Treaty and other norms of international law and in defiance of the opinion of the overwhelming majority of the world's scientists and despite the protests of the world public, the Soviet Union will use the many opportunities available to it for ensuring its security in the new conditions and preserving the rough military-strategic parity that has evolved in the world. However, considering the fact that the modern military-strategic balance is a complex macrosystem of the sides' armaments and geo-strategic factors and that its stability depends on the actions of both sides, one should bear in mind that the military-strategic balance will be restored on a higher and (through no fault of the USSR) less stable level. As a result, the risk of a disastrous nuclear war will increase. This also applies to a war breaking out through accident, for instance, through errors in an assessment of the strategic situation due to a possible self-activation of space-based components of a large-scale ABM system.

There is yet another real danger: with the development of such a system the US ruling quarters will be even more tempted to use military force as their main foreign policy instrument, for instance, by waging different kinds of wars, including nuclear wars, in an attempt to complete them "on terms advantageous for the USA," as has been repeatedly noted in the Pentagon's documents. It is not accidental that many "star wars" propagandists are also advocates and authors of various concepts of waging nuclear warfare, and claim, contrary to the findings of scientific research, that a nuclear war is "winnable." Soviet political and military leaders reject this idea. It is pointed out in unambiguous terms in the draft of an updated version of the CPSU Program that there will be neither winners nor losers in a nuclear war. It would be extremely dangerous to hope to win a nuclear war, for instance, by using strike space weapons on a mass scale.

Before the advent of the nuclear and space age the adoption by a state of unrealistic, adventurous methods based on power politics and relying on its military machine, that is, on war, might at worst mean its utter defeat in the war it unleashed. Today, the attitude to this must be changed. Statesmen and military leaders who adopt concepts disregarding the nature of strategic balance and war today, doom their country and people to certain death in the event of war. Moreover, the aggressor may drag all mankind to death with him.

All these facts show that the genuine security interests of both the USSR and the USA, and of the entire international community, urgently demand the preservation of the 1972 ABM Treaty and efforts to prevent the development and deployment of strike space weapons. Such a policy in this matter means the vital interests of international security and strategic stability. Prohibition of attack space-based weapons and preservation of limitations on ABM systems are one of the major factors facilitating progress in limiting and reducing nuclear armaments. Implementation, on the basis of a bilateral agreement, of the Soviet proposal for a 50 percent cut in the nuclear weapons of the USSR and the USA capable of reaching each other's territory could play a very important role in the efforts to free the world from nuclear weapons.

Implementation of a number of priority measures by the sides would be a major practical contribution to accomplishing this task. It is first of all necessary to suspend all work on the development, testing, and deployment of strike space weapons, to liquidate the existing antisatellite weapons, to freeze the current number of nuclear armaments with a maximal limitation of their modernization and a simultaneous cessation of the development, testing and deployment of new types of such armaments; and to halt the siting of medium-range missiles in Europe (which has already been done by the USSR). These steps could be taken even before the USSR and the USA work out an agreement on a whole range of issues pertaining to nuclear and space weapons.

The aim of stopping the nuclear arms race, and consequently averting the threat of nuclear war, can only be achieved through a general and complete ban on nuclear weapons tests. The same goal would be promoted by the adoption by the USSR and the USA of a mutual obligation to refrain from deploying any nuclear weapons on the territory of states where there are none at present, not to build up nuclear stockpiles and not to replace nuclear weapons with new ones in those countries where such weapons have already been deployed.

29

The National Pledge of Non-Participation: A Boycott of SDI Research

Editors' Note

In late 1984, the Strategic Defense Initiative Organization established the Office of Innovative Science and Technology (IST), charged with involving universities and small businesses in basic research on a wide range of exotic technologies relevant to the SDI. Although IST has been successful in attracting a number of academic researchers to work on SDI-related projects—by mid-1986 it was sponsoring research projects at seventy-three academic institutions—its efforts have generated considerable controversy within the academic community.[1]

In the spring of 1985, IST began an aggressive campaign to recruit academic scientists and engineers for SDI research which quickly brought some three thousand inquiries from the scientific community. On March 29, however, IST Director James Ionson implied to a reporter that the Massachusetts Institute of Technology was supportive of the SDI. This was followed, on April 23, by an IST press release which incorrectly announced that the California Institute of Technology had joined an IST research consortium on optical computing. These actions drew angry responses from the presidents of the two institutions, who resented the use of their schools' names to imply endorsement of the SDI. Such incidents, in conjunction with the release of an IST briefing document for potential researchers which described the academic community's reaction to SDI as "immediate and

1. The best discussion of opposition to SDI research in the academic community is Jonathan B. Tucker, "Scientists and Star Wars," Union of Concerned Scientists, *Empty Promise: The Growing Case Against Star Wars* (Beacon Press, 1986), pp. 34–61. See also Lisbeth Gronlund, John Kogut, Michael Weissman, and David Wright, "A Status Report on the Boycott of Star Wars Research by Academic Scientists and Engineers." (This unpublished report is available through Lisbeth Gronlund, Department of Physics, Clark Hall, Cornell University, Ithaca, New York 14853.)

overwhelming," served as inspiration for the National Pledge of Non-Participation.[2]

Working independently, physists at Cornell University and the University of Illinois (Champaign-Urbana) began the local circulation of boycott pledges in May 1985. The two groups became aware of each others' efforts in June and developed a loosely coordinated campaign around a joint pledge. National circulation of the pledge began on September 12, 1985. This has been done informally on the basis of professional and personal contacts among members of those departments which are eligible to receive SDI funding, e.g., physics, applied physics, chemistry, engineering, computer science, space science, and mathematics. As a result, the specific wording of the pledge may differ from campus to campus, but the common thread is a vow not to accept funds from, or to work on projects funded by, the Strategic Defense Initiative Organization.

As of September 1986, the National Pledge of Non-Participation had been signed by 3,850 faculty members and senior researchers and 2,850 graduate students and junior researchers. This includes a majority of the faculty in 110 research departments (64 physics and 46 other science and engineering research departments) and 57 percent of the combined faculty of the top twenty physics departments in the nation, as ranked by the Conference Board of the Associated Research Councils.

The pledge is currently circulating on over 120 campuses in the United States and has recently begun circulating on campuses in Great Britain, the Federal Republic of Germany, and Japan. By early December 1986, the pledge had been signed by over five hundred British and two thousand Japanese scientists. A similar pledge has been signed by over thirteen hundred Canadian university scientists and engineers.[3]

The Pledge of Non-Participation

We, the undersigned science and engineering faculty, believe that the Strategic Defense Initiative (SDI) program (commonly known as Star Wars) is ill-conceived and dangerous. Anti-ballistic missile defense of sufficient reliability to defend the population of the United States against a

2. Tucker, "Scientists and Star Wars," pp. 40–41. See also "Two Universities Say Pentagon Distorted 'Star Wars' Roles," *Chronicle of Higher Education,* June 1, 1985, p. 1; "Caltech at War on Star Wars," *Nature,* 315, 266 (1985); and "Star Wars: The Episode of the Phantom Consortium," *Science and Government Reports,* June 1, 1985, p. 1.

3. Gronlund, et al., "Status Report," p. 22. Also Lisbeth Gronlund, correspondence with the editors, September 27, 1986, and conversation with the editors, December 5, 1986.

Soviet attack is not technically feasible. A system of more limited capability will only serve to escalate the nuclear arms race by encouraging the development of both additional offensive overkill and an all-out competition in anti-ballistic missile weapons. The program will jeopardize existing arms control agreements and make arms control negotiation even more difficult than it is at present. The program is a step toward the type of weapons and strategy likely to trigger a nuclear holocaust. For these reasons, we believe that the SDI program represents, not an advance toward genuine security, but rather a major step backwards.

Accordingly, as working scientists and engineers, we pledge neither to solicit nor accept SDI funds, and to encourage others to join us in this refusal. We hope together to persuade the public and Congress not to support this deeply misguided and dangerous program.[4]

---

4. The pledge commonly appears as above. In many cases, however, the following appear as the second and third paragraphs of a four-paragraph statement:

> The likelihood that SDI funding will restrict academic freedom and blur the distinction between classified and unclassified research is greater than for other sources of funding. The structure of SDI research programs makes it likely that groups doing only unclassified research will be part of a Research Consortium and will therefore work closely with other universities and industries doing *classified* research. SDI officials openly concede that any successful unclassified project may *become* classified. Moreover, the potentially sensitive nature of the research may invoke legal restrictions required by the Export Administration Act.
>
> Participation in SDI by individual researchers would lend their institution's name to a program of dubious scientific validity, and give legitimacy to this program at a time when the involvement of prestigious research instititions is being sought to increase Congressional support. Researchers who oppose the SDI program yet choose to participate in it should therefore recognize that their participation would contribute to the political acceptance of SDI.

30

Appeal by American Scientists to Ban Space Weapons

In May 1986 the Union of Concerned Scientists sponsored a petition, calling for a ban on all space weaponry, which was circulated to the members of the National Academy of Sciences. The petition was signed by over seven hundred individuals, representing more than half of the Academy's membership. Among the signatories were fifty-seven Nobel laureates.

THE development of antisatellite weapons and space-based missile defenses would increase the risk of nuclear war and stimulate a dangerous competition in offensive nuclear arms. An arms race in space poses a great threat to the national security of the United States.

Outer space must remain free of any weapons. It should be preserved as an arena for non-threatening uses: peaceful cooperation, exploration, and scientific discovery among all nations.

We call upon the United States and the Soviet Union to negotiate a total ban on the testing and deployment of weapons in space. To create a constructive environment for the negotiations, both nations should join in a moratorium on further tests of antisatellite weapons. The Soviet Union should bring the Krasnoyarsk radar into conformity with the ABM Treaty, or dismantle it.

We ask the United States and the Soviet Union to reaffirm their commitment to the 1972 ABM Treaty, which prohibits the development, testing, and deployment of space-based ABM systems. We support the continuance of a program of research on ABM technologies in strict conformity with the provisions of the 1972 ABM Treaty.

31

The Anti-Ballistic Missile Treaty

Treaty Between the United States of America and the Union of Soviet Socialist Republics on the Limitation of Anti-Ballistic Missile Systems

THE United States of America and the Union of Soviet Socialist Republics, hereinafter referred to as the Parties,

Proceeding from the premise that nuclear war would have devastating consequences for all mankind,

Considering that effective measures to limit anti-ballistic missile systems would be a substantial factor in curbing the race in strategic offensive arms and would lead to a decrease in the risk of outbreak of war involving nuclear weapons,

Proceeding from the premise that the limitation of anti-ballistic missile systems, as well as certain agreed measures with respect to the limitation of strategic offensive arms, would contribute to the creation of more favorable conditions for further negotiations on limiting strategic arms,

Mindful of their obligations under Article VI of the Treaty on the Non-Proliferation of Nuclear Weapons,

Declaring their intention to achieve at the earliest possible date the cessation of the nuclear arms race and to take effective measures toward reductions in strategic arms, nuclear disarmament, and general and complete disarmament,

Desiring to contribute to the relaxation of international tension and strengthening of trust between States,

Have agreed as follows:

The ABM Treaty was signed in Moscow on May 26, 1972. It was ratified by the United States Senate on August 3 and signed by President Richard M. Nixon on September 30. With the exchange of instruments of ratification on October 3, 1972, the treaty entered into force.

Article I

1. Each Party undertakes to limit anti-ballistic missile (ABM) systems and to adopt other measures in accordance with the provisions of this Treaty.

2. Each Party undertakes not to deploy ABM systems for a defense of the territory of its country and not to provide a base for such a defense, and not to deploy ABM systems for defense of an individual region except as provided for in Article III of this Treaty.

Article II

1. For the purpose of this Treaty an ABM system is a system to counter strategic ballistic missiles or their elements in flight trajectory, currently consisting of:

(a) ABM interceptor missiles, which are interceptor missiles constructed and deployed for an ABM role, or of a type tested in an ABM mode;

(b) ABM launchers, which are launchers constructed and deployed for launching ABM interceptor missiles; and

(c) ABM radars, which are radars constructed and deployed for an ABM role, or of a type tested in an ABM mode.

2. The ABM system compotents listed in paragraph 1 of this Article include those which are:

(a) operational;
(b) under construction;
(c) undergoing testing;
(d) undergoing overhaul, repair or conversion; or
(e) mothballed.

Article III

Each Party undertakes not to deploy ABM systems or their components except that:

(a) within one ABM system deployment area having a radius of one hundred and fifty kilometers and centered on the Party's national capital, a Party may deploy: (1) no more than one hundred ABM launchers and no more than one hundred ABM interceptor missiles at launch sites, and (2) ABM radars within no more than six ABM radar complexes, the area of each complex being circular and having a diameter of no more than three kilometers; and

(b) within one ABM system deployment area having a radius of one hundred and fifty kilometers and containing ICBM silo launchers, a Party may deploy: (1) no more than one hundred ABM launchers and no more than one hundred ABM interceptor missiles at launch sites, (2) two large phased-array ABM radars comparable in potential to corresponding ABM radars operational or under construction on the date of signature of the Treaty in an ABM system deployment area containing ICBM silo launchers, and (3) no more than eighteen ABM radars each having a potential less than the potential of the smaller of the above-mentioned two large phased-array ABM radars.

Article IV

The limitations provided for in Article III shall not apply to ABM systems or their components used for development or testing, and located within current or additionally agreed test ranges. Each Party may have no more than a total of fifteen ABM launchers at test ranges.

Article V

1. Each Party undertakes not to develop, test, or deploy ABM systems or components which are sea-based, air-based, space-based, or mobile land-based.

2. Each Party undertakes not to develop, test, or deploy ABM launchers for launching more than one ABM interceptor missile at a time from each launcher, not to modify deployed launchers to provide them with such a capability, not to develop, test, or deploy automatic or semi-automatic or other similar systems for rapid reload of ABM launchers.

Article VI

To enhance assurance of the effectiveness of the limitations on ABM systems and their components provided by the Treaty, each Party undertakes:

(a) not to give missiles, launchers, or radars, other than ABM interceptor missiles, ABM launchers, or ABM radars, capabilities to counter strategic ballistic missiles or their elements in flight trajectory, and not to test them in an ABM mode; and

(b) not to deploy in the future radars for early warning of strategic ballistic missile attack except at locations along the periphery of its national territory and oriented outward.

Article VII

Subject to the provisions of this Treaty, modernization and replacement of ABM systems or their components may be carried out.

Article VIII

ABM systems or their components in excess of the numbers or outside the areas specified in this Treaty, as well as ABM systems or their components prohibited by this Treaty, shall be destroyed or dismantled under agreed procedures within the shortest possible agreed period of time.

Article IX

To assure the viability and effectiveness of this Treaty, each Party undertakes not to transfer to other States, and not to deploy outside its national territory, ABM systems or their components limited by this Treaty.

Article X

Each Party undertakes not to assume any international obligations which would conflict with this Treaty.

Article XI

The Parties undertake to continue active negotiations for limitations on strategic offensive arms.

Article XII

1. For the purpose of providing assurance of compliance with the provisions of this Treaty, each Party shall use national technical means of verification at its disposal in a manner consistent with generally recognized principles of international law.
2. Each Party undertakes not to interfere with the national technical means of verification of the other Party operating in accordance with paragraph 1 of this Article.
3. Each Party undertakes not to use deliberate concealment measures which impede verification by national technical means of compliance with the provisions of this Treaty. This obligation shall not require changes in current construction, assembly, conversion, or overhaul practices.

Article XIII

1. To promote the objectives and implementation of the provisions of this Treaty, the Parties shall establish promptly a Standing Consultative Commission, within the framework of which they will:

   (a) consider questions concerning compliance with the obligations assumed and related situations which may be considered ambiguous;

   (b) provide on a voluntary basis such information as either Party considers necessary to assure confidence in compliance with the obligations assumed;

   (c) consider questions involving unintended interference with national technical means of verification;

(d) consider possible changes in the strategic situation which have a bearing on the provisions of this Treaty;

(e) agree upon procedures and dates for destruction or dismantling of ABM systems or their components in cases provided for by the provisions of this Treaty;

(f) consider, as appropriate, possible proposals for further increasing the viability of this Treaty; including proposals for amendments in accordance with the provisions of this Treaty;

(g) consider, as appropriate, proposals for further measures aimed at limiting strategic arms.

2. The Parties through consultation shall establish, and may amend as appropriate, Regulations for the Standing Consultative Commission governing procedures, composition and other relevant matters.

Article XIV

1. Each Party may propose amendments to this Treaty. Agreed amendments shall enter into force in accordance with the procedures governing the entry into force of this Treaty.
2. Five years after entry into force of this Treaty, and at five-year intervals thereafter, the Parties shall together conduct a review of this Treaty.

Article XV

1. This Treaty shall be of unlimited duration.
2. Each Party shall, in exercising its national sovereignty, have the right to withdraw from this Treaty if it decides that extraordinary events related to the subject matter of this Treaty have jeopardized its supreme interests. It shall give notice of its decision to the other Party six months prior to withdrawal from the Treaty. Such notice shall include a statement of the extraordinary events the notifying Party regards as having jeopardized its supreme interests.

Article XVI

1. This Treaty shall be subject to ratification in accordance with the constitutional procedures of each Party. The Treaty shall enter into force on the day of the exchange of instruments of ratification.
2. This Treaty shall be registered pursuant to article 102 of the Charter of the United Nations.

DONE at Moscow on May 26, 1972, in two copies, each in the English and Russian languages, both texts being equally authentic.

Agreed Statements, Common Understandings, and Unilateral Statements Regarding the Treaty Between the United States of America and the Union of Soviet Socialist Republics on the Limitation of Anti-Ballistic Missiles

1. Agreed Statements

The document set forth below was agreed upon and initialed by the Heads of the Delegations on May 26, 1972 [letter designations added];

AGREED STATEMENTS REGARDING THE TREATY BETWEEN THE UNITED STATES OF AMERICA AND THE UNION OF SOVIET SOCIALIST REPUBLICS ON THE LIMITATION OF ANTI-BALLISTIC MISSILE SYSTEMS

[A]

The Parties understand that, in addition to the ABM radars which may be deployed in accordance with subparagraph (a) of Article III of the Treaty, those non-phased-array ABM radars operational on the date of signature of the Treaty within the ABM system deployment area for defense of the national capital may be retained.

[B]

The Parties understand that the potential (the product of mean emitted power in watts and antenna area in square meters) of the smaller of the two large phased-array ABM radars referred to in subparagraph (b) of Article III of the Treaty is considered for purposes of the Treaty to be three million.

[C]

The Parties understand that the center of the ABM system deployment area centered on the national capital and the center of the ABM system deployment area containing ICBM silo launchers for each Party shall be separated by no less than thirteen hundred kilometers.

[D]

In order to secure fulfillment of the obligation not to deploy ABM systems and their components except as provided in Article III of the Treaty, the Parties agree that in the event ABM systems based on other physical principles and including components capable of substituting for ABM interceptor missiles, ABM launchers, or ABM raders are created in the future, specific limitations on such systems and their components would be subject to discussion in accordance with Article XIII and agreement in accordance with Article XIV of the Treaty.

[E]

The Parties understand that Article V of the Treaty includes obligations not to develop, test or deploy ABM interceptor missiles for the delivery by each ABM interceptor missile of more than one independently guided warhead.

[F]

The Parties agree not to deploy phased-array radars having a potential (the product of mean emitted power in watts and antenna area in square meters) exceeding three million, except as provided for in Articles III, IV and VI of the Treaty, or except for the purposes of tracking objects in outer space or for use as national technical means of verification.

[G]

The Parties understand that Article IX of the Treaty includes the obligation of the US and the USSR not to provide to other States technical descriptions or blue prints specially worked out for the construction of ABM systems and their components limited by the Treaty.

2. Common Understandings

Common understanding of the Parties on the following matters was reached during the negotiations:

A. Location of ICBM Defenses

The U.S. Delegation made the following statement on May 26, 1972:

> Article III of the ABM Treaty provides for each side one ABM system deployment area centered on its national capital and one ABM system deployment area containing ICBM silo launchers. The two sides have registered agreement on the following statement: "The Parties understand that the center of the ABM system deployment area centered on the national capital and the center of the ABM system deployment area containing ICBM silo launchers for each Party shall be separated by no less than thirteen hundred kilometers." In this connection, the U.S. side notes that its ABM system deployment area for defense of ICBM silo launchers, located

west of the Mississippi River, will be centered in the Grand Forks ICBM silo launcher deployment area. (See Agreed Statement [C].)

B. ABM Test Ranges

The U.S. Delegation made the following statement on April 26, 1972:

> Article IV of the ABM Treaty provides that "the limitations provided for in Article III shall not apply to ABM systems or their components used for development or testing, and located within current or additionally agreed test ranges." We believe it would be useful to assure that there is no misunderstanding as to current ABM test ranges. It is our understanding that ABM test ranges encompass the area within which ABM components are located for test purposes. The current U.S. ABM test ranges are at White Sands, New Mexico, and at Kwajalein Atoll, and the current Soviet ABM test range is near Sary Shagan in Kazakhstan. We consider that non-phased array radars of types used for range safety or instrumentation purposes may be located outside of ABM test ranges. We interpret the reference in Article IV to "additionally agreed test ranges" to mean that ABM components will not be located at any other test ranges without prior agreement between our Governments that there will be such additional ABM test ranges.

On May 5, 1972, the Soviet Delegation stated that there was a common understanding on what ABM test ranges were, that the use of the types of non-ABM radars for range safety or instrumentation was not limited under the Treaty, that the reference in Article IV to "additionally agreed" test ranges was sufficiently clear, and that national means permitted identifying current test ranges.

C. Mobile ABM Systems

On April 29, 1972, the U.S. Delegation made the following statement:

> Article V(1) of the Joint Draft Text of the ABM Treaty includes an undertaking not to develop, test, or deploy mobile land-based ABM systems and their components. On May 5, 1971, the U.S. side indicated that, in its view, a prohibition on deployment of mobile ABM systems and components would rule out the deployment of ABM launchers and radars which were not permanent fixed types. At that time, we asked for the Soviet view of this interpretation. Does the Soviet side agree with the U.S. side's interpretation put forward on May 5, 1971?

On April 13, 1972, the Soviet Delegation said there is a general common understanding on this matter.

D. Standing Consultative Commission

Ambassador Smith made the following statement on May 22, 1972:

> The United States proposes that the sides agree that, with regard to initial implementation of the ABM Treaty's Article XIII on the Standing Consultative Commission (SCC) and of the consultation Articles to the Interim Agreement on offensive arms and the Accidents Agreement,[1] agreement establishing the SCC will be worked out early in the follow-on SALT negotiations; until that is completed, the following arrangements will prevail: when SALT is in session, any consultation desired by either side under these Articles can be carried out by the two SALT Delegations; when SALT is not in session, *ad hoc* arrangements for any desired consultations under these Articles may be made through diplomatic channels.

Minister Semenov replied that, on an *ad referendum* basis, he could agree that the U.S. statement corresponded to the Soviet understanding.

E. Standstill

On May 6, 1972, Minister Semenov made the following statement:

> In an effort to accommodate the wishes of the U.S. side, the Soviet Delegation is prepared to proceed on the basis that the two sides will in fact observe the obligations of both the Interim Agreement and the ABM Treaty beginning from the day of signature of these two documents.

In reply, the U.S. Delegation made the following statement on May 20, 1972:

> The U.S. agrees in principle with the Soviet statement made on May 6 concerning observance of obligations beginning from date of signature but we would like to make clear our understanding that this means that, pending ratification and acceptance, neither side would take any action prohibited by the agreements after they had entered into force. This understanding would continue to apply in the absence of notification by either signatory of its intention not to proceed with ratification or approval.

The Soviet Delegation indicated agreement with the U.S. statement.

3. Unilateral Statements

The following noteworthy unilateral statements were made during the negotiations by the United States Delegation:

---

1. See Article 7 of Agreement to Reduce the Risk of Outbreak of Nuclear War Between the United States of America and the Union of Soviet Socialist Republics, signed September 30, 1971.

A. Withdrawal from the ABM Treaty

On May 9, 1972, Ambassador Smith made the following statement:

The U.S. Delegation has stressed the importance the U.S. Government attaches to achieving agreement on more complete limitations on strategic offensive arms., following agreement on an ABM Treaty and on an Interim Agreement on certain measures with respect to the limitation of strategic offensive arms. The U.S. Delegation believes that an objective of the follow-on negotiations should be to constrain and reduce on a long-term basis threats to the survivability of our respective strategic retaliatory forces. The U.S.S.R. Delegation has also indicated that the objectives of SALT would remain unfulfilled without the achievement of an agreement providing for more complete limitations on strategic offensive arms. Both sides recognize that the initial agreements would be steps toward the achievement of more complete limitations on strategic arms. If an agreement providing for more complete strategic offensive arms limitations were not achieved within five years, U.S. supreme interests could be jeopardized. Should that occur, it would constitute a basis for withdrawal from the ABM Treaty. The U.S. does not wish to see such a situation occur, nor do we believe that the U.S.S.R. does. It is because we wish to prevent such a situation that we emphasize the importance the U.S. Government attaches to achievement of more complete limitations on strategic offensive arms. The U.S. Executive will inform the Congress, in connection with Congressional consideration of the ABM Treaty and the Interim Agreement, of this statement of the U.S. position.

B. Tested in ABM Mode

On April 7, 1972, the U.S. Delegation made the following statement:

Article II of the Joint Text Draft uses the term "tested in an ABM mode," in defining ABM components, and Article VI includes certain obligations concerning such testing. We believe that the sides should have a common understanding of this phrase. First, we would note that the testing provisions of the ABM Treaty are intended to apply to testing which occurs after the date of signature of the Treaty, and not to any testing which may have occurred in the past. Next, we would amplify the remarks we have made on this subject during the previous Helsinki phase by setting forth the objectivès which govern the U.S. view on the subject, namely, while prohibiting testing of non-ABM components for ABM purposes: not to prevent testing of ABM components, and not to prevent testing of non-ABM components for non-ABM purposes. To clarify our interpretation of "tested in an ABM mode," we note that we would consider a launcher, missile or radar to be "tested in an ABM mode" if, for example, any of the following events occur: (1) a launcher is used to launch an ABM interceptor missile, (2) an interceptor missile is flight tested against a target vehicle which has a flight trajectory with characteristics of a strategic ballistic missile flight trajectory, or is flight tested in conjunction with the test of an ABM interceptor missile or an ABM radar at the same test range, or is flight tested to an altitude inconsistent with interception of targets against which air defenses are deployed, (3) a radar makes measurements on a cooperative target vehicle of the kind referred to in item (2) above during the reentry portion of its trajectory or makes measurements in conjunction with the test of an ABM interceptor missile or an ABM radar at the same test range. Radars used for purposes such as range safety or instrumentation would be exempt from application of these criteria.

C. No-Transfer Article of ABM Treaty

On April 18, 1972, the U.S. Delegation made the following statement:

> In regard to this Article [IX], I have a brief and I believe self-explanatory statement to make. The U.S. side wishes to make clear that the provisions of this Article do not set a precedent for whatever provision may be considered for a Treaty on Limiting Strategic Offensive Arms. The question of transfer of strategic offensive arms is a far more complex issue, which may require a different solution.

D. No Increase in Defense of Early Warning Radars

On July 28, 1970, the U.S. Delegation made the following statement:

> Since Hen House radars [Soviet ballistic missile early warning radars] can detect and track ballistic missile warheads at great distances, they have a significant ABM potential. Accordingly, the U.S. would regard any increase in the defenses of such radars by surface-to-air missiles as inconsistent with an agreement.

Glossary

AA = Antiaircraft
AAM = Air-to-Air Missile
ABM = Antiballistic Missile
ADI = Air Defense Initiative
ADMP = Air Defense Master Plan
AFB = Air Force Base
AGARD = Advisory Group for Aerospace Research
AI = Artificial intelligence
ALB = Air-Land Battle
ALCM = Air-Launched Cruise Missile
ALL = Airborne Laser Laboratory
ANG = Air National Guard
ANZUS = Australia–New Zealand–United States alliance
ARPA = Advanced Research Projects Agency
ASAT = Antisatellite weapons
ASW = Antisubmarine Warfare
ATB = Advanced Technology Bomber
ATBM = Antitactical Ballistic Missile
ATM = Antitactical Missile
AWACS = Airborne Warning and Control System
AWST = *Aviation Week and Space Technology* (journal)
BAMBI = Ballistic Missile Booster Interceptor
BDI = Federation of German Industries
BMD = Ballistic Missile Defense
BMEWS = Ballistic Missile Early Warning System
C^3I = Command, Control, Communications, and Intelligence
CA-90 = Counter-Air 90
CDU = Christian Democratic Union (FRG)
CEP = Circular Error Probable
CIA = Central Intelligence Agency
CINCNORAD = Commander in Chief, NORAD
CIT = California Institute of Technology
CPSU = Communist Party of the Soviet Union
CRS = Congressional Research Service

CSIS = Center for Strategic and International Studies
CSU = Christian Social Union (FRG)
DARPA = Defense Advanced Research Projects Agency
DDRE = Department of Defense Research and Engineering
DEFCON = Defense Condition
DEW = Directed Energy Weapons
DEW Line = Distant Early Warning Line
DIA = Defense Intelligence Agency
DOB = Dispersed Operating Base
DOD = Department of Defense
DOS = Department of State
ECM = Electronic Countermeasures
EDI = European Defense Initiative
EKD = Protestant Church Council of Germany
EMP = Electromagnetic Pulse
ERW = Enhanded-Radiation Warhead
EUREKA = European Research Coordinating Program
FAMBDS = Field Army Ballistic Missile Defense System
FBM = Fleet Ballistic Missile
FBS = Forward-Based Systems
FDP = Free Democratic Party (FRG)
FOFA = Follow-On Forces Attack
FOL = Forward Operating Location
FRG = Federal Republic of Germany
FY = Fiscal Year
GAPA = Ground-to-Air Pilotless Aircraft Program
GDR = German Democratic Republic
GLCM = Ground-Launched Cruise Missile
HE = High Explosive
HOE = Homing Overlay Experiment
ICBM = Intercontinental Ballistic Missile
IISS = International Institute for Strategic Studies
INF = Intermediate-Range Nuclear Forces
IR = Infrared
IRBM = Intermediate-Range Ballistic Missile
IST = Office of Innovative Science and Technology
JCS = Joint Chiefs of Staff
KAL = Korean Air Lines
KEW = Kinetic Energy Weapons
Kt = Kiloton
LANL = Los Alamos National Laboratory
LLNL = Lawrence Livermore National Laboratory
LoADS = Low Altitude Defense System
LTBT = Limited Test Ban Treaty

MAD = Mutual Assured Destruction
MARV = Maneuvering Reentry Vehicle
MAS = Mutual Assured Survival
MHV = Miniature Homing Vehicle
MIRV = Multiple Independently-Targetable Reentry Vehicle
MIT = Massachusetts Institute of Technology
MOD = Ministry of Defense
MOU = Memorandum of Understanding
MRBM = Medium Range Ballistic Missile
MRV = Maneuvering Reentry Vehicle
Mt = Megaton
MX = Missile Experimental
NASA = National Aeronautics and Space Administration
NATO = North Atlantic Treaty Organization
NBC = Nuclear, Biological, and Chemical Weapons
NDP = New Democratic Party (Canada)
NORAD = North American Aerospace (formerly Air) Defense Command
NPT = Non-Proliferation Treaty
NPG = Nuclear Planning Group
NSA = National Security Agency
NSC = National Security Council
NSDD = National Security Decision Directive
NWS = North Warning System
OSA = Office of Systems Analysis
OSD = Office of the Secretary of Defense
OTA = Office of Technological Assessment
OTH-B = Over-the-Horizon, Backscatter Radar
PII = Pershing II Missile
PAR = Phased-Array Radar
PBV = Post-Boost Vehicle
PBDV = Post-Boost Delivery Vehicle
PC = Progressive Conservative Party (Canada)
PD = Presidential Directive
PDM = Presidential Decision Memorandum
PRC = People's Republic of China
R&D = Research and Development
RDT&E = Research, Development, Test, and Evaluation
RV = Reentry Vehicle
SAC = Strategic Air Command
SACEUR = Supreme Allied Commander in Europe
SAGE = Semi-Autonomous Ground Environment
SAINT = Satellite Interceptor, Project
SALT = Strategic Arms Limitation Talks
SAM = Surface-to-Air Missile

SBAM = Space-Based Antimissile System
SCC = Standing Consultative Commission
SDA = Strategic Defense Architecture
SDI = Strategic Defense Initiative
SDIO = Strategic Defense Initiative Organization
SICBM = Small Intercontinental Ballistic Missile
SIOP = Single Integrated Operational Plan
SIPRI = Stockholm International Peace Research Institute
SLBM = Submarine-Launched Ballistic Missile
SLCM = Sea-Launched Cruise Missile
SPD = Social Democratic Party of Germany (FRG)
SRAM = Short-Range Attack Missile
SRBM = Short-Range Ballistic Missile
SSBN = Nuclear-Powered Ballistic Missile Submarine
SSM = Surface-to-Surface Missile
SSN = Nuclear-powered submarine
START = Strategic Arms Reduction Talks
TBM = Tactical Ballistic Missile
TNF = Theater Nuclear Forces
TOW = Tube-launched, Optically-tracked, Wire-guided antitank missile
UCS = Union of Concerned Scientists
UK = United Kingdom
USAFSPACECOM = United States Air Force Space Command
USSPACECOM = United States Space Command
USSR = Union of Soviet Socialist Republics
USA = United States of America
USA = United States Army
USAAF = United States Army Air Force
USAF = United States Air Force
USN = United States Navy
WEU = Western European Union

Index